Isotopes in Palaeoenvironmental Research

Developments in Paleoenvironmental Research

VOLUME 10

Isotopes in Palaeoenvironmental Research

Edited by

Melanie J. Leng

*NERC Isotope Geosciences Laboratory, British Geological Survey,
Keyworth, Nottingham, UK
and School of Geography, University of Nottingham, Nottingham, UK*

A C.I.P. Catalogue record for this book is available from the Library of Congress.

ISBN-13 978-90-481-6674-9
ISBN-10 1-4020-2504-1 (e-book)
ISBN-13 978-1-4020-2504-4 (e-book)

Published by Springer,
P.O. Box 17, 3300 AA Dordrecht, The Netherlands.

www.springeronline.com

Printed on acid-free paper

TABLE OF CONTENTS

PREFACE

This volume is intended to show how stable isotopes can be applied to understanding the palaeoenvironment. There are chapters on the interpretation of isotopes in water, tree rings, bones and teeth, lake sediments, speleothems and marine sediments. Crucial to the understanding of the environmental signal contained within the isotope composition of different materials is to gain more information about how rainfall isotope compositions are determined by climate. Chapter 1 (Darling et al.) describes O, H and C stable isotope compositions in the modern day water and aqueous carbon cycles to provide a framework for the interpretation of these isotopes in the past. The chapter on the water cycle divides naturally into a number of sections. The starting point, precipitation, is especially important because it is the precursor to which most O and H isotope proxy studies are attempting to relate. While much is understood about the isotope systematics of precipitation, largely owing to the existence of the IAEA–WMO Global Network for Isotopes in Precipitation (GNIP), important questions remain to be answered in relation to the isotope-temperature gradients of past climatic conditions. The chapter describes the three reservoirs of water sustaining all terrestrial proxies; soil and vadose zone moisture, groundwater, and surface waters. In each reservoir isotope effects intervene to modify to a greater or lesser extent the isotope signature of antecedent precipitation; groundwaters are least affected and surface waters the most. Since lake sediments are an important archive of the palaeoenvironment, the continuing development of isotope mass-balance modelling techniques is playing a role of considerable importance in interpreting isotope records from lake sediments (see also Leng et al., Chapter 4). The carbon cycle is far more complex than the water cycle, and is mentioned in various guises in all the other chapters. However, carbon is still of use, especially where the various carbon sources can be constrained and can be linked to the overall geochemical conceptual model.

Trees provide a good archive of recent and earlier Holocene climate and environmental change (McCarroll and Loader, Chapter 2). Trees have the advantage of being widespread and can produce a near continuous archive of environmental information. The water isotopes provide two separate signals; the isotope composition of the source water and evaporation from leaves. The former is only a proxy for the isotope composition of precipitation if the trees are accessing relatively young and unmixed water. In other trees evaporative enrichment may be the predominant signal in the tree rings and this is controlled largely by air humidity. Carbon isotopes can complement the water data, notably in dry environments, where the main signals are air humidity and soil moisture content, and in moist environments where photosynthetic rate dominates and the signals are summer temperature and solar radiation.

Perhaps one of the most significant advances in tree ring research is the development of on-line techniques which allow the analyses of increased numbers of tree rings, and these are reviewed. The ability to analyse smaller samples enables individual trees to provide data, while combining the results from several trees allows the quantification of variability, at different temporal frequencies, and thus allows statistically defined confidence limits to be placed around mean values. This way temporal frequencies, including the low-frequency, long-term changes that are so difficult to capture using more traditional methods, are able to be investigated.

The third chapter deals with the isotope composition of bone and teeth (Hedges et al.) which are dominated by biological and ecological controls. This is even true for such isotopes as strontium that are adventitiously taken up according to local geology, and so potentially record animal movements within their ecological setting. However, there is a need to understand the biological features in the subject of study (such as food web structure) in order to get back to environmental information which cannot be done using isotopes alone. One of the problems with using bones as an environmental archive is the sporadic and acontextual deposition of bone, rarely enabling its data to be directly integrated with complementary data from the same site, apart from on a general scale. However, bone and teeth contain many different isotope signals which can be very sensitive - the non-linear change in C3/C4 plant abundance (and therefore with $\delta^{13}C$ in the bone collagen of herbivorous grazers) with local climate for example. Overall this chapter provides the background to understanding how isotope changes in bone and teeth are caused, and shown how these changes may be understood, though not always, in terms of environmental effects.

Chapter 4 on lakes (Leng et al.) demonstrates that the oxygen, carbon, nitrogen and hydrogen isotope composition of lacustrine sedimentary materials can yield a wide range of useful palaeoclimate information, although a full interpretation of isotope data from lake sediments requires a detailed knowledge of the processes that control and modify the signal. For example, Chapter 1 describes the various factors that can change the isotope composition of precipitation, which is an important consideration in palaeolimnology. Processes that can affect the isotope composition of lacustrine sedimentary materials must be largely determined for an individual lake system, although typical responses are often assumed. Oxygen isotopes are the main isotopes used in palaeolimnology and can be obtained from a large range of inorganic and organic materials. However, even within the most established materials the interpretation is not easy. For example, a change in temperature will produce a shift in the equilibrium oxygen isotope composition of carbonate forming in lake water, but will also affect the isotope composition of the rainfall and rates of evaporation. Recent research has shown that this might be resolved by the analysis of the oxygen isotope composition of lacustrine organic materials, for example aquatic cellulose, which is thought to have $\delta^{18}O$ values that are independent of water temperature.

Other palaeoenvironmental information can be achieved from organic materials in lakes (Chapter 4) from their carbon and nitrogen ratios, although they will also contain information on productivity and changes in nutrient supply.

Chapter 5 describes the stable isotope composition of speleothems (McDermott et al.). Despite the generally acknowledged lack of precise isotope equilibrium between cave drip-waters and their carbonate precipitates, speleothems are usually deposited sufficiently close to isotope equilibrium to retain useful information about climate-driven changes in the isotope composition of meteoric water that falls on a cave site. As with all archives the quantitative interpretation of stable isotope shifts in speleothems is seldom straightforward, and a detailed knowledge of the meteorological variables that control these archives in the modern cave system is essential to provide a sound interpretation of past climates (see Chapter 1). Where possible, present-day monitoring studies of cave drip waters are essential to understand fully the site-specific relationships between speleothem $\delta^{18}O$ and meteorological variables at the surface.

Temporal changes in speleothem $\delta^{13}C$ can offer considerable potential to reconstruct climate-driven changes in both the nature of vegetation (e.g., C3 vs. C4 type) and the intensity of vegetation above a cave site. One particular requirement in speleothem research is that carbon isotope data should be accompanied, where possible, by appropriate elemental data (e.g., Ca/Mg ratios) in order to evaluate other possible causes for temporal changes in $\delta^{13}C$. For example, strongly correlated carbon isotope and Mg/Ca ratios may point to a role for partial degassing of cave drip-waters and calcite precipitation in the hydrological flow-path above the cave (so-called 'prior calcite precipitation') that could produce elevated $\delta^{13}C$ values unrelated to changes in vegetation type or intensity.

One of the distinct advantages of using speleothems, is the ability to use U-series dating. The possibility of using well-dated speleothems records to refine the chronology for the high-latitude ice core records is particularly exciting. In the future it seems likely that data from speleothem fluid inclusions will provide a more robust basis for palaeoclimate interpretations, in particular enabling the relative effects of air temperature and atmospheric circulation changes to be unravelled.

Measurement of oxygen and carbon isotopes in foraminifera have been central to the development and establishment of stable isotopes as one of the most important sets of proxies within palaeoceanography. In Chapter 6, Maslin and Swann, demonstrate how oxygen isotopes records can provide a stratigraphy while enabling the reconstruction of such things as: past global ice volume, ocean temperatures, relative sea level, ocean circulation and structure, surface water salinity, iceberg melting, river discharge, and monsoonal intensity. Carbon isotope records from both carbonate and organic matter found within marine sediments can provide information on the past carbon cycle on a range of time-scales, enabling an insight into past marine productivity, ocean circulation, past surface water, atmospheric CO_2, storage and exchange of carbon on both the local and global scale. However, the interpretation of carbon isotope records is complicated by different sources of carbon entering the marine environment with an additional number of influences affecting the isotope values of each source. Recent years have witnessed the development of a variety of other isotope proxies, in particular $\delta^{15}N$ and $\delta^{11}B$. While these are still not measured routinely, great steps forward have been achieved in our understanding of them, although significant difficulties remain in ensuring robust and accurate interpretations from the generated data. There is potential for both isotopes, however, to provide a more significant insight into past oceanic conditions and palaeo-pCO_2 levels. As such, their use are likely to expand over the next decade as a greater understanding of the marine nitrogen cycle, controls on $\delta^{15}N$ and the errors associated with $\delta^{11}B$ based pCO_2 reconstructions are achieved.

Outside foraminiferal isotope based studies, there is great scope to generate isotope data from other biological sources. A current major limitation with foraminifera studies is the void of data from regions depleted in $CaCO_3$ preservation, particularly in the Southern Ocean and other high latitude regions. Future development and use of $\delta^{18}O_{(diatom)}$ and $\delta^{30}Si_{(diatom)}$ alongside $\delta^{13}C$ and $\delta^{15}N$ analysis of intrinsic organic matter within diatom frustules will ultimately open the gateway for the extension of detailed, isotope based, palaeoceanographic reconstructions to more of the world's oceans.

In summary, all the chapters presented here have several similar messages. Isotopes can be extremely powerful palaeoenvironmental tools. However, as with all archives, it is always desirable to carry out a calibration exercise to investigate the basic systematics of isotope variation in the modern environment to establish the relationship between the measured signal and the isotope composition of the host. A robust calibration is seldom easy as the materials may not occur in the contemporary environment and the site may be in an isolated geographical region, making a rigorous contemporary study impossible. Where such a calibration is not possible assumptions have to be made, but these should be based on evidence from a multi-proxy approach using isotope signals from different materials or combined with other palaeoenvironmental techniques.

Acknowledgements

Many people have helped with the planning, and production of this volume. Bill Last and John Smol (DPER series editors), Judith Terpos and Gert-Jan Geraeds (Springer) have provided technical assistance as well as advice and encouragement. Each chapter benefited from two external referees (many of whom remain anonymous), the help of all those who generously gave their time is most gratefully acknowledged by all the authors. Finally, thanks are due to colleagues at the NERC Isotope Geosciences Laboratory, especially Carol Arrowsmith, Angela Lamb and Joanne Green for administrative and organisational efforts. The chapter authors have contributed to this book because of their enthusiasm and belief in the usefulness of isotope techniques. We hope that it may encourage young scientists to enter the field and that isotopes will be a valuable source of evidence for those who in the future continue to address questions in palaeoenvironmental research.

Many of the authors wish to express personal thanks, in particular Danny McCarroll and Neil Loader thank Margaret Barbour, David Beerling, Keith Briffa, Tony Carter, Mary Gagen, Mike Hall, Gerd Helle, Debbie Hemming, Risto Jalkanen, Andreas Kirchhefer, Frank Pawellek, Iain Robertson, Roy Switsur and John Waterhouse for their assistance and many stimulating discussions. In addition their friends in FOREST, PINE and ISONET for help and support. Isotope dendroclimatology in Swansea has been supported though grants from NERC, the Royal Society, The Leverhulme Trust, and the European Union.

Frank McDermott acknowledges the help of numerous colleagues at the Open University (Mabs Gilmour, Peter van Calsteren), Bristol University (Chris Hawkesworth, Dave Richards and Siobhan McGarry), Birmingham University (Ian Fairchild and Andy Baker) and at UCD (James and Lisa Baldini). Financial support from Enterprise Ireland and Science Foundation Ireland is also acknowledged. Henry Schwarcz acknowledges the long and continued collaboration and co-operation of his past and present students and colleagues: Chas Yonge, Jim O'Neil, Russ Harmon, Peter Thompson, Mel Gascoyne, Tim Atkinson, Peter Rowe and, lately, Ren Zhang and Trish Beddows. Above all, he thanks his friend and colleague Derek Ford for many years of collaboration and encouragement in these studies. Martin Knyf has given invaluable assistance in all aspects of these analyses. Peter Rowe acknowledges the help and stimulating support of Tim Atkinson and Paul Dennis, and countless fruitful discussions with many members of the speleothem and stable isotope communities. The authors are grateful for constructive reviews by Mira Bar-Matthews and an anonymous reviewer. Informal reviews by Ian Fairchild and Lisa Baldini greatly improved the chapter.

The marine isotope chapter was supported by a NERC PhD CASE studentship award to GEAS and NERC grants to MAM. The authors would like to thank Eelco Rohling and two anonymous reviewers for their critical insights, constructive comments and help in improving the quality of the manuscript. Additional thanks are given to the Drawing office, Department of Geography, University College London for assistance in the preparation of the figures.

The following people generously provided images for the front cover: M. Jones (lake sediment), G. Swann (diatom), P. Langdon (chironomid), P. Shand (straw), N. Loader (tree ring), J. Evans (skull), P. Witney (rain drop), J. Pike and A.E.S. Kemp (marine sediment) and D. Genty (speleothem).

Notation and standardisation

To prevent repetition regarding notation and standardisation, a short summary is given here. However, fuller descriptions can be found in Bowen (1988), Coleman and Fry (1991), Clarke and Fritz (1997), Hoefs (1997), Criss (1999) and de Groot (2004). Isotope ratios (e.g., $^{18}O/^{16}O$, $^{13}C/^{12}C$) are expressed in terms of delta values (δ), because the isotope ratios are more easily measured as relative differences, rather than absolute values. We refer to the delta value (δ) and this is measured in units of per mille (‰). The δ value is defined as:

$$\delta = (R_{sample}/R_{standard}) - 1. \ 10^3$$

Where R = the measured ratio of the sample and standard respectively. Since a sample's ratio may be either higher or lower than that of the standard, δ values can be positive or negative. The δ value is dimensionless, so where comparisons are made between samples (e.g., where $\delta_A < \delta_B$) the δ value of A, is said to be 'lower' than that of B (and B 'higher' than A). Where reference is made to absolute ratios A may be said to be 'depleted' in the heavier isotope compared to B (and B 'enriched' compared to A). In the laboratory it is necessary to use 'working standards' with values calibrated against recognised standard materials, thus all values are quoted relative to the latter. For waters (oxygen and hydrogen) and silicates (oxygen) we use VSMOW (Vienna Standard Mean Oceanic Water) an average ocean water, for carbonate and organic material we use VPDB (Vienna Pee Dee Belemnite), for nitrogen the convention is atmospheric nitrogen (described as AIR), for for BoronNIST (boric acid SRM 951) and silicon NBS28 (IAEA quartz sand). See discussions in Bowen (1988), Coleman and Fry (1991), Clarke and Fritz (1997), Hoefs (1997), and Criss (1999). The data are presented as per mille (‰) deviations from the relevant international standard (e.g., ‰ VPDB).

References
Bowen G.J. and Wilkinson B. 2002. Spatial distribution of $\delta^{18}O$ in meteoric precipitation. Geology 30: 315-318.
Clark I. and Fritz I. 1997. Environmental Isotopes in Hydrogeology. Lewis Publishers, Boca Raton, New York, 328 pp.
Coleman D.C. and Fry B. 1991. Carbon Isotope Techniques. Academic Press, London, 274 pp.
Criss R.E. 1999. Principals of Stable Isotope Distribution, Oxford University Press, New York, 264 pp.
De Groot P.A. 2004. Handbook of Stable Isotope Analytical Techniques. Elsevier, 1234 pp.
Hoefs J. 1997. Stable Isotope Geochemistry. Springer, 201 pp.

LIST OF CONTRIBUTORS

Carol Arrowsmith, NERC Isotope Geosciences Laboratory, British Geological Survey, Nottingham NG12 5GG, UK.
e-mail: caar@bgs.ac.uk

Adrian H. Bath, Intellisci Ltd.,Loughborough, LE12 6SZ, UK.
e-mail: abath@intellisci.co.uk

W. George Darling, British Geological Survey, Wallingford, OX10 8BB, UK.
e-mail: wgd@bgs.ac.uk

John J. Gibson, National Water Research Institute, University of Victoria, Victoria BC, V8W 2Y2, Canada.
e-mail: john.gibson@ec.gc.ca

Timothy H.E. Heaton, NERC Isotope Geosciences Laboratory, British Geological Survey, Nottingham NG12 5GG, UK.
e-mail: theh@bgs.ac.uk

Robert E.M. Hedges, Research Laboratory for Archaeology and the History of Art, University of Oxford, Oxford OX1 3QJ, UK.
e-mail: robert.hedges@rlaha.ox.ac.uk

Jonathan A. Holmes, Department of Geography, University College London, London WC1H OAP, UK.
e-mail: j.holmes@ucl.ac.uk

Matthew D. Jones, School of Geography, University of Nottingham, Nottingham NG7 2RD, UK.
e-mail: matthew.jones@nottingham.ac.uk

Paul L. Koch, Department of Earth Sciences, University of California Santa Cruz, California 95064, USA.
e-mail: pkoch@es.ucsc.edu

Angela L. Lamb, NERC Isotope Geosciences Laboratory, British Geological Survey, Nottingham NG12 5GG, UK.
e-mail: alla@bgs.ac.uk

Melanie J. Leng, NERC Isotope Geosciences Laboratory, British Geological Survey, Nottingham NG12 5GG, UK
and
School of Geography, University of Nottingham, Nottingham NG7 2RD, UK.
e-mail: mjl@bgs.ac.uk

Neil J. Loader, Department of Geography, University of Wales, Swansea SA2 8PP, UK.
e-mail: n.j.loader@swansea.ac.uk

James D. Marshall, Department of Earth Sciences, University of Liverpool, Liverpool L69 3GP, UK.
e-mail: isotopes@liverpool.ac.uk

Mark A. Maslin, Department of Geography, University College London, London WC1H 0AP, UK.
e-mail: mmaslin@geog.ucl.ac.uk

Danny McCarroll, Department of Geography, University of Wales, Swansea SA2 8PP, UK.
e-mail: d.mccarroll@swansea.ac.uk

Frank McDermott, Department of Geology, University College Dublin, Belfield, Dublin 4, Ireland.
e-mail: frank.mcdermott@ucd.ie

Peter J. Rowe, School of Environmental Sciences, University of East Anglia, Norwich NR4 7TJ, UK.
e-mail: P.Rowe@uea.ac.uk

Kazimierz Rozanski, Faculty of Physics and Applied Computer Science, AGH University of Science and Technology, 30–059 Krakow, Poland.
e-mail: rozanski@novell.ftj.agh.edu.pl

Henry Schwarcz, Department of Geography and Geology, McMaster University, Hamilton, Ontario L8S 4M1, Canada.
e-mail: schwarcz@mcmaster.ca

Rhiannon E. Stevens, School of Geography, University of Nottingham, Nottingham NG7 2RD, UK.
and
NERC Isotope Geosciences Laboratory, British Geological Survey, Nottingham NG12 5GG, UK
e-mail: rhiannon.stevens@nottingham.ac.uk

George A. Swann, Department of Geography, University College London, London WC1H 0AP, UK.
e-mail: g.swann@ucl.ac.uk

Brent B. Wolfe, Department of Geography and Environmental Studies, Wilfrid Laurier University, Waterloo Ontario N2L 3C5, Canada.
e-mail: bwolfe@wlu.ca

PUBLISHED AND FORTHCOMING TITLES IN THE
DEVELOPMENTS IN PALEOENVIRONMENTAL RESEARCH BOOK SERIES

Series Editors:
John P. Smol,
Department of Biology, Queen's University Kingston, Ontario, Canada
William M. Last,
Department of Geological Sciences, University of Manitoba, Winnipeg, Manitoba, Canada

---

For more information on this series, please visit:

www.springeronline.com

http://home.cc.umanitoba.ca/~mlast/paleolim/dper.html

1. ISOTOPES IN WATER

W. GEORGE DARLING (wgd@bgs.ac.uk)
British Geological Survey
Wallingford OX10 8BB, UK

ADRIAN H. BATH (abath@intellisci.co.uk)
Intellisci Ltd.
Loughborough LE12 6SZ, UK

JOHN J. GIBSON (john.gibson@ec.gc.ca)
National Water Research Institute
University of Victoria
Victoria V8W 2Y2, Canada

KAZIMIERZ ROZANSKI (rozanski@novell.ftj.agh.edu.pl)
Faculty of Physics and Applied Computer Science
AGH University of Science and Technology
30–059 Krakow, Poland

Key words: Stable isotopes, water cycle, carbon cycle, precipitation, surface waters, groundwater, isotope mass balance, proxies.

Introduction

Scope of this chapter

The study of isotopic proxies has two different though related basic aims: an understanding of the way in which the water cycle is linked to alterations in climate, and how the carbon cycle has responded to these changes. This necessarily requires the study of oxygen, hydrogen and carbon stable isotope ratios. Water plays a vital role in the growth or formation of all proxies whether animal, vegetable or mineral; while water molecules consist of oxygen and hydrogen isotopes alone, they are also a solvent for several different forms of carbon. There are of course stable isotopes of other elements encountered in dissolved form, for example boron, chlorine, nitrogen and sulphur, but none of these has yet achieved more than a niche role in the study of proxies. This chapter therefore concentrates exclusively on the well-established trio of O, H and C. Note that the dating of water by carbon-14 or other radio-isotope methods is not included as direct dating of proxies provides much higher resolution. Water dating and allied information is extensively reviewed in books by Clark and Fritz (1997) and Cook and Herczeg (1999).

1

M.J. Leng (ed.), 2005. *Isotopes in Palaeoenvironmental Research.* Springer. Printed in The Netherlands.

Three main reservoirs of water may contribute to the isotopic compositions of proxy indicators: surface waters, soil and unsaturated (vadose) zone moisture, and groundwater. One important aim of studying proxy indicators is to arrive back at the isotopic composition of the antecedent rainfall during the period of growth. While the ways in which the proxies may isotopically fractionate the oxygen and hydrogen in their reservoir water during growth or formation are considered in the following chapters in this volume, it would at least be helpful to establish the typical extent to which the isotopic composition of rainfall may be modified in the three reservoirs at the present day. The largely physical nature of these fractionation effects may vary under different climatic regimes, so the hydrological parts of this chapter will consider the available evidence from a range of climate types worldwide.

Characterising the evolution of the carbon isotope content of waters prior to proxy formation is perhaps a greater challenge, involving as it does both organically and inorganically mediated interactions between water, rock and gases. The almost infinite variety of environmental settings makes it much more difficult to be prescriptive with carbon than with oxygen and hydrogen, but nevertheless we consider it worthwhile to establish the "ground rules" for dissolved carbon. Some of the individual fractionation processes are as climate-dependent as for oxygen and hydrogen isotope ratios, so climatic differences will be considered, albeit in rather less detail than for O and H.

The development of stable isotope hydrology

Research on the technical aspects of O, H and C stable isotope measurement began in the early 1950s (e.g., Epstein and Mayeda (1953); Friedman (1953); Craig (1953)), but widespread hydrological usage concentrating on $\delta^{18}O$ and $\delta^{2}H$ did not arrive until the 1970s. Although studies carried out during the 1960s on rainfall by Craig (1961) and Dansgaard (1964) paved the way for thinking about isotopes in the water cycle, attention was to a large extent diverted by tritium (^{3}H) in the wake of the atmospheric thermonuclear testing of the 1963–65 period. However, as interest in ^{3}H gradually waned, the usefulness of $\delta^{18}O$ and $\delta^{2}H$ began to be appreciated once more, leading to the widespread development of stable isotope laboratories during the 1970s and early 1980s.

It was early realised that isotope methods were not restricted to the study of water and ice, but could also be applied to steam (e.g., Giggenbach (1971)). Furthermore, it became apparent that isotopes could be instructively applied to 'hidden' forms of water — for example soil moisture (e.g., Zimmerman et al. (1967)), water of crystallisation (e.g., Sofer (1978)) and fluid inclusions (e.g., Roedder (1984)).

It also became apparent that groundwaters from the confined parts of aquifers were often isotopically depleted relative to modern waters. During the 1970s, the development of routine methods for measuring radiocarbon (^{14}C) in groundwaters led to a more quantitative understanding of the magnitude and timing of climate change away from polar regions.

Over the last few decades the proliferation of laboratories routinely using O and H isotopes to understand the water cycle, present and past, means that a fair degree of maturity has been reached. While there remains much data to be collected to fill various gaps in knowledge, the subject area is understood well enough (e.g., Clark and Fritz (1997)) to constrain the interpretation of proxy data.

The involvement of C isotopes in hydrology and environmental studies has been more cryptic. While it was known as early as the 1930s that there were differences in $^{13}C/^{12}C$ between organic and inorganic mineral carbon (Nier and Gulbransen 1939), research in the 1950s and 60s was focused on the use of $\delta^{13}C$ for correcting radiocarbon ages of groundwaters for 'dilution' by inorganic carbonate that is dead to radiocarbon (e.g., Vogel and Ehhalt (1963); Pearson and Hanshaw (1970)) and on the study of reaction mechanisms for carbon in plants, for example the C_3 and C_4 photosynthetic pathways (Bender 1968; Deines 1980). Early work on carbonates centred around the use of $\delta^{18}O$ rather than $\delta^{13}C$, and it was not until the mid-1960s that much attention started to be paid to carbon in freshwater fauna and lake deposits (e.g., Keith and Weber (1964); Stuiver (1970)). Likewise, speleothems were initially valued more for their $\delta^{18}O$ and dating potential (summarised in Schwarcz (1986)) and it was only since the work of Hendy (1971) that the significance of variations in $\delta^{13}C$ has led to it being more routinely measured and interpreted. The carbon in travertine-type deposits and other potential freshwater proxies may have derived from a multiplicity of sources so that their $\delta^{13}C$ composition often reveals more about mode of formation than palaeoenvironment (Turi 1986). Nevertheless, interest is developing in the interpretation of carbon isotopes in freshwater tufas and allied cyanobacterial deposits (e.g., Pazdur et al. (1988); Andrews et al. (1993)).

The water and carbon cycles

Figure 1 depicts the global water (or hydrological) cycle. While the concept of the water cycle may be simple in essence — water evaporates from the sea, falls as rain over land, and eventually returns to the sea mainly via river and groundwater discharge — there are a number of complicating factors, at least where stable isotopes are concerned. Much of the focus of this chapter lies in considering how these factors affect the more active parts of the water cycle. The physical processes governing the isotopic composition of precipitation are a natural starting point. Following on from this is a consideration of the transformation of precipitation into the basic 'terrestrial reservoirs' of soil-zone moisture, groundwater, and surface waters. Processes such as seasonal freeze-thaw, plant transpiration and soil evaporation can each cause isotopic changes, but perhaps of most significance is evaporation from surface waters. Lake sediments are an important archive for climate change, and it is therefore vital to understand as far as possible the often complex isotopic relationship between individual lakes and regional precipitation. Accordingly this chapter considers isotope mass-balance modelling in some detail.

The situation with regard to carbon is more complicated, because the C cycle cannot be summarised in a simple climate-related manner. While C stable isotopes are crucial to palaeo-environmental reconstruction, as other chapters in this volume demonstrate (McCarroll and Loader; Hedges et al.; Leng et al.; McDermott et al. all this volume), their incorporation in proxies is affected by many factors, including hydrology, that may be more or less related to climate. For instance, the mass budgets and isotopic compositions of carbon in land animal diet and plant photosynthetic pathways are not

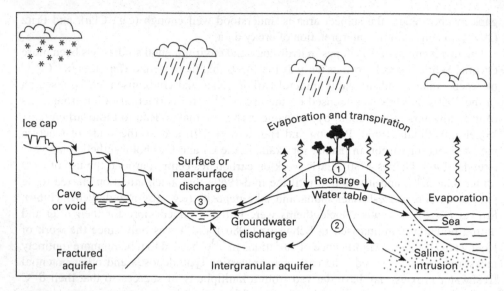

Figure 1. Schematic of the global water cycle. Evaporation of seawater leads to cloud formation with progressive rainout as temperature decreases. The moisture will eventually return to the ocean mainly via surface runoff or subsurface discharge, though it may be delayed by recycling via evaporation from lakes or transpiration from vegetation. Also shown on the figure are the three 'terrestrial reservoirs' of water that are of importance to proxy formation: 1 – soil and unsaturated zone moisture, 2 – groundwater, 3 – surface waters.

controlled simply by the carbon contents of environmental waters. However, many proxies growing in aquatic environments will have carbon isotope compositions that can be related back in various ways to the original water signal.

The underlying aim of this chapter is to demonstrate that changes associated with the water cycle have implications for understanding the climatic information recorded by the O, H and C stable isotope compositions of proxies. Constraints on the isotopic approach are discussed in the final section of the chapter.

Oxygen and hydrogen stable isotopes in precipitation

The global atmosphere provides a dynamic link between the major reservoirs of water on Earth (cf. discussion above). Atmospheric water vapour is the primary vehicle through which this link is accomplished. The mass of water vapour in the contemporary atmosphere is estimated to be around 1.3×10^{16} kg, of which about 85% is located in the lower troposphere, below the altitude of 5 km. The amount of water stored in the atmosphere as water vapour is negligible when compared with the world's oceans or continental ice sheets, the corresponding mass ratios being approximately 10^{-5} and 4×10^{-4}, respectively (Berner and Berner 1987). The atmosphere is only partially saturated with water, with the globally averaged relative humidity of the lower troposphere below the planetary boundary level being around 85%. This lack of saturation stems from the existing thermal structure and large-scale circulation patterns of the troposphere.

Distribution of water vapour in the lower atmosphere is highly inhomogeneous. It is controlled by existing thermal gradients imposing phase changes (evaporation-condensation-freezing, sublimation-resublimation), as well as by the distribution of major water sources on the Earth's surface. In the stratosphere, photochemical reactions such as oxidation of methane constitute the principal mechanism controlling spatial distribution of water vapour. Expressed as a mixing ratio (kg H_2O per kg of dry air), water vapour content decreases from approximately 10^{-3} in the lower troposphere close to the planetary boundary level, to around 3×10^{-6} in the lower stratosphere and then increases again to approximately 10^{-4} at the altitude of 30 km. Close to the Earth's surface, it varies between approximately 1.5×10^{-2} over tropical oceans to 5×10^{-3} over dry regions in the interior of continents and less than 10^{-3} over ice sheets at high latitudes.

The reservoir of atmospheric water vapour is highly dynamic, quickly responding to changes of external conditions such as the flux of solar energy reaching the Earth's surface and/or changes of temperature lapse rate in the troposphere. The mean turnover time is a useful quantity characterising dynamic behaviour of water vapour in the global atmosphere. It is defined as the ratio of the total mass of water in the atmosphere to the flux of water leaving the atmospheric reservoir as precipitation, and is equal to around 10 days.

Although being a trace constituent of the global atmosphere, atmospheric moisture plays a crucial role in the global ecosystem behaviour. Firstly, it is the most important greenhouse gas, responsible for approximately 60% of the natural greenhouse effect. Secondly, atmospheric moisture serves as an important regulator of heat fluxes in the atmosphere, reducing thermal gradients between low and high latitudes. Thirdly, atmospheric water vapour serves as the primary source material from which precipitation is formed.

Isotopic composition of atmospheric water vapour

Interest in the isotopic composition of atmospheric water vapour goes back to the beginning of isotope hydrology. In the early 1960s Craig and Gordon (1965) measured the isotopic composition of atmospheric water vapour over the North Pacific. During the years 1966–67 and 1971–73, Ehhalt (1974) obtained a number of vertical profiles of the ^2H content in tropospheric water vapour up to the tropopause region over several locations in the continental USA. In Europe, the ^2H and ^{18}O content in tropospheric water vapour up to an altitude of 5 km was measured during 1967 and 1968 (Taylor 1972). These two sets of measurements remain to date the major source of information about the vertical distribution of ^2H and ^{18}O in water vapour in the lower troposphere, the region where most precipitation is formed.

A common feature observed was a gradual depletion in heavy isotopes with increasing altitude up to the tropopause region, with a reversed trend detected within the stratosphere. The generalised vertical profile of δ^2H in atmospheric water vapour is shown in Figure 2. The ^2H content decreases gradually with altitude from a δ^2H value of $-150 \pm 50‰$ close to the surface to $-650 \pm 30‰$ in the tropopause region, then increases up to $-500 \pm 50‰$ at about 40 km altitude (Rozanski and Sonntag 1982; Johnson et al. 2001). Changes of $\delta^{18}O$ with altitude reveal similar characteristics.

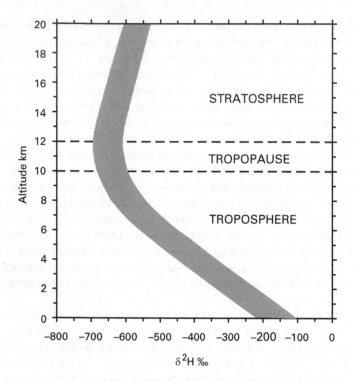

Figure 2. Schematic representation of the vertical distribution of deuterium content in atmospheric water vapour in the troposphere and the lower stratosphere.

The observed distribution of δ^2H and $\delta^{18}O$ with altitude is currently explained as a consequence of the following processes: (i) isotope fractionation (both equilibrium and kinetic), associated with gradual condensation of atmospheric water vapour during cloud formation and subsequent rainout processes, leading to preferential removal of heavy isotopes from the reservoir of atmospheric moisture; (ii) turbulent mixing of different air masses containing water vapour with contrasting isotope compositions; (iii) addition of isotopically heavy moisture, originating in the evapotranspiration process; (iv) lofting (upward advection) and subsequent evaporation of ice crystals; and (v) oxidation of methane. The first two processes are thought to be major mechanisms controlling vertical distribution of δ^2H and $\delta^{18}O$ throughout the troposphere. Transpired moisture modifies the isotopic composition of atmospheric water vapour close to the Earth's surface (Jacob and Sonntag 1991; Bariac et al. 1991). Lofting of ice crystals across the tropopause region was suggested as an important phenomenon influencing ^{2}H and [18]O content in the lowermost stratosphere (Keith 2000). Photochemical oxidation of methane in the middle and upper stratosphere is responsible for the observed increase of water vapour content with altitude as well as the gradual increase of ^{2}H and [18]O content in this region (Johnson et al. 2001).

Systematic studies aimed at characterising the isotopic composition of near-ground water vapour on a regional scale were launched in the early 1980s in Europe. The isotopic composition of daily composite samples of atmospheric moisture was monitored by a network of several stations across the European continent between 1980 and 1984 (Schoch-Fischer et al. 1984). One site (Heidelberg, Germany) has continued to do this up to the present day (Jacob and Sonntag 1991; Agemar 2001) and has accumulated a unique, high-resolution record of the ^{18}O and ^{2}H content of atmospheric water vapour spanning more than two decades. During the 1990s the isotopic composition of atmospheric moisture near the ground was further investigated, although these studies were usually local in character and limited in time. The past several years have seen a renaissance of interest in isotopic studies of near-ground atmospheric moisture, stimulated mostly by the need for better understanding of the isotopic signatures of precipitation and the links between the water and carbon cycles within the soil-plant-atmosphere domain.

Most observations of the isotopic composition of atmospheric water vapour carried out to date have been performed at ground level, with moisture being sampled within the planetary boundary layer. Very large short-term variability of the isotopic composition of near-ground water vapour has been detected; day-to-day variations of ^{2}H and ^{18}O content recorded at mid-latitude continental sites are often of the same magnitude as seasonal changes, and are tightly linked to weather changes, reflecting the passage of air masses with different rainout histories. This variability is particularly pronounced during the winter months. The isotopic composition of atmospheric water vapour recorded near the ground during the summer months reflects a significant contribution of transpired moisture, with the isotopic signature reflecting the mixture of summer and winter precipitation. This has a damping effect on the short-term isotope variability induced by synoptic changes during summer.

Isotopic composition of precipitation
Clearly the reservoir of atmospheric water vapour constitutes the primary source material from which precipitation is formed. The available data suggest that the isotopic composition of monthly precipitation, which is the basic type of data being collected worldwide, properly reflects the mean isotopic status of the atmospheric water vapour reservoir within the planetary boundary layer during these time intervals. Whilst this holds for moderate climates, it may not be the case under semi-arid and arid conditions. In contrast to the limited number of observations relating to the isotopic composition of atmospheric water vapour, a sizeable amount of data on the isotopic composition of precipitation is now available. Over more than four decades precipitation samples have been measured through various research projects and monitoring networks at temporal scales ranging from single minutes (fraction of individual rain event) to monthly composite samples. Among the various initiatives to characterise the isotopic variability in precipitation at global and regional scales, the Global Network for Isotopes in Precipitation (GNIP), jointly operated by the International Atomic Energy Agency (IAEA) and the World Meteorological Organization (WMO), is by far the largest undertaking (http://isohis.iaea.org). The network was launched in 1960 with the primary aim of providing basic information about the isotopic composition of precipitation on the global scale.

Figure 3 summarises the major processes and mechanisms controlling the isotopic composition of atmospheric water vapour and precipitation in the troposphere. They

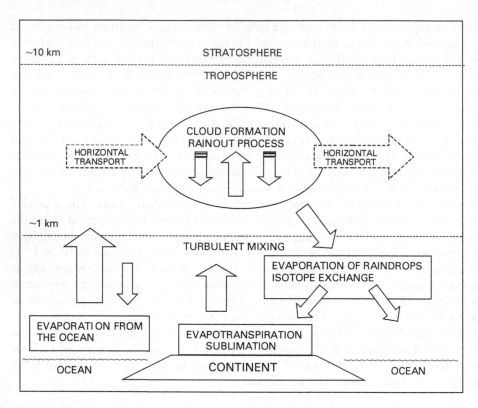

Figure 3. Schematic representation of the processes controlling the isotopic composition of atmospheric water vapour and precipitation.

constitute a physical basis for understanding the observed spatial and temporal variability of heavy isotope ratios in monthly precipitation to which the majority of available data are related.

The first step in the chain of processes depicted in Figure 3 is evaporation from the ocean. Isotopic effects associated with this process are quantified in the framework of a mechanistic model proposed by Craig and Gordon in 1965 (cf. the mass balance modelling section of this chapter). The major controlling parameters are: isotopic composition of the ocean surface, sea-surface temperature, relative humidity of the atmosphere, and wind regime. Water vapour leaving the surface of the ocean is isotopically depleted in comparison with oceanic water.

Marine moisture is transported, both vertically and horizontally, until saturation conditions are reached and cloud formation starts. Partial condensation of water vapour associated with this transport and cloud formation leads to further isotopic depletion of the vapour, even without moisture loss via rainout, since the heavy water molecules $^1H^2H^{16}O$ and $^1H^1H^{18}O$ are preferentially converted into water droplets or ice crystals. In-cloud processes may involve fast advective upward transport with entrainment of moist air from below, phase changes with formation of rain drops and/or snow/hail particles, isotopic exchange between liquid and gaseous phases in the cloud, kinetic isotope effects during snow formation, removal of precipitation elements at the cloud base and dissipation of dry air with isotopically depleted moisture at cloud tops.

The isotopic composition of rainfall collected at the ground surface appears to be close to isotopic equilibrium with near-ground water vapour. This results from the fact that raindrops leaving the cloud base continuously re-equilibrate isotopically with the surrounding moisture on their way to the surface, the degree of this re-equilibration being controlled by the size of the raindrops, the actual height of the cloud base and the relative humidity of the atmosphere beneath the cloud. Solid precipitation (snow, hail) is usually more depleted in heavy isotopes, reflecting equilibrium conditions at much lower in-cloud temperatures. At sufficiently low condensation temperatures, the isotopic composition of snow is influenced by an additional kinetic effect, linked to supersaturation conditions around the forming snow flakes (Jouzel and Merlivat 1984).

The first comprehensive review of isotopic data gathered by the GNIP network, carried out by Dansgaard (1964), resulted in the formulation of a number of empirical relationships between the observed isotopic composition of monthly precipitation and environmental parameters, such as surface air temperature, amount of precipitation, latitude, altitude, or distance from the coast. Subsequent reviews of the GNIP database (Yurtsever and Gat 1981; Rozanski et al. 1993) have largely confirmed the early findings of Dansgaard. It has become apparent that these empirical relationships, often called 'effects', can be viewed as a measure of the average degree of rainout of moist air masses transported from the major vapour source regions (mainly intertropical ocean) to the site of precipitation.

Spatial variability in the isotopic composition of precipitation
The spatial variability in the isotopic composition of precipitation is shown on global maps of $\delta^{18}O$ and δ^2H on continents presented in Figure 4. The maps have been constructed using the GNIP database as well as other available data and drawn using multivariate interpolation techniques (Bowen and Wilkinson 2002; Bowen and Revenaugh 2003; http://es.ucsc.edu/~gbowen/Isomaps.html).

The gradual depletion in the ^2H and ^{18}O content of precipitation from the tropics towards mid- and high-latitudes, seen in Figure 4, can be understood as a result of the progressive removal of water from the moist air masses being transported from the tropical areas towards the poles. Typical $\delta^{18}O$ values in coastal areas of Antarctica are around −15 to −20‰ and close to the South Pole they reach about −50‰ (Ciais et al. 1995). Very negative $\delta^{18}O$ values of around −20 to −26‰ are also found in the northern part of Canada and in eastern Siberia (IAEA 2002; Kurita et al. 2004). The highest $\delta^{18}O$ and δ^2H values are observed in eastern tropical Africa and the Arabian Peninsula.

Precipitation on oceanic islands and coastal, low-elevation continental areas in the intertropical regions shows δ^2H and $\delta^{18}O$ values close to those of the ocean (typical $\delta^{18}O$ values are from −2 to −4‰) (Figure 4). These values usually represent the first portion of condensate from undisturbed marine moisture. More negative δ-values are found in equatorial Southeast Asia and the West Pacific, where high precipitation rates may locally lead to a relatively high degree of rainout of the given air mass induced by displacement of the Intertropical Convergence Zone (Rozanski and Araguas-Araguas 1995; Araguas-Araguas et al. 1998).

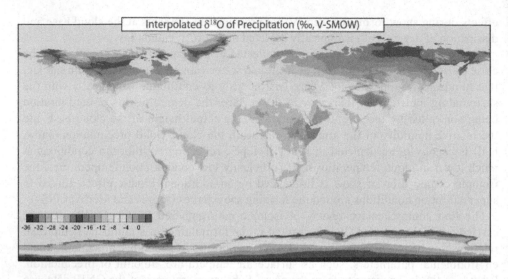

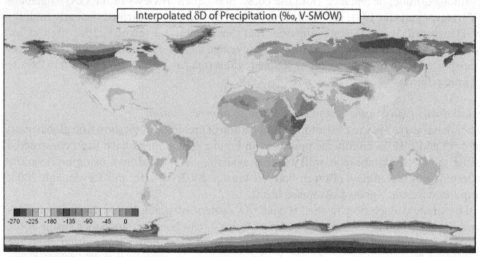

Figure 4. The global maps of $\delta^{18}O$ and δ^2H in precipitation on continents, constructed using the GNIP database as well as other available data and drawn using multivariate interpolation techniques (http://es.ucsc.edu/~gbowen/Isomaps.html).

The lowering of temperature with increasing elevation in mountainous regions usually leads to enhanced condensation and therefore to a progressive depletion in the heavy isotopes of precipitation with altitude, known as the 'altitude effect'. Such an effect is well seen in Figure 4, along the Andes in South America and on the Tibetan Plateau. The vertical isotope gradients in precipitation vary between −0.15 and −0.50‰ per 100 m for ^{18}O and −1 to −4‰ per 100 m for ^2H (e.g., Siegenthaler and Oeschger (1980); Holdsworth et al. (1991)).

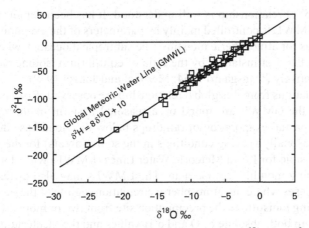

Figure 5. The relationship between long-term annual means of $\delta^{18}O$ and δ^2H in precipitation, derived for stations of the GNIP network.

The 'amount effect' is a well-known feature in oceanic islands and in coastal areas of the tropical regions, where seasonal variations of temperature are minimal. It manifests itself as an apparent correlation between the heavy isotope composition and the amount of rainfall. It can be understood, at least partially, as a consequence of the extent of the rainout process of deep convective clouds producing rainfall in these regions. Prominent examples of this effect can be found in Southeast Asia and intertropical islands in the Pacific (Rozanski et al. 1993; Araguas-Araguas et al. 1998).

In mid-latitudes, a progressive depletion of precipitation in heavy isotopes with increasing distance from the coast has been observed. It is interpreted as a consequence of the gradual rainout of the air masses entering the continents (Salati et al. 1979; Sonntag et al. 1983; Rozanski et al. 1993). The continental gradient in ^{18}O content observed in present-day European precipitation is around $-2‰$ $\delta^{18}O$ per 1000 km (Rozanski et al. 1982; Rozanski et al. 1993). A substantially lower isotope gradient has been found in the Amazon Basin in South America. It has been attributed to intense recycling of moisture within the basin by evapotranspiration (Salati et al. 1979; Gat and Matsui 1991).

The $\delta^2H – \delta^{18}O$ relationship

It is apparent from the isotopic maps shown in Figure 4 that the ^2H and ^{18}O contents of precipitation are closely correlated. This close relationship between δ^2H and $\delta^{18}O$ is observed on both regional and global scales. Craig (1961), in his survey of the ^2H and ^{18}O contents of fresh waters world-wide, suggested the best fit line of his data points:

$$\delta^2H = 8\,\delta^{18}O + 10 \tag{1}$$

be termed the 'Global Meteoric Water Line' (GMWL, also sometimes known simply as the Meteoric Water Line, MWL, or the World Meteoric Line, WML). This relationship is illustrated by Figure 5 which shows the long-term annual mean δ^2H and $\delta^{18}O$ values of precipitation collected at the stations belonging to the GNIP network.

The global δ^2H–$\delta^{18}O$ relationship is well understood. It has been demonstrated that the intercept of the GMWL is controlled mainly by parameters of the evaporation process in major source areas for atmospheric moisture (the subtropical oceans), whereas the slope is determined, in the first instance, by the ratio of equilibrium isotope enrichments for ^2H and ^{18}O respectively (Dansgaard 1964; Merlivat and Jouzel 1979).

Significant deviations from this global relationship are observed (Rozanski et al. 1993). Deviations from the GMWL are found on a seasonal basis in many regions, mainly due to enhanced partial evaporation of raindrops below the cloud base during summer months and/or seasonally varying conditions in the source area(s) for the vapour. These effects are responsible for Local Meteoric Water Lines (LMWL) with slopes lower than 8, constructed using monthly isotope data. The LMWLs may also differ substantially from GMWL in cases where the atmospheric circulation regime of the given area varies seasonally, bringing moisture to the precipitation site from two or more different sources which may differ in both absolute $\delta^{18}O$ and δ^2H values and the 'deuterium excess' value d, which is defined as (Dansgaard 1964):

$$d = \delta^2H - 8\delta^{18}O \qquad (2)$$

Slopes of LMWLs larger than 8 are found in areas where the air masses responsible for the relatively enriched δ-values in rainfall are characterised by high d-values and the isotopically depleted rains are below the GMWL (IAEA 1992).

Temporal variability in the isotopic composition of precipitation
Numerous studies have demonstrated or inferred that the isotopic composition of precipitation may vary on wide temporal scales, ranging from single minutes to hundreds of thousands of years. Isotopic studies of single rain events have revealed that in-storm variations can be relatively large, reaching in some cases 10 to 12‰ in $\delta^{18}O$ (e.g., Ambach et al. (1975); Rindsberger et al. (1990)). The general conclusion of these studies is that the isotopic composition of precipitation sample(s), representing the given rainfall event in whole or part, depends strongly on the meteorological history of the air and specific conditions in which the precipitation is produced, as well as the isotopic composition of moist air through which it falls. Precipitation samples collected on a per-event basis reveal a strong linkage between their isotopic signature and the storm's track, structure and evolution (e.g., Gedzelmann and Lawrence (1990); Smith (1992)).

Regular seasonal variations of $\delta^{18}O$ and δ^2H in precipitation are observed at mid- and high latitudes where seasonal variations of temperature are well developed. These seasonal changes of $\delta^{18}O$ and δ^2H may result from the interplay of several factors: (i) seasonal changes of the temperature at the precipitation sites leading to substantial changes of the total precipitable water in the atmosphere with season, when compared to the source areas; (ii) seasonal changes in the evapotranspiration flux on the continents, amplifying seasonal differences in total precipitable water; (iii) seasonal changes of prevailing circulation patterns changing the origin of moisture. In the tropical regions, where changes in air temperature are minimal, the seasonality in the ^2H and ^{18}O content of precipitation may have its origin in seasonal displacement of the Intertropical Convergence Zone (ITCZ) which controls the movement of moist air masses and induces

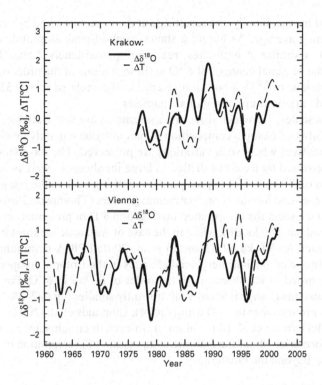

Figure 6. The relationship between long-term trend curves of $\delta^{18}O$ and surface air temperature for two stations in central Europe (Vienna and Krakow), with the seasonal variation component removed (see text for details of this). With a value of approximately 2.5‰, the long-term amplitude in $\delta^{18}O$ is only about one-third of the seasonal variation for both stations. The degree of correlation between the stations suggests regional variations in climate. Based on GNIP data.

rainy period(s) at a particular site. In many cases, for the given amount of rainfall and almost constant monthly temperature, large differences in the isotopic composition of precipitation have been reported, depending on the season. This has been attributed to the differing origins of the moisture that produces precipitation over the course of a year.

Apart from well-established seasonal variations, the isotopic signature of precipitation also exhibits some changes on longer timescales. The available instrumental record of $\delta^{18}O$ and δ^2H in precipitation is still relatively short, reaching several decades for a few sites only. Analysis of these records reveals that $\delta^{18}O$ and δ^2H fluctuate also on a decadal timescale, although amplitudes of these changes are significantly smaller than those observed for shorter timescales. Figure 6 is an illustration of this variability. The trend curves of $\Delta\delta^{18}O$ and ΔT shown have been derived for two continental sites in central Europe (Vienna and Krakow) from the available time series of monthly $\delta^{18}O$ and surface temperature data. First, the seasonal component was removed by calculating a 12-month running average of $\delta^{18}O$ and temperature. Then, the differences $\Delta\delta^{18}O$ and ΔT were calculated for both stations by subtracting the running average curves from the long-term annual means. Finally, the resulting curves were smoothed by again applying a 12-month running average. As Figure 6 shows, peak-to-peak amplitude of long-term changes of $\delta^{18}O$ is similar at both sites, reaching approximately 2.5‰. This may be compared with the seasonal changes of $\delta^{18}O$ at those stations of the order of 7–8‰. The long-term fluctuations of $\delta^{18}O$ at both stations are well correlated ($r^2 = 0.52$) suggesting that the observed variations are of regional character.

On longer timescales, for which direct measurements are not available, information about past variability of isotopic composition of precipitation can only be obtained from environmental archives where such variations are preserved. The most straightforward information is provided by ice cores drilled in large ice sheets as well as in continental glaciers. Records of $\delta^{18}O$ and δ^2H in precipitation reaching back to the last glaciation are now available for several low-latitude continental glaciers (Thompson 2001). In the case of the Greenland ice sheet the information on $\delta^{18}O$ and δ^2H of precipitation goes back to the last interglacial (ca. 240 ka BP), while in the case of Antarctic ice sheet it has recently been extended back to 740 ka BP (Johnsen et al. 2001; EPICA community members 2004). The amplitude of glacial/interglacial changes in the isotopic composition of past precipitation, recorded in ice sheets, is of the order of 6–7‰ in $\delta^{18}O$. For low-altitude continental regions this change is usually significantly smaller (up to 2–3‰), recorded in archives such as groundwater (e.g., Darling (2004); Edmunds et al. (2004)) or carbonate deposits (e.g., McDermott et al. (this volume)). In certain circumstances such archives may provide information about the isotopic composition of precipitation on even longer timescales, going beyond the Quaternary.

Links between climatic parameters and isotopic composition of precipitation

Stable isotope ratios of oxygen and hydrogen in water have long been considered powerful tools in palaeoclimatology. The attractiveness of these tools stems from the fact that strong apparent links exist between some climatically-relevant meteorological parameters, such as surface air temperature or amount of rainfall, and the distribution patterns of ^2H and ^{18}O in precipitation observed for present-day climatic conditions. Based on this link, numerous attempts have been made to reconstruct past climatic changes from records of the isotopic composition of ancient precipitation preserved, often via proxies, in various environmental archives (glacier ice, sediments, groundwater, organic matter, and others). However, quantitative reconstruction of past climatic changes from isotopic records preserved in continental archives requires that the isotope palaeothermometer or isotope palaeopluviometer be adequately calibrated for the timescales of interest.

Most important for palaeoclimatic reconstructions is an apparent link between the isotopic composition of precipitation and surface air temperature (e.g., Fricke and O'Neil (1999)). Dansgaard (1964) presented the empirical relationship between the annual mean values of $\delta^{18}O$ and δ^2H of precipitation and local surface air temperature, derived from the data gathered during the first three years of operation of the GNIP network. The relationship was developed for mid- and high northern latitude coastal stations and is characterised by the isotope–temperature coefficient equal to 0.69‰ per °C for $\delta^{18}O$ and 5.6‰ per °C for δ^2H (Rozanski et al. 1993). Although this relationship was frequently used in the past for isotope-aided palaeoclimatic reconstructions, for the wide range of climates and timescales involved, doubts have often arisen as to whether spatial relations between the isotopic composition of precipitation and climatic variables, derived from present-day conditions, can be used with confidence to interpret isotopic records of past precipitation preserved in various environmental archives.

The available instrumental records of the ^{18}O and ^2H content of precipitation permit the link between the stable isotope signature of precipitation waters and surface air temperature ('δ–T') to be considered in three different ways:

i) A spatial relationship between the long-term (annual) averages of $δ^{18}O$ ($δ^2H$) of precipitation and surface air temperature for different stations. The slope of spatial relationship for $δ^{18}O$ varies from about 0.8–0.9‰ per °C for high latitude areas to virtually zero for tropical regions, where a strong correlation between $δ^{18}O$ ($δ^2H$) and the amount of precipitation is usually observed. For the European continent this slope is equal to 0.58‰ per °C (Rozanski et al. 1992).

ii) A temporal relationship between short-term (seasonal) changes of $δ^{18}O$ ($δ^2H$) of precipitation and surface air temperature for a single station or a group of stations in a given region. For mid-latitude stations of the GNIP network (40°N to 60°N) this relationship has a slope of 0.32‰ per °C for $δ^{18}O$ (Rozanski et al. 1993). For tropical regions this slope is close to zero or even negative (the monsoon-type climate). It has been argued that the δ–T coefficients derived from seasonal cycles of $δ^{18}O$ ($δ^2H$) and temperature for mid- and high-latitude regions might be useful for palaeoclimatic applications because they sample a wide range of different 'climates' (e.g., Siegenthaler and Oeschger (1980); van Ommen and Morgan (1997)).

iii) A temporal relationship between long-term (interannual) changes of $δ^{18}O$ ($δ^2H$) and temperature for a fixed location or a region. Analysis of long-term trends of $δ^{18}O$ and temperature at a number of European stations resulted in an average $δ^{18}O$–T coefficient of 0.63 ± 0.04‰ per °C (Rozanski et al. 1992). For the trend curves shown in Figure 6 the resulting $δ^{18}O$–T slopes are equal to 0.60 and 0.58‰ per °C, for the Vienna and Krakow stations respectively. Validity of the $δ^{18}O$–T coefficient derived from precipitation records has been confirmed for the last 200 years through study of the $δ^{18}O$ of ostracodes preserved in the sediments of Ammersee Lake in Germany (von Grafenstein et al. 1996).

The δ–T coefficients based on interannual changes in $δ^{18}O$ ($δ^2H$) and surface air temperature are probably the most relevant as far as palaeoclimatic applications of stable isotopes are concerned, as they reflect the long-term linkage between the isotopic composition of precipitation and climate for a given area. However, the numerical values of these coefficients have been derived from relatively short instrumental records (several decades) and the question arises to what extent they are valid for longer periods, including major climatic shifts. Moreover, they most probably vary from region to region (Rozanski et al. 1993).

Proper assessment of the temporal changes of long-term δ–T coefficients is a vital problem for calibration of the isotope palaeothermometer (Jouzel 1999; Jouzel et al. 2000). Several processes have been identified which may lead to long-term changes of this relationship over time on a given area: (i) changes in the source region of moisture (e.g., Boyle (1997)); (ii) global or regional changes of the isotopic composition of the ocean (e.g., Shackleton (1987); Werner et al. (2000a)); (iii) changes of atmospheric circulation patterns shifting the sources and routes of air masses precipitating over the given site (Charles et al. 1994); (iv) changes in seasonality of precipitation (Werner et al. 2000b).

Some insight into temporal variations in the δ–T relationship which occur on timescales exceeding instrumental records has been gained through studies of ice cores. The $\delta^{18}O$ time series available for ice cores drilled in central Greenland has been compared with the measured borehole temperatures. This comparison in the case of the GRIP ice core yielded $\delta^{18}O$–T gradients of around 0.6‰ per °C for the Holocene and only 0.23‰ per °C for the last glacial transition (Johnsen et al. 1995). Somewhat different values were obtained for the GISP2 ice core: from 0.25 to 0.46‰ per °C for the Holocene (the higher values being for recent centuries) and 0.33‰ per °C for the last glacial transition (Cuffey et al. 1995). It has been argued that, unlike for Greenland, present-day spatial isotope gradients can be used as a surrogate of the temporal gradients to interpret glacial–interglacial isotopic changes at sites such as Vostok and Dome C in central Antarctica (Jouzel et al. 2003).

Long-term changes of temporal δ–T relationships can be addressed also through studies of groundwater systems. While the isotopic composition of groundwater yields information about the isotopic composition of past precipitation, the concentration of noble gases provides information about recharge temperature which is directly linked to the mean surface air temperature of the given area (e.g., Bath et al. (1979); Andrews and Lee (1979)). Although there are some issues regarding the dispersion of isotopic signals in groundwater (e.g., Stute and Schlosser (2000)), the groundwater archives allow the assessment of variability in the stable isotope composition of precipitation as a function of temperature from the perspective of major climatic changes during the last glacial–interglacial cycle.

The link between the isotopic composition of precipitation and changes in climate on timescales exceeding direct observations can also be addressed by General Circulation Models (GCMs). They can be employed to simulate isotopic water cycles ($^1H^2H^{16}O$, $^1H^1H^{18}O$) on different spatial and temporal scales, including periods of major climatic shifts (e.g., Jouzel et al. (1991, 1994); Hoffmann et al. (1998)). The simulated changes in the global or regional isotope fields in precipitation can be compared with predicted changes of climate (temperature). The GCM approach is being continuously refined and has been applied successfully in a number of studies aimed at better understanding of the isotopic records preserved in continental archives. However, the available GCMs still lack adequate resolution and computing power to provide continuous reconstructions of changes in isotopic water cycles across timescales addressed in the framework of palaeoclimatic reconstructions.

Outlook

It is apparent from the above discussion that the calibration of isotope palaeothermometers for terrestrial environments remains to be fully achieved. It appears that long-term δ–T gradients derived from analysis of instrumental records of the isotopic composition of precipitation and temperature can be used, with care, to interpret isotopic records preserved in environmental archives located at mid- and high latitudes, under a Holocene climate. In the absence of sufficiently long instrumental records, spatial relationships derived for a given area can be used as a first-order surrogate for temporal isotope gradients. However, this approach may not be generally valid when the interpreted isotope records cover periods of major climatic shifts, such as the last deglaciation period.

From precipitation to terrestrial water

The purpose of this section is to cover the fate of O and H isotopes in water in the three terrestrial reservoirs described in connection with Figure 1 (with the exception of lakes, which are considered in detail in the next section of this chapter). The necessity for looking at transformation effects on isotopes is clear from the previous section on precipitation, where it was shown that an alteration of 1‰ in $\delta^{18}O$ can under certain conditions represent a temperature change of as much as 3°C, a significant amount in palaeoclimatic terms.

Processes affecting infiltration

It is necessary for the discussion that follows to distinguish between infiltration (water which enters the soil) and recharge (water which eventually reaches the water table of an aquifer) as they are not usually synonymous, owing to the processes described below.

Soil physics
Under temperate climatic conditions, soil moisture is at a saturated state in winter but not in summer, leading to the concept of 'soil moisture deficit' (SMD). Where there is an SMD, rainfall may be absorbed at the soil surface but is unable to move down against pore pressure, and is therefore liable to be removed by evaporation or transpiration. Allied to this is the concept of the 'zero flux plane' (ZFP), the notional division between water percolating downwards and water in liquid or vapour form travelling upwards. When the SMD is fully satisfied, the ZFP will be at ground surface. In contrast, towards the end of a dry summer the ZFP could be some meters below the surface.

Moisture in the layers of soil above the ZFP is most prone to isotopic effects because of vaporisation. Under temperate climates it may be difficult to use stable isotope ratios to distinguish evaporative enrichment from transient storage of individual rainfall events in the upper soil layers, though where both O and H isotope data are available the position of the data on a δ-plot may provide a resolution. It is also worth bearing in mind that different soil-water collection methods in the unsaturated zone may be sampling "relatively more and less mobile components of soil water that may have different histories and pathways" (Landon et al. 1999).

Effects of vegetation cover
Even before infiltration can take place, interception by leaves can lead to a certain amount of re-evaporation and hence isotopic fractionation (the 'canopy effect': Gat and Tzur (1967); DeWalle and Swistock (1994)). Re-evaporation is not the only effect of trees; branches and trunks can be very effective at conducting rainfall as 'stem flow' via root channels to the underlying unsaturated zone without a prolonged residence in the shallow soil zone where evaporation could affect it (Dawson 1993). While it is likely that much

of this infiltration will be recycled by transpiration from the tree itself, the moisture may subsequently be re-distributed in the soil by 'siphoning' (Smith et al. 1999), possibly leading to complex isotopic changes in the soil moisture profile (Adar et al. 1995).

Although it may vary with the type of precipitation from event to event, the magnitude of canopy interception effects rarely seems to be extreme. The amount of evaporative enrichment in throughfall, where directly measured, appears remarkably consistent under different climatic conditions: 0.3‰ in $\delta^{18}O$ for pine forest in southern Sweden (Saxena 1986) and Pennsylvania (DeWalle and Swistock 1994), 0.5‰ in $\delta^{18}O$ for chestnut, oak and cork in the south of France (Pichon et al. 1996), and of the same order for tropical montane rainforest in Ecuador (Goller et al. 2005).

Indirectly-measured effects inferred from soil water measurements are of the same order. Brodersen et al. (2000) showed that trees in the Black Forest cause enrichments of just under 0.4‰ in $\delta^{18}O$, irrespective of whether they are coniferous or deciduous, while Darling (2004) reported a case where 'ventilation' (position of a tree at the forest edge) apparently contributed to a 5‰ increase in δ^2H compared to soil moisture elsewhere in a wood in southern England, equivalent to perhaps 0.7‰ in $\delta^{18}O$.

A corollary of the isotopically-enriching vegetation canopy effect is that tree or scrub clearance should introduce a corresponding isotope depletion in soil moisture compositions. This was found to be the case in semi-arid South Australia, where Allison et al. (1985) measured soil profiles beneath dunes with and without native vegetation. Midwood et al. (1993) noted a similar effect in southern Texas, but occurring in reverse order as grassland was replaced by thorn woodland.

All forms of vegetation, whether plants or trees, must of course transpire to survive. The balance of evidence suggests that under normal growth conditions the process of transpiration is to a great extent non-fractionating (Zimmerman et al. 1967; Allison et al. 1984), so it might be thought to be of little importance in terms of isotopic modification of the infiltration which will become recharge. However, where rainfall has a strong seasonal variation in isotopic composition, transpiration might be subtracting a component of potential recharge. This would result in a divergence between the composition of recharge and transpiration water, neither of which will be fully representative of bulk rainfall. This needs to be considered in the interpretation of isotopic proxies such as δ^2H in tree rings (Epstein et al. 1976; White et al. 1985), and $\delta^{18}O$ in leaves (Iacumin and Longinelli 2002).

Evaporation from the soil
In arid or semi-arid conditions vegetation may be sufficiently sparse that interception and transpiration are only very minor facets of the water balance. However, evaporation from the soil surface layers then emerges as an important process in the isotopic modification of any rainfall which manages to infiltrate (e.g., Dincer et al. (1974)). Obviously, the warmer and drier the climate the more evaporation, and therefore fractionation, will occur. The depth to which soil moisture is significantly affected is rarely more than 1–2 m below ground level (Barnes and Allison 1988), but within that range can give rise to extreme evaporative enrichment. These enrichments will usually show some seasonal changes as a result of the moisture balance. Figure 7 shows an example of this from sequential soil profiles taken in the Negev desert of Israel (Sonntag et al. 1985).

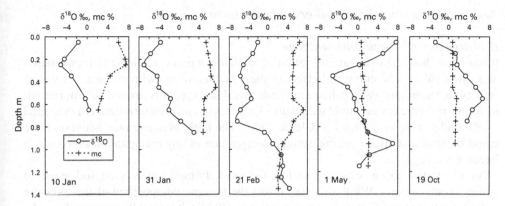

Figure 7. Sequential soil profiles of $\delta^{18}O$ and soil moisture content (mc) taken in the Negev desert of Israel. The profiles in January and February show infiltration from relatively-depleted winter rainfall moving downwards in the profile. By early summer, evaporation is driving the residual moisture towards isotopic enrichment. By the autumn, infiltration is beginning again. Data from Sonntag et al. (1985).

The profiles in January and February show infiltration from relatively-depleted winter rainfall moving downwards in the profile, with the ZFP not far from the surface. By early summer, however, the dry conditions and increasing temperature mean that the ZFP is moving down, while above it evaporation is driving the residual moisture towards isotopic enrichment. Similar seasonal changes were observed by Liu et al. (1995) in the desert of southern Arizona. There must therefore be a large margin of uncertainty over the bulk isotope input to proxies relying on soil moisture under arid conditions. This appears to apply as much to clay soils as sands, despite their higher moisture contents (e.g., Mathieu and Bariac (1996)).

Some desert areas are zones of groundwater discharge rather than recharge. If the outflow of water is great enough, these will be manifested as lake or wetland oases and the usual surface-water evaporation dynamics will apply (see next section). However, where outflows are low the production of 'dry' lakes is common. Cores taken in the lake beds show the same extreme evaporative enrichment in near-surface moisture as found in recharge zones, but below this there is a progressive depletion until the isotopic value of the underlying groundwater is reached (e.g., Allison and Barnes (1985)). Because there is no infiltration from rainfall events, there are no significant seasonal fluctuations in isotope values at any particular level, though seasonal changes in atmospheric humidity etc. may cause some variation in the maximum isotope enrichment.

In cases of both recharge and discharge, soil moisture from the shallowest depths in the soil zone (say <25 mm) may actually show major isotope depletion (Barnes and Allison 1988). The reason for this is that any soil moisture at such depths consists largely of condensed water vapour, which must be depleted to preserve isotope mass balance with the enriched water immediately below from which it originated.

Aquifer recharge and the rainfall-groundwater isotope balance

Recharge through the unsaturated zone
It has been shown above that infiltration of rainfall is a process subject to transpiration and evaporation. The combined effect of these factors varies with climate and soil or rock type, but simple water balance methods reveal that typically around 30% of rainfall becomes recharge under humid temperate climates, 10% in Mediterranean climates, and 0–5% under more arid climates (Gleick 2000). The high 'wastage rate' suggests there could be profound effects on the isotopic composition of any precipitation surviving to become recharge.

As illustrated above, evaporation has the most dramatic effects on soil-moisture isotope compositions. While this may affect the isotopic composition of the moisture before it is transpired by plants (e.g., Brunel et al. (1997)), it is equally apparent from the low moisture contents involved that this soil water cannot be an important factor in the overall water balance: either the rainfall event was too small to satisfy the SMD, or it was large enough to provide recharge and the isotopically enriched soil moisture is merely the remnant of percolating water. Transpiration has a greater but more insidious role in modifying recharge, in that it may remove part of the annual rainfall before mixing can occur. If the precipitation has a high seasonal variability in isotopic composition (see for example Rozanski et al. (1993)), then there is the possibility of a missing component in the isotopic balance of recharge. This may not necessarily mean summer growing-season rainfall; it depends on the depth from which roots are extracting water. For example, Robertson et al. (2001) found the best correlation for the $\delta^{18}O$ values of cellulose from oak trees in East Anglia (UK) was with rainfall of the previous December/January period, when vegetation is dormant.

However, despite evaporation and more significantly transpiration, studies have shown that the isotopic modification during the precipitation–recharge process is generally small, within perhaps ~0.5‰ in $\delta^{18}O$ of the bulk rainfall. This is almost irrespective of climate: for example it applies in many parts of North America (Yonge et al. 1985; Clark and Fritz 1997; Kendall and Coplen 2001), across Europe (Bath 1983; Eichinger et al. 1984; Deak et al. 1996; Gehrels et al. 1998; Kortelainen and Karhu 2004), Jordan (Bajjali et al. 1997), India (Krishnamurthy and Bhattacharya 1991), West Africa (Mathieu and Bariac 1996), East Africa (Nkotagu 1996), and Australia (Sharma and Hughes 1985).

Recharge by surface waters
The results of the above studies suggest that it is actually quite difficult to cause much isotopic bias or selection in rainfall-derived recharge. Mass-balance considerations demand a sizeable input of energy to cause the necessary fractionation, but it appears there is simply not enough of this energy (solar or other) available during normal areal infiltration to cause much modification via evapotranspiration, whatever the terrain or climate. Solar energy is however much more available to surface waters, and it is among the more slow-moving of these that the greatest evaporative effects are seen, usually in lakes (see next section). While terminal (endorheic) lakes in discharge areas will normally be the most affected, lakes in recharge areas are not immune and may pass on evidence of their isotopic signatures to the groundwater. In more temperate areas the

effects may persist for only a few km (Krabbenhoft et al. 1994) before they are damped by areal rainfall recharge, but in warmer, more arid areas the effects may be evident for tens of km and involve isotopic enrichments of many permil ($\delta^{18}O$) or tens of permil ($\delta^{2}H$) compared to unmodified groundwater in the vicinity (Darling et al. 1996).

It is of course possible for river as well as lake waters to infiltrate the subsurface, most easily via swallow holes in karstic terrain (Plummer et al. 1998; Perrin et al. 2003), though generally this will cause relatively little isotopic modification to the underlying aquifer. The main exception to this would be rivers flowing through arid areas, which tend to lose water by infiltration in addition to evaporation. Examples include the Colorado (Payne et al. 1979), the Niger (Fontes et al. 1991), and perhaps most-studied, the Nile (Verhagen et al. (1991); Geirnaert and Laeven (1992); see also Figure 8). These effects are seen to good advantage in the Gezira region of Sudan, immediately to the south of the confluence of the Blue and White Niles at Khartoum. The Niles have different isotopic signatures owing to the different geographical origin of their waters, and this enabled Farah et al. (2000) to detect river infiltration up to 12 km from the rivers, with the White Nile being the greater contributor.

However, rivers do not necessarily lose or gain water over any particular reach for the whole of a hydrological year. Seasonal changes in hydraulic gradient mean that groundwaters near a river can show quite large changes in isotopic composition due to reversals in flow direction. This is particularly the case with large rivers having major fluctuations in flow rate due to snowmelt: for example, Navada and Rao (1991) found seasonal changes of up to 2.5‰ in $\delta^{18}O$ for groundwaters within a few km of the Ganges.

In arid areas it is common to find lenticular aquifers underlying wadis, which are recharged during flood flow. Depending on how long the waters were exposed to the atmosphere, there may be greater or lesser amounts of evaporative modification (Darling et al. 1987; Clark and Fritz 1997). There is also the possibility of evaporation through the unsaturated zone (Aranyossy et al. 1992; Clark and Fritz 1997).

Mixing with seawater
In coastal areas the possibility exists for mixing between fresh and marine waters, with obvious implications for isotopic compositions. Where this is simple mixing of groundwater and seawater, often resulting from intensive groundwater abstraction or eustatic change (Rao et al. 1987; Edmunds et al. 2001), isotope values cannot be more enriched than approximately 0‰, the mean ocean water value for both $\delta^{18}O$ and $\delta^{2}H$. Often, however, the situation will be more complex as groundwaters, rainfall and seawater variously interact in lagoonal and sabkha settings, with evaporation frequently giving rise to positive isotope values (Gat and Levy 1978; Manzano et al. 2001; Duane et al. 2004).

Layered aquifers and stratification
Depending on geological conditions, it is possible to have two or more aquifer units underlying an area or region, each of which can have a different isotope composition owing to combinations of factors such as differences in mode of recharge, catchment altitude and residence time (e.g., Brown and Taylor (1974); Gonfiantini et al. (1974)).

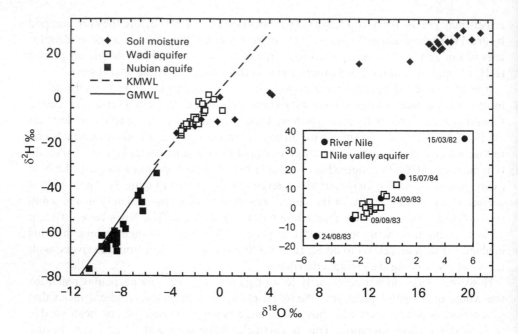

Figure 8. Large contrasts in the O and H stable isotope composition of different reservoirs of water are characteristic of arid regions. This δ-plot, based on a study carried out in the Butana region of northern Sudan, demonstrates the effects of some of the processes outlined elsewhere in this chapter. The observed isotope ranges are large: more than 30‰ in $\delta^{18}O$ and over 100‰ in δ^2H. The data in the figure can be considered in terms of the three terrestrial reservoirs between which water is distributed. *Soil moisture* (diamonds) shows extreme enrichment along the slope of ~2 typical of soil evaporation. Deeper unsaturated zone moisture is included in this category, and trends back towards the shallow groundwater composition. There can be very little direct rainfall recharge under such an arid climate, and the possibility exists that this trend is due to evaporative discharge driven by the humidity gradient. *Groundwater* (squares) in the main part of the figure divides into two populations: firstly the wadi aquifer, actively being recharged (via occasional flood events) by modern rainfall lying along the Khartoum meteoric water line (KMWL), and secondly the regional aquifer in the underlying Nubian formation, which radiocarbon measurements show was predominantly recharged during early Holocene pluvial episodes, and which tends to conform to the Global Meteoric Water Line (GMWL). For each aquifer there is evidence of isotopic modification. For example, some wells in the wadi aquifer yield water showing signs of evaporation. Whether this occurred before or after recharge is not clear; most sampling points were large-diameter dug wells where in-situ evaporation might have occurred. In the regional aquifer, the less-depleted groundwaters are evidence for mixing between old and modern waters. *Surface water* (circles, inset) is shown in the form of the Nile, sampled seasonally at Shendi. The large range in composition is due to the interplay between the strong evaporation prevailing throughout the year in Sudan, and the depleted water flowing down from the Ethiopian highlands during the annual Blue Nile flood in the late summer. Water from the Nile seeps into the shallow aquifer, where it may mix with locally-recharged groundwater, giving rise to the range of compositions measured. Data from Darling et al. (1987).

Fracturing may then lead to waters from different aquifer units mixing and emerging at the surface in springs with isotopic contents at variance with local meteoric recharge.

Age-related layering within single aquifer units has been recognised in a number of studies (e.g., Edmunds and Smedley (2000); Chen et al. (2004)), and where the age profile spans the Holocene-Pleistocene boundary there is scope for significant isotope gradients: for example, Deak et al. (1993) found a 3‰ change in $\delta^{18}O$ between the surface and a depth of 160 m in an aquifer in southeast Hungary. Natural spring discharge from such aquifers may represent various degrees of mixing.

Interception of recharge by caves and tunnels
As discussed above, once infiltration has passed through the soil into the unsaturated zone it normally becomes recharge to the underlying aquifer. However, in karstic or fractured-rock areas there is always the possibility that percolating water will find its way into cave systems or artificial excavations, giving rise to speleothem features. These have been used as proxies in two ways: by interpreting the $\delta^{18}O$ (and $\delta^{13}C$) in deposited calcite (e.g., Schwarcz (1986)), and by attempting to measure more directly the $\delta^{18}O$ and δ^2H in the original dripwater via fluid inclusions (e.g., Ayalon et al. (1998)). Both these aspects are covered in detail in McDermott et al. (this volume); the comments here are concerned with the hydrology of the dripwaters.

Two potential difficulties arise with dripwaters. Firstly, there may be seasonal selection of recharge; secondly, the altitude of recharge may not be known with certainty. With regard to the first, karst is not generally as effective as most other lithologies in mixing percolating waters. Seasonal changes or selection effects have been observed in dripwaters of the Villars cave (southwestern France) by Genty et al. (2002), in the B7 cave (western Germany) by Niggemann et al. (2003), and in Reed's Cave (South Dakota, USA) by Serefiddin et al. (2004). However, such effects are not inevitable under humid temperate climates: for example, Williams and Fowler (2002) found no evidence for either in a cave at Waitomo in New Zealand.

Warmer, drier conditions might be expected to affect isotope values in dripwaters. The complexities of karst unsaturated zone behaviour in semi-arid terrain have been demonstrated in relation to the much-studied Soreq cave of Israel (e.g., Ayalon et al. (1998); Kaufman et al. (2003)). Two types of infiltration water were identified: fast flow (residence time of weeks) discharging from fractures in the cave roof during heavy rainfall events in the wet season, and slow flow in the form of speleothem dripwater throughout the year (residence time tens of years). Generally the fast-flow waters were enriched 0.3–1.0‰ in $\delta^{18}O$ compared to the bulk rainfall on the land overlying the cave. Dripwaters tended to vary seasonally by about 1.5‰ in $\delta^{18}O$, and were themselves slightly enriched relative to the fast-flow waters; the seasonal variation has been preserved despite the lengthy percolation time. There was however little evidence of evaporative enrichment, so different selection processes for the fast and slow waters were proposed. Another Mediterranean locality showed contrasting behaviour: at the Clamouse cave in the south of France, dripwaters showed a remarkable consistency in their isotopic composition over time (Genty et al. 2002), with values similar to modern groundwater in the area (Dever et al. 2001) and to the weighted mean rainfall at the nearest GNIP station (Barcelona), therefore showing no evidence of selection.

Depending on the topography of the overlying land surface, altitude effects may play a part in affecting the interpretation of dripwaters. For example, a comprehensive study of water inflows to tunnels in Alpine crystalline massifs found that $\delta^{18}O$ values tended to be more negative than expected from the predicted catchment areas (Marechal and Etcheverry 2003). This was attributed to geological structure directing recharge from higher altitudes into the tunnel fracture systems. However, it was also noted that recharge from rivers and lakes might cause local perturbations to the overall pattern. A study on another deep tunnel within granite in the Central Alps concluded that isotopically depleted winter precipitation and glacial meltwaters were important contributors to the local groundwater flow system (Ofterdinger et al. 2004).

In some cases speleothem deposits form in caves that are associated with the discharge rather than the recharge end of a groundwater system, for example the Devil's Hole vein calcite deposit in Nevada, USA (Winograd et al. 1992; Quade 2004). In these circumstances, not only is the usual uncertainty over recharge area compounded, but residence time and the possibility of mixing between groundwaters of different provenance also have to be considered.

Fluid inclusions and water of crystallisation

Fluid inclusions have already been referred to above in connection with speleothems. However, many other origins are possible: for example during the formation of evaporites (e.g., Knauth and Roberts (1991)), or in high-temperature hydrothermal systems (e.g., Polya et al. (2000)). Typically their δ^2H and sometimes $\delta^{18}O$ compositions are used to infer hydrological conditions during periods well beyond the Quaternary, but diagenetic or other complicating factors can lead to rather imprecise conclusions about the composition of contemporaneous meteoric waters.

There tends to be a mineral-dependent fractionation between water of crystallisation and free water, but where the magnitude of this is known it is possible to draw broad conclusions about groundwater fluxes and timing from isotopic measurements of the crystallisation water. Gypsum in particular has been used for this purpose (Halas and Krouse 1982; Bath et al. 1987).

Runoff, discharge and effects on streams and rivers

Consideration of the water cycle diagram shown in Figure 1 reveals several different discharge routes for any precipitation escaping initial evapotranspiration. While some groundwater discharges directly to the sea, water also finds its way into surface watercourses via overland flow or after residence in the subsurface. In practice, overland flow of rainfall is relatively rare, relying as it does on highly impermeable conditions at the ground surface. Under most climatic and geological conditions, a whole continuum of flowpath lengths is likely to contribute to streams and rivers.

From precipitation to river water

It has been mentioned in the second section of this chapter that rainfall can show large variations in its isotopic composition even between individual events, let alone seasonally. Yet infiltration appears to be very effective at mixing what will become recharge (see above). Therefore only that fraction of rainfall which flows rapidly to a river can conceivably have a much different isotope composition. In practical terms this is not usually important. In the permeable lowland catchments of minor rivers there is little variation in isotopic composition throughout the year (e.g., Lawler (1987); Darling et al. (2003)). Even streams in relatively impermeable, 'flashy' upland catchments often fail to show much isotopic response to rainfall, whether under temperate climate (Sklash and Farvolden 1979; McDonnell et al. 1991; Durand et al. 1993) or tropical rainforest conditions (Goller et al. 2005).

Unlike for the more humid climates referred to above, arid-zone runoff can vary considerably in its composition (up to 2‰ $\delta^{18}O$) on a scale of hours (Adar et al. 1998) to days (Lange et al. 2003) because of the lack of damping in the catchment. However, the general impersistence of this wadi-type flow means that it is not on the whole important in the growth of proxies.

A major factor affecting the isotopic composition of water reaching streams and rivers in regions subject to cold winters is snowmelt. Where there is a significant accumulation of snow during the winter, a pulse of isotopically-depleted water can be detected arriving in the upper parts of catchments during the spring (Sklash and Farvolden 1979; Herrmann and Stichler 1980; Laudon et al 2004). In smaller river systems, depletions due to the snowmelt pulse are generally within the range 2 to 4‰ in $\delta^{18}O$ (Bottomley et al. 1986; Schuerch et al. 2003). In larger rivers, depletions are smaller and it can take months for the snowmelt pulse to pass down the river (Mook 1970; Rank et al. 1998) so that proxies may be affected during their growing season (e.g., Ricken et al. (2003)). There are however limits to the propagation of such pulses in recognisable form in continental-scale catchments (e.g., Winston and Criss (2003)).

The effects of evaporation on river and stream waters are difficult to distinguish from seasonal changes, particularly when a number of upstream tributaries are involved. Given that residence time and surface area are important factors in evaporation, its effect on river waters would not be expected to be as profound as on lake waters (see next section). This appears to be the case: for example, after de-convoluting the effects of tributaries, Simpson and Herczeg (1991) calculated that the maximum enrichment of water in the Murray basin (Australia) was about 5‰ in δ^2H, which would equate to 1‰ in $\delta^{18}O$ assuming a surface-water evaporation slope of 5. However, they also calculated that irrigation return water contributed an approximately equal enrichment, so anthropogenic factors can have important modifying effects. Where water is delayed in natural or artificial lakes, scope for further isotope enrichment exists: Payne et al. (1979) calculated that damming the Colorado River had resulted in evaporative enrichment of some 2.5‰ in $\delta^{18}O$. In very large river systems there may be rather complex changes as tributaries fed by snowmelt interact with lowland tributaries subject to evaporation (e.g., Kendall and Coplen (2001); Winston and Criss (2003)). However, large changes in isotope composition are not exclusively caused by snowmelt: strongly seasonal rainfall can have a similar effect, as the example of the Nile shows (Figure 8).

Mixing of river waters
An additional complication in understanding isotopic compositions in major rivers is the mixing – or lack of it – occurring when major tributaries join the main river. Differences

in morphology, flow characteristics etc. mean that the magnitude of incomplete mixing is difficult to predict. In cases that have been investigated, the downstream distance before full mixing ranged from 50 km for the Main-Rhine confluence at Mainz (Fritz 1981), to at least 120 km for the Amazon tributaries Solimoes and Negro at Manaus (Matsui 1976), and 300 km for the Liard-Mackenzie confluence in northern Canada (Krouse and McKay 1971).

Springs
It may seem slightly perverse to include springs towards the end rather than the beginning of a discussion on discharge, but in volumetric terms they are likely to be unimportant contributors to river flow, at least as far as visible manifestations are concerned. (Admittedly there is no simple definition of what constitutes a spring; much river baseflow will be the result of groundwater discharge in the river bed, whether in visible form or not.) It is apparent from evidence discussed earlier in this section that most groundwater is fairly well mixed during percolation through the unsaturated zone and/or during residence in the aquifer, and therefore little variation in spring isotope composition would be expected. The major exceptions are some springs or resurgences found in karstic terrain. These sometimes consist of little more than surface water that has spent only a brief period underground, and therefore tend to show stream-like amplitudes of isotopic variation, typically 1–2‰ in $\delta^{18}O$ (Flora and Longinelli 1989; Winston and Criss 2004). However, not all karst springs show this amplitude of variation; in particular the higher-volume outlets are often damped by mixing and dilution (Stewart and Downes 1982; Nativ et al. 1999).

In nearly all cases springs will be discharging waters in the age range of a few years to a few thousand years, simply because longer residence times are unlikely to occur in actively circulating flow systems. Such Holocene waters will generally carry the same isotope signatures as present-day recharge. By contrast in the late Pleistocene, beyond 10,000 years before present, the colder climate resulted in groundwaters generally having isotopic signatures depleted by anything from 1 to 10‰ in $\delta^{18}O$ depending on location (e.g., Darling (2004); Edmunds et al. (2004)). The vast majority of these data come from borehole samples, and not all the aquifers are still being recharged and therefore actively circulating. Nevertheless 'palaeowater' springs have been identified, in major aquifer systems like the Nubian Sandstone of North Africa (Sturchio et al. 1996; Sultan et al. 1997) and the Great Artesian Basin of Australia (Mudd 2000; Love et al. 2000).

There are equally rare cases where spring waters contributing to streams and rivers may have been isotopically modified in the subsurface. A prime example would be discharges of thermal water which may be enriched in $\delta^{18}O$ owing to mineral exchange at high (>150°C) temperatures (the so-called 'oxygen shift') after passing through geothermal systems (e.g., Fontes and Zuppi (1976)). Superimposed on this shift may be some evaporative enrichment due to steam separation at the surface (Giggenbach et al. 1983). However, high-temperature springs are not normally important in volumetric terms, often discharging into non-thermal rivers which dilute away any unusual isotopic characteristics. There may however be local effects within those rivers, for example as observed in the Firehole River of Yellowstone National Park (Woodward et al. 2000).

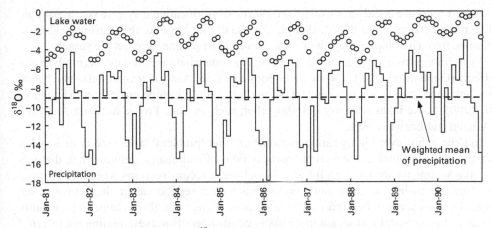

Figure 9. A decade-long record of $\delta^{18}O$ in the Neusiedlersee, a steppe lake situated on the Pannonian plain in eastern Austria, showing the sizeable displacement between local precipitation and lake water isotope compositions caused by evaporative enrichment. After Rank et al. (1984), with further data courtesy of D. Rank.

Wetlands

No discussion of isotopes in discharge would be complete without a mention of wetlands: these can be important habitats for proxy species (White et al. 2001; Lambs and Berthelot 2002; Menot-Combes et al. 2002). However, it is difficult to be prescriptive about isotopic effects since wetlands vary greatly in their hydrology — indeed not all wetlands are discharge areas. There is often a complex balance between groundwater, river water and rainfall, and stable isotopes can at least be useful in unravelling it (Hunt et al. 1998).

Lake waters and mass balance modelling

There is an almost infinite variety in the form and size of lakes studied for proxy purposes, from upland tarns to the great lakes of tropical Africa and North America. Hydrological context also varies, the balance between volume, surface area and residence time revealing itself as a continuum between simple inflow-outflow systems and true terminal lakes. The most profound isotope effects in lakes are due to the enrichment produced by evaporation. Typically the residual waters lie along a slope of 5 on a δ-plot. The precise gradient of the evaporation slope in a particular area is largely a function of local humidity, while the amount of evaporation determines the amount of displacement along the slope (Gonfiantini 1986).

Lake water compositions may change fairly predictably on a seasonal basis, particularly in temperate or boreal regions, or much less predictably in semi-arid areas subjected to periodic droughts and floods. The magnitude of these changes typically ranges from 1–2‰ in $\delta^{18}O$ in temperate regions, up to 4–5‰ where snowmelt is involved, and >10‰ for some Mediterranean and tropical lakes. Even at their most depleted, lake waters are often several permil enriched over local bulk rainfall or river inputs (an example is shown in Figure 9).

Lake waters are not always well mixed. Vertical stratification may give rise to isotopic differences (Gonfiantini 1986), though not usually as large as the seasonal changes. Plumes from river inputs may show an isotopic contrast depending on the season.

Evaporation from very saline lakes is actually somewhat inhibited by thermodynamic effects (Gonfiantini 1986). Thus a water body like Lake Asal, a terminal lake well below sea level and mainly fed by seawater, shows rather little isotope enrichment (Fontes et al. 1980). Some lakes may dry out altogether, such as Lake Frome in South Australia (Allison and Barnes 1985).

In general however highly saline, periodic or other 'problem' lakes tend to be avoided for the study of proxies: the isotopic compositions of carbonates, cellulose and diatoms in lake sediments are most useful to palaeoclimatic reconstructions where they record fluctuations in lake water isotope composition on a regular rather than event basis. Smaller lakes seem to be preferred to large ones (Leng et al. this volume), presumably because hydrological factors are more easily constrained. However, relating the inferred lake water isotope values back to the original rainfall composition is not necessarily straightforward, as earlier discussion suggests. The application of isotope modelling techniques can help to resolve this.

Isotopic tracing of evaporation losses to the atmosphere

For an evaporating water body, isotopic fractionation occurs at the air–water interface owing to slightly lower vapour pressures and attenuated rates of molecular diffusion in air of $^1H^1H^{18}O$ and $^1H^2H^{16}O$ with respect to the common, light isotope species $^1H^1H^{16}O$. This fractionation typically gives rise to evaporating moisture which is isotopically lighter (depleted in ^{18}O or ^2H) relative to the remaining liquid, an effect which has been exploited to study evaporation by several approaches, including: (i) tracing of evaporate admixtures to the atmosphere based on changes in the isotopic signature of downwind precipitation (e.g., Gat (2000)); (ii) measurement of isotopic gradients in the boundary layer above evaporating surfaces (e.g., Wang and Yakir (2000)); and, (iii) tracing of heavy-isotope buildup which occurs in the residual liquid, i.e., in water bodies (e.g., Gonfiantini (1986); Gat (1995)). The latter approach has by far the most common application owing to the ease with which liquid water can be sampled in the natural environment. In this case, isotope buildup is applied in conjunction with a mass-balance model to account for atmospheric and water balance controls on the isotopic composition of lake water. The high preservation potential of isotopic enrichment signals (mainly $\delta^{18}O$) contained within inorganic and organic fractions of lake sediments, namely bulk carbonate, molluscs, ostracodes, aquatic cellulose and biogenic silica, has also facilitated palaeohydrological and palaeoclimatic reconstructions using this technique (Talbot 1990; Talbot and Kelts 1990; Edwards 1993; Holmes 1996; Li and Ku 1997; von Grafenstein et al. 1999; Ito 2001; Wolfe et al. 2001; Schwalb 2003; Leng et al. this volume). An overview of the fractionation mechanisms and isotopic balance is provided below.

Isotopic fractionation during evaporation is practically described as the product of fractionations occurring by exchange between water molecules in liquid and vapour (in thermodynamic equilibrium), and diffusion of water molecules from the liquid to vapour phase (i.e., kinetic or transport fractionation).

For the case of a water body in thermodynamic equilibrium with the atmosphere, when there is no humidity gradient, the equilibrium fractionation factor (α^*) between liquid and vapour can be represented as:

$$\alpha^*_{\ liquid\text{-}vapour} = \frac{R_L}{R_V} \tag{3}$$

where R_L and R_V are the ratios of heavy isotope species to the common isotope species in liquid and vapour, respectively where $R = \frac{^{18}O}{^{16}O}$ or $\frac{^2H}{^1H}$, and $\alpha^* > 1$ such that the rare, heavy isotope species is more abundant in the liquid phase. Isotopic differences between coexisting phases are conventionally discussed in terms of isotope separation factors. In this case the equilibrium isotope separation between liquid and vapour (ε^*) is given by:

$$\varepsilon^* \ (‰) = 1000 \,(\alpha^* - 1) \approx 1000 \ln \alpha^* \approx \delta_L - \delta_V \tag{4}$$

and δ_L and δ_V are the δ-notation equivalents of R_L and R_V respectively.

Numerous laboratory measurements of ε^* have been conducted for oxygen and hydrogen over wide range of temperatures such that their values are reasonably well-constrained for use in water balance studies. Currently, the equilibrium separations are commonly evaluated using the empirical, experimental equations proposed by Horita and Wesolowski (1994):

$$\varepsilon^*_{18} \approx 10^3 \cdot \ln \alpha^*(^{18}O) = \ -7.685 + 6.7123 \,(10^3 /T)$$
$$-1.6664 \,(10^6 /T^2) + 0.35041 \,(10^9 /T^3) \ (‰) \tag{5}$$

$$\varepsilon^*_2 \approx 10^3 \cdot \ln \alpha^*(^2H) = 1158.8 \,(T^3 /10^9) - 1620.1 \,(T^2/10^6)$$
$$+ 794.84 \,(T /10^3) - 161.04 + 2.9992 \,(10^9 /T^3) \ (‰) \tag{6}$$

where T is the air–water interface temperature (K). These equations yield values only slightly different from earlier laboratory experiments, including Majoube (1971) and Bottinga and Craig (1969), in the range of temperatures expected for average evaporation conditions (0 to 35°C). For example, one standard deviation of at 10°C is about 0.2 and 2.0 for oxygen and hydrogen, respectively, which is similar in magnitude to analytical uncertainty. Notably, ε^* values are solely temperature-dependent, and therefore are substantially better-defined than the analogous kinetic isotope separations ε_K (described below).

For the case of evaporation into undersaturated air, the isotopic fractionation typically exceeds equilibrium values due to kinetic effects. The total isotope fractionation α in this case is:

$$\alpha = \alpha^* \, \alpha_K \tag{7}$$

where α_K is the kinetic fractionation. Alternatively, this can be expressed as total isotope separation given by:

$$\varepsilon = \varepsilon^* + \varepsilon_K \tag{8}$$

where ε_K, the kinetic isotope separation, (and also α_K) are controlled by the evaporation mechanism.

Craig and Gordon (1965) first proposed the use of a 1-D linear resistance model to describe transport of the isotopic species through the boundary layer during constant evaporation, i.e., constant vertical flux with no convergence or divergence in the air column. As summarised in Gat (1996) the model depicts transport of vapour through a series of sublayers, including in order: a saturated sub-layer above the air/water interface, where relative humidity $h = 1$ and the isotopic ratio in the saturated sub-layer according to Equation (3) is $R_V = R_L/\alpha^*$; a boundary layer consisting of a diffusive sublayer and turbulently mixed sublayer in which transport occurs by diffusion and turbulent transfer respectively; and finally, the free-atmosphere where $h = q_A/q_s$, q_A and q_s being the specific humidity in the saturated sublayer and the free atmosphere, respectively. Normally the surface sublayer humidity is assumed to be 1 so that $h = h_A$.

The Craig and Gordon model predicts the net evaporating moisture contribution to the free atmosphere to be:

$$\delta_E(‰) = \frac{\alpha^{*\,-1}(\delta_L) - h\delta_A - \varepsilon}{1 - h + \varepsilon_K \cdot 10^{-3}} \tag{9}$$

where δ_E, δ_L, and δ_A are the isotopic compositions of the evaporate, the liquid (assumed to be well-mixed), and the free atmosphere, respectively (Gat 1996).

The kinetic isotope separation depends on both the boundary layer conditions and the humidity deficit according to:

$$\varepsilon_K = \theta C_K (1 - h) \tag{10}$$

where C_K is the kinetic fractionation factor, θ is a mixing/advection index, and $(1-h)$ is the atmospheric humidity deficit. For water balance applications ε_K has often been evaluated with the assumption that $\theta = 1$, using climate or micrometeorological data to define h, and using C_K values from wind tunnel experiments (e.g., Vogt (1976)). Typically C_K is specified to be in the range of 14 to 15‰ for oxygen and 12 to 13‰ for hydrogen, representative of natural open-water bodies under fully-turbulent exchange conditions (Gonfiantini 1986, Araguas-Araguas et al. 2000). While experimental evidence has suggested that C_K can vary depending on the evaporation mechanism, increasing under smooth to static transport conditions, these are not typical of open-water bodies in a continental setting. However, two-fold higher C_K values are typical of evaporation through the unsaturated zone, a process which appears to closely mimic static transport (Allison et al. 1983). The value of ε_K for open water bodies is therefore strongly humidity dependent, varying between 2.9 and 7.2‰ for oxygen and 2.5 and 6.3‰ for hydrogen, over the relative humidity range of between 80 and 50%.

The mixing/advection parameter $\theta = (1-h')/(1-h)$, has been introduced for the special case where the evaporation flux contributes to humidity buildup in the free air, as observed for the atmosphere on the leeward side of large water bodies (see also Gat (1995)). The adjusted humidity h' is the humidity of the atmosphere following the admixture of evaporate. In principle, θ also can be used to describe any situation where humidity and sampling of vapour for isotopic analysis is conducted within the turbulently mixed sublayer, rather than in the free atmosphere. The net impact of humidity buildup from an evaporating water body is to limit the overall fractionation via reduction in ε_K and increase in δ_E, as well as the over-riding influence of reduced evaporation.

Where the concentration of dissolved solids in lake water is sufficient to affect the thermodynamic activity of water, and where ion hydration and crystallisation may lead to isotopic fractionation within the water body, it is necessary to use isotope activity ratios rather than isotopic composition and to normalise the humidity according to the thermodynamic activity of water. In general, activity of water is reduced substantially only at high concentrations of dissolved solids, as in the case of brines: for example the thermodynamic activity of seawater is only 0.98. Salinity effects may therefore be a factor in surface water studies only if using water samples or isotopic proxies from inland seas and sabkhas. Although coupling of the salinity with isotopic changes is a significant complication, use of archives such as brine inclusions in halite have been successfully applied to reconstruct palaeoclimate in high-salinity systems (Horita 1990).

These basic fractionation mechanisms, incorporated within an isotope mass balance context, are useful for prediction and interpretation of the isotopic behaviour of modern water bodies and palaeohydrological archives. An overview follows of isotope mass balance models for simple lake systems.

The isotope mass balance of lake water budgets and processes

For a shallow, well-mixed lake during the ice-free period (Figure 10), the water and isotope balance may be written respectively as:

$$dV/dt = I - Q - E \qquad (11)$$

$$\frac{V d\delta_L + \delta_L dV}{dt} = I\delta_I - Q\delta_Q - E\delta_E \qquad (12)$$

where V is the volume of the lake, t is time, dV is the change in volume over time dt, I is instantaneous inflow where $I = I_F + I_G + P$, I_F is surface inflow, I_G is groundwater inflow, and P is precipitation on the lake surface; Q is instantaneous outflow where $Q = Q_R + Q_G$, Q_R is surface outflow and Q_G is the groundwater outflow; E is evaporation; and δ_L, δ_I, δ_Q and δ_E are the isotopic compositions of the lake, inflow, outflow, and evaporative flux respectively (Figure 10a). Note that δ_Q and its subcomponents, δ_R and δ_G will be similar and approximately equal to δ_L in well-mixed lakes.

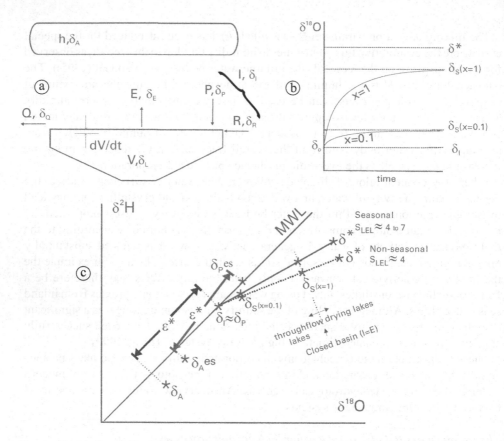

Figure 10. (a) Generalised lake balance schematic showing storage and hydrological fluxes and their isotopic compositions for a simple, well-mixed lake. Note that I is total inflow, P is precipitation, R is surface and subsurface runoff; E is evaporation, Q is the isotopic composition of outflow, V is lake volume, dV/dt is change in lake storage, and h is ambient atmospheric humidity. Note that δ values refer to the isotopic compositions of the respective components. δ_A and δ_L denote the isotopic compositions of ambient atmospheric moisture and lake water, respectively. (b) Time series plot of $\delta^{18}O$ enrichment for two lakes in the same climatic region but with differing values of x (evaporation/inflow). The scenario assumes an initial value for δ_0 that is isotopically depleted, typical of precipitation sources. δ^* is the limiting isotopic enrichment under local atmospheric conditions, δ_S is the steady-state isotope composition. (c) Plot of $\delta^{18}O$ vs δ^2H depicting major controls on isotopic enrichment in lakes. A seasonal and non-seasonal scenario both assume equilibrium between the atmosphere and precipitation. The seasonal scenario is distinct in that isotopic signatures of inflow δ_I and precipitation during the evaporation season (δ_P^{es}) are significantly offset due to the seasonality in evaporation rates, which enhances the slope of the local evaporation lines (S_{LEL}). Other influences that create disequilibrium between inflow and atmosphere could similarly alter the S_{LEL} from around 4. MWL = meteoric water line.

Substitution of the Craig and Gordon model (9) for δ_E into (12) yields:

$$V\frac{d\delta_L}{dt} + \delta_L\frac{dV}{dt} = I\delta_I - Q\delta_L - \frac{E}{1 - h + \varepsilon_K \cdot 10^{-3}}({\alpha^*}^{-1}\delta_L - h\delta_A - \varepsilon) \qquad (13)$$

which can be further simplified if the lake maintains a near-constant volume ($dV / dt = 0$):

$$\frac{d\delta_L}{dt} = - [(1 + mx)\delta_L - \delta_I - x\delta^*](I/V) \tag{14}$$

where $x = E/I$ is the fraction of lake water lost by evaporation, and $1 - x = Q/I$ is the fraction of water lost to liquid outflows. Here, δ_0 is the initial isotope composition of the reservoir, m is the temporal enrichment slope (Figure 10b; see also Welhan and Fritz (1977); Allison and Leaney (1982)), defined as:

$$m = \frac{h - 10^{-3} \cdot \varepsilon}{1 - h + 10^{-3} \cdot \varepsilon_K} \tag{15}$$

and δ^* is the limiting isotope enrichment (Gat and Levy 1978; Gat 1981) given by:

$$\delta^* = \frac{h\delta_A + \varepsilon}{h - \varepsilon \cdot 10^{-3}} \ (\text{‰}) \tag{16}$$

Integration with respect to δ_L, holding all other isotope and water balance parameters constant, yields (Gonfiantini 1986):

$$\delta_L = \delta_S - (\delta_S - \delta_0)\exp[-(1 + mx) \ (It/V)] \tag{17}$$

where the steady-state isotope composition approached by the reservoir over time is:

$$\delta_S = (\delta_I + mx\delta^*) / (1 + mx) \tag{18}$$

In the case where well-mixed lakes are not subjected to pronounced seasonal shifts in water budget, or when seasonal fluctuations converge on a relatively stable interannual balance point, a lake can be described as attaining long-term isotopic ($\delta_L \approx \delta_S$) and hydrological steady-state (i.e., $dV/dt = 0$). The isotopic composition of inflow, especially for headwater lakes is expected to be close to the isotopic composition of mean annual precipitation δ_P, plotting close to the intersection of the MWL and LEL (Local Evaporation Line) (Figure 10c). In this situation, net displacement of δ_L from the MWL along the LEL is controlled mainly by the fraction of water loss by evaporation:

$$x = \frac{E}{I} = \frac{m(\delta_L - \delta_I)}{(\delta^* - \delta_L)} \tag{19}$$

In the case of small lakes with pronounced isotope changes over one annual cycle, we can define an 'interannual steady-state' which may be reached under constant year-to-year water balance conditions. For interannual steady-state composition in cold climates, rates of evaporation E and inflow I should be defined based on the annual average values, rather than values based upon observations during the thaw season. Archives may selectively record isotopic signatures during summer which can bias the record, typically towards higher evaporation loss, especially in boreal regions.

Volume, residence effects and stratification

Although lake volume does not exert a direct influence on the isotopic composition of lake water in systems which are close to isotopic and hydrological steady state, it does play a role in determining the temporal weighting of the isotopic signals that develop in lake water and in related isotope archives. The temporal footprint of δ_L and hence δ_P is fundamentally dependent on the residence time τ of water in a lake:

$$\tau = \frac{V}{I} \ or \ x \cdot \frac{V}{E} \ (yr) \tag{20}$$

which suggests that the isotopic composition of long-residence lakes, typically higher-volume lakes, is likely to be weighted over longer time intervals as compared to short-residence lakes, and for this reason display smoother records of changes. Lakes of smaller volume will also tend to be more susceptible to short-term perturbations related to flood-drought cycles and other events. These effects may lead to significant changes in the lake volume which can complicate interpretation of the δ_L record. The rapidity of readjustment of isotopic composition following perturbation events is also dependent on the residence time of isotopes in the lake (see Horita (1990)).

In stratified lakes, or lakes with pronounced horizontal inhomogeneities, it may be necessary to account separately for epilimnion and hypolimnion volumes and exchanges, provided these have distinct isotope compositions. In modern systems, incomplete mixing can be characterised by spatial and temporal sampling to bracket potential errors to any desired level of precision, although this is not always practical. A simple approximation for systems with similar epilimnion and hypolimnion compositions is to use an average value to represent the undifferentiated lake volume. Neglecting stratification can lead to overestimation of the importance of evaporation loss if epilimnion sampling is conducted during dry, stratified periods, and underestimation of evaporation loss if sampling is conducted during wet, stratified periods. From the perspective of palaeohydrological archives, isotopic records obtained from pelagic and benthic organisms, for example, may exhibit differing isotope signatures which may be useful for examining stratification history. On the other hand, use of a single source archive may also lead to systematic bias in the palaeohydrological record. Inflow bypass or short-circuiting of the system may also reduce the effective volume of the lake during wet periods or reduce the effective input in the opposite situation. Incomplete mixing within the lake itself is also a potential source of error when applying the isotope mass-balance approach to large lakes, although such uncertainty may be reduced by using multiple records of isotopic changes in different parts of the lake. While to a large degree these propositions remain untested, some of the complexities have been considered in theoretical terms by Lewis (1979).

Catchment runoff

One important application of lake isotope balance modelling, as suggested by Gibson et al. (2002), is the tracing of long-term runoff or water yield from the catchment area. Assuming that runoff from the catchment is equal to inflow to the lake minus precipitation on the lake, then the runoff from the drainage basin (Q_{DBA}) can be calculated as:

$$Q_{DBA} = \left\{ \frac{E_L}{x} - PL \right\} \frac{LA}{DBA} \tag{21}$$

where E_L (*mm*) is the lake evaporation rate, P_L (*mm*) is the precipitation rate on the lake, and *LA/DBA* is the ratio of the lake area to the drainage basin area which contributes to the lake. This simplified model assumes a headwater lake setting whereby the isotope enrichment signal is produced entirely by lake evaporation in the water body in question, and not inherited from upstream water sources. In principle, this approach may be useful for examining historical changes in runoff, but this remains to be thoroughly tested. In more complex systems, such as non-headwater lakes or string-of-lakes drainage networks, more detailed models are required to account for multiple sources and feedbacks (see also Gat and Bowser (1991); Gibson and Edwards (2002)). Overall, it is important to note that the $\delta^{18}O$ or δ^2H signals preserved in the palaeohydrological record will depend significantly on both climatic conditions and the hydrology of the lake and watershed. A continuum of settings ranging from climatically-dominated (high throughflow, low *x*) to hydrologically-dominated (closed basin, x ≈ 1) is expected based on theoretical considerations, implying differential suitability of lakes for climatic reconstruction (Figure 10c). Note that drying lakes (x>1) where inflow takes place once or sporadically, often will not preserve a long-term continuous sedimentary record and are therefore not expected to be well-suited to regional palaeohydrological reconstructions. Nevertheless, more gradual drying or filling of lakes on climatic timescales (decades and longer) may be represented using volume-variant versions of Equation (13) or step-wise hydrologic steady state models (i.e., Equation (15)).

Practical considerations for modern and palaeoclimate applications

For quantitative isotope balance applications, model parameterisation is normally achieved by using local climate to define humidity and temperature and, in turn, applying standard temperature- and humidity-dependent equations (i.e., Equations (5), (6) and (10)) to estimate the equilibrium and kinetic fractionations under the assumption that exchange during evaporation is fully turbulent and advection is minor. For modern studies, precipitation input is normally defined using the GNIP database, other local precipitation or groundwater database, or indirectly from proxy data or by inferring the input values from the intersection of the local evaporation line and meteoric water line.

The most difficult parameter to measure or estimate is the isotopic composition of atmospheric moisture. The two common approaches for defining δ_A include evaporation-flux-weighting (see Gibson (2002)) and the best-fit approach, whereby δ_A is constrained to fall on a precipitation-vapour tie line in δ-space (see Figure 10c), and to match the slope of the local evaporation line given by:

$$S_{LEL} = \left(\delta_A - \delta_P + \varepsilon/h\right)_2 / \left(\delta_A - \delta_P + \varepsilon/h\right)_{18} \tag{22}$$

Note that the subscripts 18 and 2 refer to ^{18}O and ^2H, respectively. The latter approach, which explicitly combines the two tracers, has been found to yield similar results to evaporation-flux-weighting for northern Canada, and adequately explains regional trends in the slopes of local evaporation lines (Gibson et al. 2005). Palaeoarchives often do not have sufficient information to explicitly resolve all components of the isotopic balance, but modern analogues have helped to develop more reasonable and realistic scenarios for interpreting past conditions. For example, the application of isotope mass balance to assess regional variations in water balance across the northern treeline in Canada has provided new insight into modern water-balance variability in boreal, tundra and transition areas (Figure 11), as well as better context for understanding the possible range of conditions preserved in palaeoclimatic and palaeohydrological archives.

Dissolved carbon

Unlike oxygen isotopes in water, which represent the solvent itself and therefore have an invariably dominant influence in the oxygen mass balance of an aqueous environmental system, carbon isotopes represent a solute that is (or was) present in relatively minor quantities. Carbon may be present in solutions in two basic forms: inorganic and organic, in which there are many different species and compounds. Isotopic compositions reflect their differing sources and subsequent transformations. Typically, dissolved inorganic carbon (DIC) is present in terrestrial environmental waters (streams, lakes, groundwaters) at concentrations up to 10 millimoles per litre. Dissolved organic carbon (DOC) including dissolved volatiles, principally methane, as well as complex organic molecules, is typically present at concentrations below 1–2 millimoles per litre but may be much higher in certain environments such as peaty lakes and groundwaters. These low budgets of carbon in terrestrial aqueous systems, relative to the dominance of the aqueous oxygen budget, dictate that $^{13}C/^{12}C$ is more susceptible to external perturbations at the time of formation or deposition of proxy media. Interpretation of $^{13}C/^{12}C$ isotope signals must also consider the possibility of post-depositional exchange that would compromise the validity of climate or environmental information.

This section describes: (i) the $\delta^{13}C$ compositions of the various inorganic and organic sources of carbon that are in contact with terrestrial waters; (ii) the interactions of these C sources with water and thus the processes that relate the $\delta^{13}C$ of waters to the $\delta^{13}C$ of sources; (iii) the $\delta^{13}C$ compositions of DIC which is the dominant or partial precursor for most of the carbon-bearing proxy media such as carbonates, plant material and biogenic solids; and (iv) the $\delta^{13}C$ of DOC that may have contributed in various ways to the preserved proxy media.

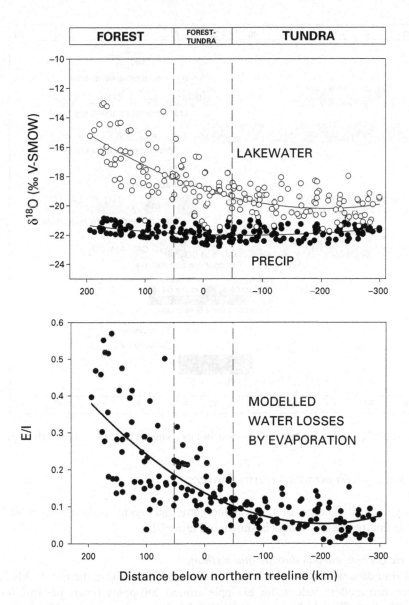

Figure 11. Gradients observed across the northern treeline in central Canada including (a) measured $\delta^{18}O$ in lake water and interpolated $\delta^{18}O$ in precipitation based on the Canadian Network for Isotopes in Precipitation database, and (b) modelled watershed losses by evaporation (expressed as evaporation/inflow ratios (E/I %) based on steady state $\delta^{18}O$ mass balance (Equation 19). Note that gradients in the isotopic composition of lake water are larger than variations in precipitation which is attributed to higher evaporation losses at lower latitudes. Gibson and Edwards (2002) applied this dataset to quantify the frequency distribution of water budgets in boreal forest, transitional and tundra regions. Evaporative enrichment signals ($\delta_L-\delta_P$) were also used in a subset of lake basins to partition evaporation and transpiration losses.

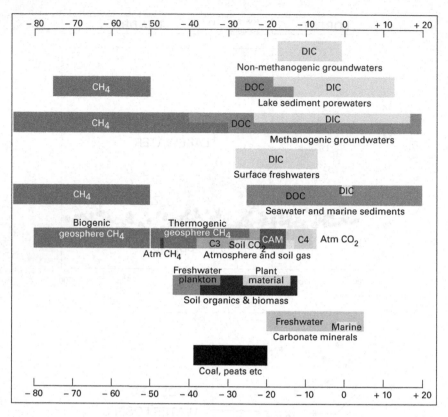

Figure 12. The $\delta^{13}C$ ranges for dissolved carbon and its sources (scale in permil with respect to VPDB).

Carbon sources in contact with terrestrial waters

A summary of the $\delta^{13}C$ ranges of potential inorganic and organic carbon sources for DIC and DOC in environmental waters is shown in Figure 12.

Atmospheric carbon: carbon dioxide and methane

Historical abundances of atmospheric carbon dioxide (CO_2) within the last 20 Ma have been lower than modern values: for example around 200 ppmv (parts per million by volume) at the Last Glacial Maximum (LGM) and 280 ppmv in the pre-industrial period (IPCC, 2003). Since the industrial revolution, atmospheric CO_2 has increased to 380 ppmv in 2004 (NOAA data, see http://www.cmdl.noaa.gov/ccgg/insitu.html), rising by 20% in the past 50 years. The $\delta^{13}C$ of atmospheric CO_2 was around −6.7‰ when first measured in the mid-1950s, and had decreased to −7.2‰ in 1979 and −8‰ in 1995 because fossil fuel burning adds isotopically light C (Keeling et al. 1979; 1980; 1995). This historical shift in $\delta^{13}C$ is correlated with a decrease in atmospheric radiocarbon (^{14}C), also reflecting the addition of fossil C (the 'Suess effect'; Keeling (1979)). Short-term seasonal fluctuations of $\delta^{13}C$ are superimposed on the long-term anthropogenic shift due to variations in the rates of CO_2 uptake and release by the biosphere (Keeling et al. 1995).

Methane (CH_4) in the atmosphere has also increased over time, from around 390 ppbv (parts per billion by volume) at the LGM to 650 ppbv in pre-industrial times and 1750 ppbv today (NOAA data, see http://www.cmdl.noaa.gov/ccgg/insitu.html). The isotopic composition is around −47‰ (Tyler 1986; Cicerone and Oremland 1988). However the atmospheric abundance is so low that, for dissolved methane to be significant for the content and isotopic composition of carbon in terrestrial waters, there must be an additional dominant terrestrial source of CH_4 such as reductive degradation of organic material.

Carbonate minerals
Carbonate minerals are in many cases the most important contributor, directly or indirectly, to DIC in terrestrial waters. The $^{13}C/^{12}C$ ratios of carbonate minerals are determined by their origin and possible post-depositional alteration. However the likely dominance of carbonate-carbon in most environments means that significant alteration of $^{13}C/^{12}C$ occurs only in extreme environments or with prolonged exchange.

Marine carbonates have $\delta^{13}C$ values of ~0‰, with a range from −2 to +3‰ depending on age, depositional environment and mineral phase. Non-marine carbonates have $\delta^{13}C$ values reflecting closely the isotopic compositions of waters (lake waters, groundwaters, streams) from which they precipitated. They may be biogenic carbonates in lake sediments, diagenetic intergranular or vein carbonates in rock formations, evaporitic deposits in arid playas, speleothems, or tufas in streams in limestone terrains. Typically, $\delta^{13}C$ values in these environments are in the range −20 to +5‰, though values outside this range are quite possible in special cases representing extremes of, for example, biological, evaporitic, hydrothermal or cryogenic activity in the hydrological environment (e.g., Stiller et al. (1985)). It may be noted that, in addition to carbon origin and temperature, pH of the depositional environment is also a factor in determining the $\delta^{13}C$ of freshwater carbonates relative to dissolved carbonate in source water because of the pH-dependence of dissolved carbonate speciation and thus of isotopic fractionation (Clark and Fritz 1997).

Organic matter including hydrocarbons
Plant material has $\delta^{13}C$ values ranging from about −38 to −8‰, dividing into narrower ranges for each of the photosynthetic pathways: −38 to −22‰ for C_3 plants, −15 to −8‰ for C_4 plants, and −30 to −13‰ for CAM plants (Deines 1980; Yeh and Wang 2001). The C_3 pathway ('Calvin cycle') dominates in terrestrial ecosystems, especially in temperate and high-latitude regions and typically produces plant material with $\delta^{13}C$ values of −27‰. The C_4 pathway ('Hatch-Slack cycle') is prominent in tropical and savannah grasses and typically produces plant material with values of −13‰. The CAM (Crassulacean acid metabolism) cycle occurs mostly in succulents in arid ecosystems and essentially involves diurnal switching between C_3 and C_4 pathways leading to intermediate $\delta^{13}C$ values.

Coals, peats and intermediate deposits have $\delta^{13}C$ values between −39 and −20‰ which are within the range found in modern-day plant material (see above). Organic material in non-marine sediments, derived from ancient plant material, also has a similar range of compositions. For example, organic carbon in sediments in a freshwater lake in New York State was measured at $-27\pm1‰$ and depth-dependent breakdown of this material by methanogenesis followed by methane oxidation was interpreted to be responsible for variations in the $\delta^{13}C$ of DIC from -16 to $+6‰$ at different locations in the water and shallow (<10cm) sediment profile (Herczeg 1988). Fossil organic carbon in marine sediments has a narrower range of $\delta^{13}C$ from -27 to $-19‰$ (Hoefs 1973; Deines 1980). Liquid hydrocarbons, i.e., bulk oils, have $\delta^{13}C$ values in the approximate range -32 to $-21‰$, mostly around $-28‰$ (Deines 1980). Volatile hydrocarbons, principally methane, have a wide range of $\delta^{13}C$ values from -80 to $-25‰$ reflecting the diversity of sources and pathways of formation (Deines 1980). Essentially these divide into biogenic methane in shallow geosphere environments and young sediments (-80 to $-50‰$ $\delta^{13}C$) and thermogenic methane associated directly or indirectly with oils or coals at greater depths (-50 to $-25‰$ $\delta^{13}C$) (Schoell 1980; 1984).

Interactions of carbon sources with water

Geochemical reactions and exchanges
Carbon dioxide in a gaseous phase dissolves in water according to Henry's Law partitioning, i.e., $C_{soln} = K_H P_{CO2}$, where K_H is the Henry's Law constant and P_{CO2} is the partial pressure of CO_2. Dissolved CO_2 then becomes part of the system in which DIC is distributed among pH-dependent species HCO_3^- and CO_3^{2-}:

$$CO_2 \text{ (g)} \leftrightarrow CO_2 \text{ (aq)}$$

$$CO_2 \text{ (aq)} + H_2O \leftrightarrow HCO_3^- + H^+$$

$$HCO_3^- \leftrightarrow CO_3^{2-} + H^+$$

In general, it can be assumed that chemical exchange between the DIC species is rapid relative to the rate of dissolution and exchange between gaseous and dissolved CO_2. Exchange between gaseous CO_2 and dissolved CO_2 may be significantly slower, for example at the air–water interface of the ocean, a lake or a stream (e.g., Zhang et al. (1995)). The rate and spatial scale of equilibration may be controlled by rates of advection or diffusion of CO_2 towards the interface and/or across the interface. Thus a perturbed or 'rough' interface may exchange CO_2 more efficiently than a calm, undisturbed interface.

Reaction between the DIC system in solution and solid carbonate phases normally occurs until carbonate dissolution-precipitation equilibrium is achieved, for example with respect to calcite:

$$CaCO_3 \leftrightarrow Ca^{2+} + CO_3^{2-}$$

The forward reaction, calcite dissolution, is dominated under normal environmental conditions by a reaction involving dissolved carbon dioxide:

$$CaCO_3 + CO_2 + H_2O \rightarrow Ca^{2+} + 2HCO_3^-$$

whilst the reverse reaction, calcite precipitation, involves loss of CO_2. The stoichiometry of these reactions in the DIC system is the background for understanding the consequent $^{13}C/^{12}C$ fractionations (Deines et al. 1974; Plummer 1977).

Geochemical equilibrium and reaction models calculate DIC speciation and simulate carbon isotope ratios in DIC based on isotopic fractionation and mass balance in relation to mineral dissolution and secondary mineral precipitation (Reardon and Fritz 1978; Wigley et al. 1978; Fontes and Garnier 1979). Although these were aimed primarily at using $\delta^{13}C$ data to estimate the extent of radiocarbon dilution by ^{14}C-dead, carbonate-derived C, they provide fairly rigorous numerical descriptions of the various pathways for evolution of DIC, taking account of equilibrium or non-equilibrium and open-system or closed-system behaviours. The widely-used geochemical modelling program PHREEQCI uses carbon isotope mass balance as a constraint on inverse modelling of water compositions (Parkhurst and Appelo 1999).

Interactions between water and CH_4, or other organic C sources, vary from simple dissolution to oxidation to CO_2. DIC in groundwaters in unconfined aquifers in glacial sediments in Canada has occasional anomalously high $\delta^{13}C$ values because of a small but significant contribution of CO_2 produced by microbial fermentation of organic material to CH_4 and CO_2 (Fritz et al. 1978; Barker and Fritz 1981). Dissolved CH_4 may have this type of provenance or be the product of microbial reduction of DIC. Higher organic compounds in water are closely related to their parent solid organics and thus are similar is isotopic composition.

Isotopic fractionations
Knowledge of isotopic fractionations is essential for relating $\delta^{13}C$ in a proxy indicator of palaeoclimate or palaeoenvironment to the precursor carbon source. Values of $\delta^{13}C$ in DIC or DOC and of a proxy that has grown from these aqueous carbon reservoirs tend to be dominated by the compositions of non-aqueous reservoirs modified by fractionations between them. This happens because aqueous C budgets are generally much lower than those in co-existing solid and/or gaseous phases.

The magnitudes of isotopic fractionations between the various chemical species of dissolved or gaseous carbon are generally determined by carbon bond strengths and phase transformations, so that gaseous or aqueous CO_2 in equilibrium with dissolved carbonate ions gives rise to the greatest fractionation towards light isotope ratios. Solid calcium carbonate is isotopically heavier than dissolved ionic carbonate ions with which it is in equilibrium, though the magnitude of fractionation is much less. Isotopic fractionations between carbon species may be at equilibrium or they may be affected by non-equilibrium kinetic processes. Kinetic effects are more likely in irreversible reactions, for example between solids and dissolved forms, or where microbial intervention occurs.

Figure 13 illustrates the equilibrium fractionations between the primary solid, dissolved and gaseous forms of carbon. Isotopic fractionation between $CO_2(g)$ and $CO_2(aq)$ is about $-1‰$, so the $\delta^{13}C$ value of dissolved CO_2 in equilibrium with atmospheric CO_2 ($-8‰$) is expected to be around $-9‰$. Bicarbonate ion, HCO_3^-, in isotopic equilibrium with dissolved CO_2 at $-9‰$ is around $+1$ to $+2‰$ $\delta^{13}C$. Values of $\delta^{13}C$ for DIC in surface

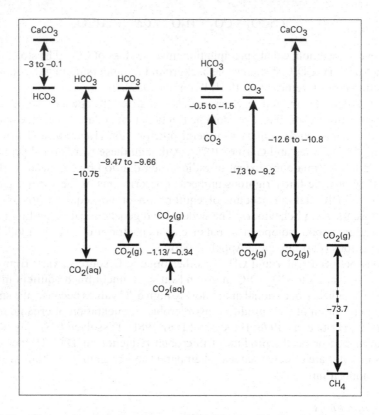

Figure 13. Equilibrium fractionations between the primary solid, dissolved and gaseous forms of carbon, based on data sources as set out in Table 1.

waters are not controlled simply by equilibrium fractionation with atmospheric CO_2 and are usually more negative than these values. There are other sources of DIC in surface waters, such as respiration of soil organic matter and dissolution of carbonates, that are quantitatively of more significance than atmospheric CO_2.

Precipitation of carbonate minerals from solution may take place rapidly, far from chemical equilibrium, introducing the possibility of isotopic disequilibrium between DIC and precipitated carbonate. The extent to which $^{13}C/^{12}C$ fractionation is affected by carbonate precipitation kinetics varies from case to case and between experimental observations and interpretations of natural systems. In some experiments, kinetic fractionation factors decrease as the precipitation rate increases (e.g., Turner (1982)) and in others fractionation is independent of precipitation rate (Romanek et al. 1992; Jimenez-Lopez et al. 2001).

Another possible cause of kinetic isotope fractionation during calcite precipitation is outgassing of CO_2, though reported deviations from equilibrium fractionation are small in normal hydrological environments. Carbon isotope studies of modern speleothems in Barbados show that most calcite is close to or at isotopic equilibrium with DIC (Mickler et al. 2004). Deviations from equilibrium fractionation are around $\pm 1‰$ $\delta^{13}C$ and were attributed to both precipitation kinetics and degassing of CO_2. Similarly, calcite precipitation from a spring in the French Pyrenees was found to occur with virtually no isotope fractionation (Dandurand et al. 1982). Kinetic fractionations that are even greater than the maximum theoretical values may occur in extreme environments. Outgassing of CO_2 as evaporating waters evolve towards brines in the Dead Sea is associated with non-equilibrium isotope fractionation. Experimental $^{13}C/^{12}C$ fractionation between $CO_2(g)$ and DIC in Dead Sea brine has been measured at –16.2 to –18.2‰, which is greater than the maximum theoretical kinetic fractionation of –11.2‰ for diffusion-controlled degassing and the fractionation of around –11‰ that is typical for $CO_2(g)$-DIC in ocean water (Stiller et al. 1985; Inoue and Sugimura 1985; Zhang et al. 1995; Barkan et al. 2001).

In many natural systems, some or all of the dissolved carbon has entered solution by non-equilibrium processes that involve kinetic isotope fractionations or irreversible stoichiometric dissolution (Dandurand et al. 1982; Usdowski and Hoefs 1986; Clark et al. 1992; Zhang et al. 1995; Skidmore et al. 2004). Simple dissolution of calcium carbonate usually proceeds without any significant isotope fractionation because it is a stoichiometric and irreversible mass transfer from solid to solution. Kinetic fractionation up to –17‰ was found in experimental dissolution of fine-grained carbonate from glacial sediments (Skidmore et al. 2004). The fractionation was found to vary with the water: solids ratio and to decline with increasing contact time.

Complex biogeochemical pathways and microbial transformations can result in substantial fractionations in their contributions to DIC and resulting biomass and/or carbonate deposits. For example, microbial mats that encrust travertine deposits at hot springs in Yellowstone, USA have variable $\delta^{13}C$ composition of total biomass from –16 to –28‰. Within this material, specific lipids have $\delta^{13}C$ values from –12 to –37‰. The carbon source for all of this material is CO_2, from the atmosphere (–7 to –8‰) and the geothermal source (–5‰), so large fractionations are occurring during the growth of biomass (Zhang et al. 2004).

Dissolution of minor water-soluble components such as humic and fulvic acids from complex organic sedimentary material is likely to involve little if any isotopic fractionation unless extensive microbiological mediation is involved (Wassenaar et al. 1991). Thus DOC is likely to inherit a $\delta^{13}C$ composition similar to its parent organic material, i.e., close to –25‰.

Microbial reduction of inorganic carbon or of higher molecular weight organics to CH_4 produces a large negative fractionation of $^{13}C/^{12}C$. Equilibrium fractionation between CH_4 and CO_2 is around –74‰ based on theoretical calculation (Bottinga 1969). Actual fractionation may be larger or smaller for microbial pathways from inorganic or organic sources (Schoell 1980; Deines 1980). Isotopic fractionation during production of CH_4 leaves the source inorganic carbon isotopically enriched which may have a significant effect in the inorganic carbon system depending on the mass balance between product and source compounds (Thorstenson et al. 1979; Barker and Fritz 1981; Aravena and Wassenaar 1993).

Typical $^{13}C/^{12}C$ fractionations for environmentally-relevant processes involving inorganic and organic carbon in aqueous, gaseous and solid phases are compiled in Table 1.

Table 1. Compilation of typical $^{13}C/^{12}C$ fractionations for environmentally-relevant processes involving inorganic and organic carbon in aqueous, gaseous and solid phases. Fractionation values are tabulated as $\delta^{13}C$ differences or experimental fractionation factors ε, depending on how reported in the data source.

Process	Typical fractionation (‰)	Examples
Precipitation of calcite at neutral pH and in isotopic equilibrium	+3 to +0.1 ($CaCO_3$–HCO_3) +2.3 to +1.8 ($CaCO_3$–HCO_3, 25°C)	Salomons and Mook 1986; Emrich et al. 1970; Wendt 1971; Turner 1982
Precipitation of calcite at alkaline pH and in isotopic equilibrium	+1.0 ± 0.2 and +0.94 ± 0.06 ($CaCO_3$–HCO_3, 25°C) +9.0 and +8.9 ($CaCO_3$–CO_2, 25°C)	Romanek et al. 1992; Jimenez-Lopez et al. 2001
Precipitation of calcite and aragonite in isotopic equilibrium	−1.4 (calcite–aragonite, 25°C)	Turner 1982
Precipitation of aragonite at alkaline pH and in isotopic equilibrium	+2.7 ± 0.6 ($CaCO_3$–HCO_3, 25°C) +10.6 ($CaCO_3$–CO_2, 25°C)	Romanek et al. 1992
Precipitation of siderite at neutral pH and in isotopic equilibrium	+17.5 ($FeCO_3$–$CO_{2\,gas}$) +10 ($FeCO_3$–$CO_{2\,gas}$, 30°C) +7.5,+13.3 to +14.5 (biogenic $FeCO_3$–$CO_{2\,gas}$, 45/30°C)	Golyshev et al. 1981; Carothers et al. 1988; Zhang et al. 2001; Romanek et al. 2003
Precipitation of siderite at alkaline pH and in isotopic equilibrium	+0.5 ± 0.2 ($FeCO_3$–HCO_3, 25°C) +8.5 ± 0.2 ($FeCO_3$–CO_2, 25°C)	Jimenez-Lopez and Romanek 2004
Outgassing/dissolution/ exchange of CO_2 in carbonate solution at neutral pH and in isotopic equilibrium	−10.2, −9.4 to −9.7 ($CO_{2\,gas}$–HCO_3) +0.3 to +1.1 ($CO_{2\,gas}$–$CO_{2\,aq}$)	Wendt 1968; Emrich et al. 1970; Mook et al. 1974; Friedman and O'Neil 1977; Szaran 1997; Zhang et al 1995; Vogel et al. 1970
Outgassing/exchange of CO_2 from carbonate solution at high pH and in isotopic equilibrium	−7.3 to −9.2 ($CO_{2\,gas}$–CO_3) −5.0 ± 0.2 ($CO_{2\,gas}$–CO_3, 25°C)	Deines et al. 1974; Thode et al. 1965; Halas et al. 1997
Diffusive exch. between atmos-pheric CO_2 and DIC in seawater	−11 ($CO_{2\,gas}$–DIC)	Zhang et al. 1995
Uptake of CO_2 into hyperalkaline water	∼−20, −19.5 18°C (CO_3–$CO_{2\,gas}$)	Clark et al. 1992; Usdowski and Hoefs 1986

Table 1 continued

Process	Typical fractionation (‰)	Examples
Movement of respired CO_2 through soil	$\sim +4$ ($CO_{2\,in\,soil} - CO_{2\,flux\,at\,soil\,surface}$)	Cerling et al. 1991; Davidson 1995
Precipitation of calcite with CO_2 outgassing not in isotopic equilibrium	$\sim 0 \pm 1$ ($CaCO_3$–DIC) +0.3 to +3.4 with decreasing precipitation rate (25°C)	Mickler et al. 2004; Dandurand et al. 1982; Turner 1982
Precipitation of calcite from evaporating brine	-16.2 to -18.2 ($CO_{2\,gas}$–DIC)	Stiller et al. 1985; Barkan et al. 2001
Precipitation of $CaCO_3$ and outgassing of CO_2 from freezing seawater	-2 to -20 ($C_{lost} - C_{remaining}$) (-2 for outgassing + precipitation, -20 for outgassing only)	Papadimitriou et al. 2004
Dissolution of calcite in a non-reversible reaction	~ 0 (DIC–$CaCO_3$)	—
Rapid dissolution of very fine-grained calcite	0 to -17 (DIC–$CaCO_3$)	Skidmore et al. 2004
Reduction of CO_2 to CH_4 at isotopic equilibrium	-73.7 (CH_4–CO_2)	Bottinga 1969
Biogenic CH_4 co-existing with DIC in groundwater but not in isotopic equilibrium	$-68, -75$ ($CH_{4\,aq}$–DIC)	Barker and Fritz 1981
Growth of microbial biomass associated with travertine	~ -4 to -32 (C_{org}–CO_2)	Zhang et al. 2004
Dissolution of water-soluble humics from solid organic material	~ 0 ($C_{org\,diss} - C_{org\,solid}$)	Wassenaar et al. 1991

Potential climatic and temporal effects

The potential climatic and temporal influences on the interaction of C sources with water involve several factors: (i) temperature effects on geochemical reactions involving carbon transfers between solid sources and solutions; (ii) direct long-term metabolic pathways, and variations in soil and other surficial compositions in terms of environmental changes such as the $^{13}C/^{12}C$ isotope ratio of atmospheric CO_2, variations of types of organic material such as relative proportions of vegetation with C_3 and C_4 balances of organic and inorganic (i.e., carbonate) C sources; (iii) residence times of dissolved C in the available carbon reservoirs with consequences for the extent of isotopic exchange and whether equilibrium has been achieved in exchange reactions having slow kinetics.

The temperature effects on $^{13}C/^{12}C$ fractionations between gaseous atmospheric CO_2, the various DIC species and solid carbonate are rather low compared to the temperature effects that are propagated into the $\delta^{18}O$ of freshwater carbonates. The temperature coefficients for $\delta^{13}C$ fractionation between calcite and DIC and between calcite and $CO_2(g)$ are only +0.05‰ and −0.16‰ per °C respectively (Bottinga 1968; Salomons and Mook 1986). The temperature coefficient for $\delta^{18}O$ fractionation between calcite and water is around −0.23‰ per °C and the temperature-dependent variation of $\delta^{18}O$ in meteoric water due to atmospheric liquid–vapour fractionation is in the range 0.6 to 0.2‰ per °C (O'Neil et al. 1969; Rozanski et al. 1993). Therefore the direct sensitivity of $\delta^{13}C$ as a proxy for environmental temperature is relatively low. Other temporally variable environmental factors in addition to temperature affect isotopic fractionation of $^{13}C/^{12}C$, for example pH-dependent speciation of DIC, and thus cause additional uncertainties in temperature dependence.

Groundwaters in karstic and other carbonate aquifer rocks may undergo isotopic exchange between DIC and solid carbonate after the establishment of chemical equilibrium (Smith et al. 1975; Edmunds et al. 1987). It is likely that dissolution and precipitation reactions at a microscopic local scale on the solid-solution interface account for isotopic exchange in chalk which is a fine-grained calcite that retains its micritic structure. Using $\delta^{13}C$ and ^{14}C activity, the rate of isotopic exchange in a Libyan confined karst aquifer has been estimated to be equivalent to a half-life of 11,000 years (Gonfiantini and Zuppi 2003). Based on this measurement and on surface areas for water-rock exchange, half-lives from 4 ka to 40 ka for ^{13}C exchange in other aquifers were estimated.

Case studies

Selected case studies are summarised here to illustrate the potential complexity of carbon isotope systematics in water-gas-mineral-(organics) systems, especially in surface or near-surface environments.

The $\delta^{13}C$ composition of DIC in surface waters and shallow groundwaters is influenced variably by input from soil organics and carbonates, and by exchange with atmospheric CO_2. Input from soil organics and carbonates is dominant in tributaries of the Ottawa River in Canada, where $\delta^{13}C$ of DIC varies from −8 to −16‰ in lowland carbonate catchments and upland silicate weathering terrains respectively (Telmer and Veizer 1999). This is the case also in the Rhône River in France where $\delta^{13}C$ of DIC in headwaters and downstream flow varies from −5 to −11‰ and is interpreted as being dominated by carbonate inputs with only minor influence from respiration of soil organics (Aucour et al. 1999). In the upper St Lawrence River in Canada, $\delta^{13}C$ varies from +2.2 to −13.7‰, with the larger variations towards negative values being associated with the more active carbon cycle of near-shore ecosystems such as are found in creeks and marshes (Barth and Veizer 1999). Smaller catchments with lower flows and situated on non-carbonate rock may have different $\delta^{13}C$ characteristics. One such catchment in the Vosges mountains, France, has seasonal variations of $\delta^{13}C$ from −24 to −20‰ in shallow groundwaters that feed stream flow, whereas the stream at the bottom of the catchment has $\delta^{13}C$ around −12‰ (Amiotte-Suchet et al. 1999). These data show that input of CO_2 from respiration of soil organics dominates the isotopic composition of DIC in shallow groundwater, whereas isotopic exchange with atmospheric CO_2 has taken over in the stream.

The $\delta^{13}C$ of DIC in groundwaters in carbonate terrain may indicate carbon origins that are more complex than stoichiometric reaction of soil CO_2 with calcium carbonate. DIC in a karst spring water was found to be derived from the mixing of unsaturated zone and saturated zone water sources (Emblanch et al. 2003). The $\delta^{13}C$ of each component water was characteristic, the unsaturated zone water being open to exchange with soil CO_2 and the saturated zone water being closed to exchange. This example, though taken from a specific type of karstic groundwater system with a thick unsaturated zone and deep water table, is probably representative of many shallow groundwaters in which seasonal or other temporal fluctuations of recharge and water table location cause variations in the degree of open- or closed-system behaviour with respect to CO_2 and thus variation in $\delta^{13}C$ of the DIC. Distinct $\delta^{13}C$ evolutionary pathways for unsaturated zone water according to seasonal hydrological differences have been observed in shallow chalk (Dever et al. 1982). In winter, the soil CO_2 system is isolated from exchange (closed system) with the atmosphere, and consequently soil water evolves its DIC content in isotopic equilibrium with soil CO_2. In contrast, in summer the soil is drier, the CO_2 content is higher, and DIC evolves out of isotopic equilibrium with the soil CO_2 which is exchanging with the atmosphere (open system).

The complexity of reaction pathways by which the $\delta^{13}C$ of DIC in groundwaters can evolve is illustrated by case studies from many aquifers. A single example serves to illustrate the multiplicity of processes that may be involved in addition to stoichiometric dissolution of carbonate minerals. A geochemical model of the carbon sources and sinks in a confined aquifer in the USA simulated the oxidation of organic matter, reduction of sulphate and the precipitation of calcite (McMahon and Chapelle 1991). The number of alternative models was constrained by consideration of $\delta^{13}C$ data which suggested that the CO_2 involved in DIC production in the aquifer is dominantly from biogenic fermentation rather than microbial respiration. CO_2 with the $\delta^{13}C$ signature of microbial fermentation along with organic acids and sulphate is most likely coming from confining beds that contain organics and are also cemented by carbonate. For the present purpose, however, groundwaters in confined aquifers are not of direct relevance unless discharges from them to surface have some influence on the isotopic compositions of potential proxy indicators.

There are rare cases where the $\delta^{13}C$ of shallow groundwaters feeding springs is totally controlled by dissolution of carbonate by a non-atmospheric source of CO_2. These generally occur in geothermal or tectonically-active areas. For example, CO_2 of deep-seated origin is transmitted through faults into shallow groundwaters in Sichuan, China and dissolves carbonate rock aggressively prior to feeding streams that deposit tufa (Yoshimura et al. 2004). The $\delta^{13}C$ of DIC in the outflow is a mixture between the isotopic compositions of CO_2 with a value of −3‰ and carbonate rock with +3‰. Further isotope fractionation occurs as CO_2 outgases from the stream water and tufa precipitates.

Evaporation from soils of arid regions causes enrichment of $\delta^{18}O$ in residual unsaturated zone waters (see Figures 7 and 8); the corresponding effects on $\delta^{13}C$ in soil waters of evaporation and of short-term and long-term changes in aridity are potentially quite complex. Isotopic compositions of calcrete ('caliche') in a desert area of the USA indicated that evaporated soil waters develop enriched $\delta^{13}C$ values (Knauth et al. 2003). This enrichment was tentatively attributed to either fractionation during photosynthetic loss of water from vegetation or to kinetic isotope effects due to direct evaporation. In studying signals of long-term climate change in Bangladesh, Alam et al. (1997) considered $\delta^{13}C$ enrichment in pedogenic carbonate nodules to indicate that DIC in soil waters had changed over time due to a climate-related decrease in soil CO_2 productivity and a greater input from atmospheric CO_2.

Implications for proxy growth and interpretation

Dissolved inorganic carbon
There are various sources for DIC: atmospheric, geological (e.g., carbonate minerals) and organic/biogenic. Atmospheric CO_2, soil CO_2 and DIC itself (e.g., HCO_3^- in lake waters as an indicator of productivity in the C cycle) are the main potential proxies for past climate and other environmental conditions. Therefore any proposed use of $\delta^{13}C$ in inorganic materials as a proxy for past environmental or climatic conditions should have a conceptual geochemical model that can be tested or otherwise evaluated independently. The contributions of various sources of C and the processes that link them with DIC may be dependent on environmental conditions. For example, temperature and aridity are likely to have affected the balance between the alternative metabolic pathways for C_3 and C_4 vegetation.

Low sensitivity to temperature of $^{13}C/^{12}C$ fractionation between the various DIC species and solid phases and the potential complexity of various reaction paths and speciations means that DIC geochemistry introduces large uncertainties into the interpretation of variations of $\delta^{13}C$ in carbonates as palaeoclimate and palaeoenvironment proxies.

Dissolved organic carbon
The $\delta^{13}C$ of DOC is potentially useful as a proxy for past climate and environmental conditions where the organics have been immobilised and preserved in solid matrices. Specific organic compounds such as humic and fulvic acids, specific biomarkers, and methane are likely to be of greatest interest, though uncertainties about trapping and preservation and likely very low or zero abundances make these studies somewhat speculative and vulnerable to contamination by anthropogenic carbon sources.

DOC may be derived from parent organic material, for example plant material and soil humus, via intermediate biochemical processes that alter $\delta^{13}C$ values in unpredictable ways. CO_2 loss during microbial respiration would alter or fractionate $^{13}C/^{12}C$, usually to heavier ratios. Dissolved organic carbon may have complex origins including 'dead' fossil organic material that has a geological age much greater than that of the proxy sought.

Methane is unlikely to be preserved as a useful proxy of past environmental conditions in view of its reactivity in oxidising conditions and mobility as a gas phase or as a dissolved gas. However the evidence for CH_4 intervention in past processes, for example anomalously light $\delta^{13}C$ in carbonate precipitates, is an important indicator of a strongly reducing or anoxic/anaerobic environment in which methanogenic bacteria were viable. Some preliminary geochemical analyses of organics trapped as solutes or as colloids in speleothem carbonate have been carried out, but stable isotope analyses of the organics have not yet been achieved. Trace organic carbon compounds that are found in old deep groundwaters also have the potential to provide proxy information about environmental conditions at the time of groundwater recharge. However, an origin at the surface and the possibility of mixing with organic compounds from geological sources, for instance fossil hydrocarbons, have to be investigated. Dissolved or colloidal organic carbon is itself now a medium for ^{14}C dating of old groundwaters and has been compared in various studies with ^{14}C dates inferred from co-existing dissolved inorganic carbon, with varying degrees of correlation that are indicative of the complexities of geochemical evolutions of dissolved inorganic and/or organic carbon in the subsurface.

From proxy to climate – constraints on interpretation

In attempting to reconstruct climatic conditions from proxy isotope values, we have to consider the various steps in the following basic chain: proxy→growth water→ground/soil/surface water→precipitation→climate. None is necessarily insurmountable, but the present state of knowledge in some areas may be insufficient to give a confident answer. The complete sequence applies mainly to O and H isotopes; C isotopes effectively by-pass the precipitation stage. (It may not always be possible to reconstruct the O and H isotope composition of precipitation either, but from the inferred isotopic signature of soil/surface water we may still be able to draw some conclusions about climate, for example the role of evaporation.)

Proxy to growth water

The first step in the chain lies beyond the scope of this chapter; there are so many different proxies, usually with their own particular fractionation factors for O, H and C isotopes, that it is difficult to discuss in terms of a few universally-applicable caveats. In any case, as Chapters 2 to 5 show (McCarroll and Loader, Hedges et al., Leng et al., McDermott et al., all this volume), the practitioners of the mainstream proxy research areas are well aware of the constraints and difficulties in making this step.

Growth water to ground/soil/surface water

On the assumption that a realistic composition for growth water has been achieved, what are the main constraints to making the next step? Again, there are many proxies, but if we take the main themes covered by Chapters 2 to 5 (tree rings, bones, lake sediments, speleothems), certain caveats are apparent. For trees, different rooting depths make it difficult to know even approximately what section of the year's rainfall infiltration is represented in the rings; the only solution would appear to be control experiments on the nearest modern analogues in terms of species, soil conditions and likely climate regime. For bones it is essential to have a good knowledge of water and food sources for O and C isotopes respectively, though evaporation effects on drinking water sources seem likely to remain a problem area. For lake sediment proxies there is one overriding problem: determining the temperature of proxy growth. As with tree rings, studies of modern analogues may be the only answer. For speleothems the main difficulty lies with separating seasonal from altitude effects on O and H isotopes, and the only way of doing this convincingly is by monitoring in the cave itself, assuming of course that deposition is still active.

Ground/soil/surface water to precipitation

If the previous steps have been accomplished successfully, the following are now known (where relevant): the isotopic composition of the growth water, depth of rooting in the soil (trees), predominant source of food and water (bones), temperature/season of growth in water (sediments), season and/or ultimate source of recharge (speleothems). These data now have to be linked to hydrology. For soils, how does the soil moisture profile vary seasonally in O and H isotope content? For bones, what is the likely bulk O isotope composition of the drinking water source(s), and bulk C isotope content of diet? (In the case of water, factors such as animal hibernation mean that the annual mean composition of a particular reservoir may not be appropriate.) For lakes, what is the scale of variation in O, H and C isotopes; is it only seasonal, or are there longer-term cycles? How do isotopic compositions relate to temperature and rainfall inputs, and what are the tolerances on lake water composition modelling? For speleothems, how secure is the interpretation of water source: if there is a seasonal isotope variation, does it reflect similar variation in rainfall or is there some selection mechanism?

Precipitation to climate

By this stage the O and H proxy data have been converted to a notional rainfall isotope composition, representing either the annual mean or a known portion of the annual cycle. So far this chapter has focused mainly on present-day processes; the important question now is, to what degree is the present the key to the past? There are two important elements in this: possible changes in the balance between the three terrestrial reservoirs of water referred to earlier, and possible changes to the isotope–temperature (δ–T) relationship.

Regarding the water balance, for earlier Holocene moderate climatic conditions it is reasonable to assume that present-day isotope hydrological relationships remain a good guide, with the proviso that particular climatic regimes with their associated rainfall cycles and carbon productivity may have waxed or waned in area as the various cold or warm periods occurred. Uncertainty over water balance (for example, was there more river runoff and/or isotopic selection of recharge) arises under conditions of widespread glaciation, for which relatively little present-day information exists. Therefore it becomes an issue with regard to proxies of Pleistocene age. It is not necessarily straightforward to disentangle the various factors affecting water balance. An example of this would be the fairly ubiquitous decline in groundwater recharge during the latter part of the last glaciation. Reduced rainfall seems likely to have played a major part: this is supported by (for example) pollen evidence and low speleothem/travertine growth even in non-glaciated areas. A permafrost seal in periglacial areas is a less likely contender since modern permafrost is usually far from totally impermeable. Surface ice cover would appear to preclude recharge altogether, but areas with basal melting might actively promote it. The best that can be said is that currently there is rather little evidence for major isotope selection effects in high-latitude water balances (see discussion in Darling (2004)).

Assuming such factors can be evaluated to arrive at a notional rainfall isotope composition, how easily can this be converted to a temperature? We have seen earlier in this chapter that δ–T relationships can be viewed in three different ways: a spatial relationship based on long-term annual averages, a temporal relationship based on seasonal variability, and a temporal relationship based on long-term changes in the isotopic signature of rainfall and the temperature. The most appropriate for palaeoclimatic reconstruction is undoubtedly the last of these, but it has hitherto been established with confidence for only few sites/regions. The important question which has to be addressed, particularly when interpreting proxies deposited under different climatic regimes such as glacial climate, is to what extent the δ–T relationship established for present-day climatic conditions is also valid for past climates. One factor which will clearly influence this relationship is the change of the isotopic composition of the global ocean over time. The δ–T is sometimes resolvable by measurement: if groundwater of the right age is available, recharge temperatures can be calculated from noble gas contents, against which isotope values can be compared. However, the dispersion effects in groundwater referred to earlier render this approach less satisfactory during periods of rapid climate change, an important focus of palaeoclimate studies.

The carbon isotope effects of past climatic change are less easily approached in a quantitative way. There are several possible effects: for example the CO_2 productivity of soil tends to drop with temperature, the balance between C_3 and C_4 plants may change, and pluvial episodes may promote the development of peat-type deposits rich in reduced carbon. All these processes carry isotopic implications, as the previous section of this chapter has shown. Nevertheless if the hydrology can be determined from O and H isotopes, this should provide a more secure framework for a C-isotope interpretation of environmental conditions.

Summary

The O, H and C stable isotope compositions of proxies reflect conditions in the water and carbon cycles at the time of formation. To provide a framework for the interpretation of such proxy data now and in the past, isotopic effects in these cycles at the present day need to be understood and characterised.

The water cycle divides naturally into a number of compartments which have been studied in varying detail over the past 40 years since the birth of isotope hydrology. The starting point, precipitation, is especially important because it is the precursor back to which most O and H isotope proxy studies are attempting to relate. While much is understood about the isotope systematics of precipitation, largely owing to the existence of the IAEA–WMO Global Network for Isotopes in Precipitation (GNIP), important questions remain to be answered in relation to the isotope–temperature gradients of past climatic conditions.

The three reservoirs of water sustaining all terrestrial proxies are soil and unsaturated (vadose) zone moisture, groundwater, and surface waters. These can be viewed broadly as the infiltration, recharge, and discharge stages of the land-based portion of the water cycle. In each case isotopic effects intervene to modify to a greater or lesser extent the stable isotope signature of antecedent precipitation. In general terms these effects vary in severity from low for groundwaters to high for surface waters. Since lake sediments of one type or another form important and accessible archives of the palaeoenvironment, the continuing development of isotope mass-balance modelling techniques is playing a significant role in interpreting the proxy isotope values.

The carbon cycle is far more complex than the water cycle even if, as here, the focus is restricted to dissolved organic and inorganic C. This is because there are several sources of both, in contrast to the single source of water (precipitation). Dissolved carbon stable isotopes are unlikely ever to feature as the 'palaeothermometers' and 'palaeopluviometers' that O and H isotopes are becoming, yet they have great potential in the interpretation of past environmental conditions where the overall geochemical conceptual model can be suitably constrained.

The conversion of proxy data into the stable isotope composition of precursor rainfall or carbon source(s), and by way of this to the characterisation of palaeoclimate and palaeoenvironment, is a multi-step process that has tended to receive most attention at the proxy end. This review shows the need to subject each step to careful scrutiny.

References

Adar E.M., Gev I., Lipp J., Yakir D., Gat J. and Cohen Y. 1995. Utilization of oxygen-18 and deuterium in stem flow for the identification of transpiration sources: soil water versus groundwater in sand dune terrain. In: Applied Tracers in Arid Zone Hydrology, IAHS Publ. No. 232, pp. 329–338.

Adar E.M., Dodi A., Geyh M., Yair A., Yakirevich A. and Issar A. 1998. Distribution of stable isotopes in arid storms: I. Relation between the distribution of isotopic composition in rainfall and in the consequent runoff. Hydrogeol. J. 6:50–65.

Agemar T. 2001. Temporal variations of δ^2H, $\delta^{18}O$ and deuterium excess in atmospheric moisture and improvements in the CO_2-H_2O equilibration technique for the $^{18}O/^{16}O$ ratio analysis. PhD thesis, University of Heidelberg, Germany.

Alam M.S., Keppens E. and Paepe R. 1997. The use of oxygen and carbon isotope composition of pedogenic carbonates from Pleistocene palaeosols in NW Bangladesh, as palaeoclimatic indicators. Quaternary Sci. Rev. 16: 161–168.

Allison G.B. and Barnes C.J. 1985. Estimation of evaporation from the normally "dry" Lake Frome in South Australia. J. Hydrol. 78: 229–242.

Allison G.B. and Leaney F.W. 1982. Estimation of isotopic exchange parameters, using constant-feed pans. J. Hydrol. 55: 151–161.

Allison G.B., Barnes C.J. and Hughes M.W. 1983. The distribution of deuterium and 18O in dry soils. J. Hydrol. 64: 377–397.

Allison G.B., Barnes C.J., Hughes M.W. and Leaney F.W.J. 1984. Effect of climate and vegetation on oxygen-18 and deuterium profiles in soils. In: Isotope Hydrology 1983. IAEA, Vienna, pp. 105–122.

Allison G.B., Stone W.J. and Hughes M.W. 1985. Recharge in karst and dune elements of a semi-arid landscape as indicated by natural isotopes and chloride. J. Hydrol. 76: 1–25.

Ambach W., Elsasser M., Moser H., Rauert W., Stichler W. and Trimborn P. 1975. Variationen des gehaltes an Deuterium, Sauerstoff-18 und Tritium während einzelner Niederschläge. Wasser und Leben 27: 186–192.

Amiotte-Suchet P., Aubert D., Probst J.L., Gauthier-Lafaye F., Probst A., Andreux F. and Viville D. 1999. δ^{13}C pattern of dissolved inorganic carbon in a small granitic catchment: the Strengbach case study (Vosges Mountains, France). Chem. Geol. 159: 129–145.

Andrews J.N. and Lee D.J. 1979. Inert gases in groundwater from the Bunter Sandstone of England as indicators of age and palaeoclimatic trends. J. Hydrol. 41: 233–252.

Andrews J.E., Riding R. and Dennis P.F. 1993. Stable isotopic composition of Recent freshwater cyanobacterial carbonates from the British Isles: local and regional environmental controls, Sedimentology 40: 303–314.

Araguas-Araguas L., Froehlich K. and Rozanski K. 1998. Stable isotope composition of precipitation over Southeast Asia. J. Geophys. Res. 103, D22: 28,721–28,742.

Araguas-Araguas L., Froehlich K. and Rozanski K. 2000. Deuterium and oxygen-18 isotope composition of precipitation and atmospheric moisture. Hydrol. Process. 16: 1341–1355.

Aranyossy J.F., Filly A., Tandia A.A., Louvat D., Ousmanc B., Joseph A. and Fontes J.C. 1992. Estimation des flux d'evaporation diffuse sous couvert sableux en climat hyperaride, Erg de Bilma, Niger. In: Isotope Techniques in Water Resources Development 1991. IAEA, Vienna, pp. 309–324.

Aravena R. and Wassenaar L.I. 1993. Dissolved organic carbon and methane in a regional confined aquifer, southern Ontario, Canada: Carbon isotope evidence for associated subsurface sources. Appl. Geochem. 8: 483–493.

Aucour A.M., Sheppard S.M.F., Guyomar O. and Wattelet J. 1999. Use of ^{13}C to trace origin and cycling of inorganic carbon in the Rhône river system. Chem. Geol. 159: 87–105.

Ayalon A., Bar-Matthews M. and Sass E. 1998. Rainfall-recharge relationships within a karstic terrain in the eastern Mediterranean semi-arid region, Israel: δ^{18}O and δD characteristics. J. Hydrol. 207: 8–31.

Bajjali W., Clark I.D. and Fritz P. 1997. The artesian thermal groundwaters of northern Jordan: insights to their recharge history and age. J. Hydrol. 192: 355–382.

Bariac T., Deleens E., Gerbaud A., Andre M. and Mariotti A. 1991. La composition isotopique (^{18}O, ^{2}H) de la vapeur d'eau transpiree: Etude en conditions asservies. Geochim. Cosmochim. Ac. 55: 3391–3402.

Barkan E., Luz B. and Lazar B. 2001 Dynamics of the carbon dioxide system in the Dead Sea. Geochim. Cosmochim. Ac. 65: 355–368.

Barker J.F. and Fritz P. 1981. The occurrence and origin of methane in some groundwater flow systems. Can. J. Earth Sci. 18: 1802–1816.

Barnes C.J. and Allison G.B. 1988. Tracing of water movement in the unsaturated zone using stable isotopes of hydrogen and oxygen. J. Hydrol. 100: 143–76.

Barth J.A.C. and Veizer J. 1999. Carbon cycle in St. Lawrence aquatic ecosystems at Cornwall (Ontario), Canada: seasonal and spatial variations. Chem. Geol. 159: 107–128.

Bath A.H. 1983. Stable isotope evidence for palaeo-recharge conditions of groundwater. In: Palaeoclimates and Palaeowaters: a Collection of Environmental Isotope Studies. IAEA, Vienna, pp. 169–186.

Bath A.H., Edmunds W.M. and Andrews J.N. 1979. Palaeoclimatic trends deduced from the hydrochemistry of a Triassic sandstone aquifer, United Kingdom. In: Isotope Hydrology 1978, Vol II. IAEA, Vienna, pp. 545–566.

Bath A.H., Darling W.G., George I.A. and Milodowski A.E. 1987. ^{18}O/^{16}O and ^{2}H/^{1}H changes during progressive hydration of a Zechstein anhydrite formation. Geochim. Cosmochim. Ac. 51: 3113–3118.

Bender M.M. 1968. Mass spectrometric studies of carbon-13 variations in corn and other grasses. Radiocarbon 10: 468–472.

Berner E.K. and Berner R.A. 1987. The Global Water Cycle, Geochemistry and Environment. Prentice Hall, Inc, 398 pp.

Bottinga Y. 1968. Calculation of fractionation factors for carbon and oxygen in the system calcite-carbon dioxide-water. J. Phys. Chem. 72: 800–808.

Bottinga Y. 1969. Calculated fractionation factors for carbon and hydrogen isotope exchange in the system calcite-CO_2-graphite-methane hydrogen and water vapour. Geochim. Cosmochim. Ac. 33: 49–64.

Bottinga Y. and Craig H. 1969. Oxygen isotope fractionation between CO_2 and water and the isotopic composition of the marine atmosphere. Earth Planet. Sc. Lett. 5: 285–295.

Bottomley D.J., Craig D. and Johnson L.M. 1986. Oxygen-18 studies of snowmelt runoff in a small Precambrian shield watershed: Implications for streamwater acidification in acid sensitive terrain. J. Hydrol. 88: 213–234.

Bowen G.J. and Wilkinson B. 2002. Spatial distribution of $\delta^{18}O$ in meteoric precipitation. Geology 30: 315–318.

Bowen G.J. and Revenaugh J. 2003. Interpolating the isotopic composition of modern meteoric precipitation. Water Resour. Res. 39: 1299. doi:10.1029/2003WR002086.

Boyle E.A. 1997. Cool tropical temperature shift the global $\delta 18O$–T relationship: An explanation for the ice core $\delta^{18}O$-borehole thermometry conflict. Geophys. Res. Lett. 24: 273–276.

Brodersen C., Pohl S., Lindenlaub M., Leibendgut C. and von Wilpert K. 2000. Influence of vegetation structure on isotope content of throughfall and soil water. Hydrol. Process. 14: 1439–1448.

Brown L.J. and Taylor C.B. 1974. Geohydrology of the Kaikoura Plain, Marlbourough, New Zealand. In: Isotope Techniques in Groundwater Hydrology, Vol I. IAEA, Vienna, 169–188.

Brunel J.P., Walker G.R., Dighton J.C. and Moteny B. 1997. Use of stable isotopes of water to determine the origin of water used by the vegetation and to partition evapotranspiration: A case study from HAPEX-Sahel. J. Hydrol. 188-189: 466–481.

Carothers W.W. Adami L.H. and Rosenbauer R.J. 1988. Experimental oxygen isotope fractionation between siderite-water and phosphoric acid liberated CO_2-siderite. Geochim. Cosmochim. Ac. 52: 2445–2450.

Cerling T.E., Solomon D.K., Quade J. and Bowman J.R. 1991. On the isotopic composition of carbon in soil carbon dioxide. Geochim. Cosmochim. Ac. 55: 3403–3405.

Charles C.D., Rind D., Jouzel J., Koster R.D. and Fairbanks R.G. 1994. Glacial-interglacial changes in moisture sources for Greenland: Influence on the ice core record of climate. Science 263: 508–511.

Chen J., Tang C., Sakura Y., Kondoh A., Yu J., Shimada J. and Tanaka T. 2004. Spatial geochemical and isotopic characteristics associated with groundwater flow in the North China Plain. Hydrol. Process. 18: 3133–3146.

Ciais P., White J.W.C., Jouzel J. and Petit J.R. 1995. The origin of present-day Antarctic precipitation from surface snow deuterium excess data. J. Geophys. Res. 100(D9): 18,917–18,927.

Cicerone R.J. and Oremland R.S. 1988. Biogeochemical aspects of atmospheric methane. Global Biogeochem. Cy. 2: 299–327

Clark I.D. and Fritz P. 1997. Environmental Isotopes in Hydrogeology, Lewis Publishers, Boca Raton, 328 pp.

Clark I.D., Fontes J.C. and Fritz P. 1992. Stable isotope disequilibria in travertine from high pH waters: Laboratory investigations and field observations from Oman. Geochim. Cosmochim. Ac. 56: 2041–2050.

Cook P.G. and Herczeg A.L. 1999. Environmental Tracers in Subsurface Hydrology. Kluwer Academic Press, Boston, 529 pp.

Craig H. 1953. The geochemistry of the stable carbon isotopes. Geochim. Cosmochim. Ac. 3: 53–92.

Craig H. 1961. Isotopic variations in meteoric waters. Science 133: 1702–1703.

Craig H. and Gordon L.I. 1965. Deuterium and oxygen-18 variations in the ocean and marine atmosphere. In Tongiorgi E. (ed.), Stable Isotopes in Oceanographic Studies and Paleotemperatures, Pisa: Cons. Naz. Rich. Lab. Geol. Nucl., pp. 9–130.

Cuffey K.M., Clow G.D., Alley R.B., Stuiver M., Waddington E.D. and Saltus R.W. 1995. Large Arctic temperature change at the Wisconsin-Holocene glacial transition. Science 270: 455–458.

Dandurand J.L., Gout R., Hoefs J., Menschell G., Schott J. and Usdowski E. 1982. Kinetically controlled variations of major components and carbon and oxygen isotopes in a calcite-precipitating spring. Chem. Geol. 36: 299–315.

Dansgaard W. 1964. Stable isotopes in precipitation. Tellus 16: 436–468.

Darling W.G. 2004. Hydrological factors in the interpretation of stable isotopic proxy data present and past: a European perspective. Quaternary Sci. Rev. 23: 743–770.

Darling W.G., Gizaw B. and Arusei M.K. 1996. Lake-groundwater relationships and fluid-rock interaction in the East African Rift Valley: Isotopic evidence. J. Afr. Earth Sci. 22: 423–431.

Darling W.G., Bath A.H. and Talbot J.C. 2003. The O and H stable isotopic composition of fresh waters in the British Isles. 2. Surface waters and groundwater. Hydrol. Earth Syst. Sc. 7: 183–195.

Darling W.G., Edmunds W.M., Kinniburgh D.G. and Kotoub S. 1987. Sources of recharge to the Basal Nubian Sandstone aquifer, Butana Region, Sudan. In: Isotope Techniques in Water Resources Development. IAEA, Vienna, pp. 205–224.

Davidson G.R. 1995. The stable isotopic composition and measurement of carbon in soil CO_2. Geochim. Cosmochim. Ac. 59: 2485–2489.

Dawson T.E. 1993. Water sources of plants as determined from xylem-water isotopic composition: perspectives on plant competition, distribution, and water relations. In: Ehleringer J.R., Hall A.E. and Farquhar G.D. (eds), Stable Isotopes and Plant Carbon–Water Relations. Academic Press, San Diego, CA, pp. 465–496.

Deak J., Deseo E., Bohlke J.K. and Revesz K. 1996. Isotope hydrology studies in the Szigetkoz region, northwest Hungary. In: Isotopes in Water Resources Management, Vol I. IAEA, Vienna, pp. 419-432.

Deak J., Forizs I., Deseo E. and Hertelendi E. 1993. Origin of groundwater and dissolved ammonium in SE Hungary: evaluation by environmental isotopes. In: Tracers in Hydrology, IAHS Publ. No. 215, pp. 117–124.

Deines P. 1980. The isotopic composition of reduced organic carbon. In: Fritz P. and Fontes J.C. (eds), Handbook of Environmental Isotope Geochemistry, Vol. 1, The Terrestrial Environment, A. Elsevier, Amsterdam, pp. 329–406.

Deines P., Langmuir D. and Harmon R.S. 1974. Stable carbon isotope ratios and the existence of a gas phase in the evaluation of carbonate groundwaters. Geochim. Cosmochim. Ac. 38: 1147–1164.

Dever L., Durand R., Fontes J.C. and Vachier P. 1982. Geochimie et teneurs isotopiques des systèmes saisonniers de dissolution de la calcite dans un sol sur craie. Geochim. Cosmochim. Ac. 46: 1947–1956.

Dever L., Travi Y., Barbecot F., Marlin C. and Gibert E. 2001. Evidence for palaeowaters in the coastal aquifers of France. In: Edmunds W.M. and Milne C.J. (eds), Palaeowaters in Coastal Europe. Geol. Soc. Lond. Spec. Publ. No. 189, pp. 93–106.

DeWalle D.R. and Swistock B.R. 1994. Differences in oxygen-18 content of throughfall and rainfall in hardwood and coniferous forests. Hydrol. Process. 8: 75–82.

Dincer T., Al-Mugrin A. and Zimmerman U. 1974. Study of the infiltration and recharge through the sand dunes in arid zones with special reference to the stable isotopes and thermonuclear tritium. J. Hydrol. 23:79–109.

Duane M.J., Al-Zamel A. and Eastoe C.J. 2004 Stable isotope (chlorine, hydrogen and oxygen), geochemical and field evidence for continental fluid flow vectors in the Al-Khiran sabkha. J. Afr. Earth Sci. 40: 49–60.

Durand P., Neal M. and Neal C. 1993. Variations in stable oxygen isotope and solute concentrations in small submediterranean montane streams. J. Hydrol. 144: 283–290.

Edmunds W.M. and Smedley P.L. 2000. Residence time indicators in groundwaters: The East Midlands Triassic Sandstone aquifer. Appl. Geochem. 15: 737–752.

Edmunds W.M., Buckley D.K., Darling W.G., Milne C.J., Smedley P.L. and Williams A.T. 2001. In: Edmunds W.M. and Milne C.J. (eds), Palaeowaters in Coastal Europe. Geol. Soc. Lond. Spec. Publ. No. 189, pp. 71–92.

Edmunds W.M., Cook J.M., Darling W.G., Kinniburgh D.G., Miles D.L., Bath A.H., Morgan-Jones M. and Andrews J.N. 1987. Baseline geochemical conditions in the Chalk aquifer, Berkshire, U.K.: A basis for groundwater quality management. Appl. Geochem. 2: 251–274.

Edmunds W.M., Dodo A., Djoret D., Gasse F., Gaye C.B., Goni I.B., Travi Y., Zouari K. and Zuppi G.M. 2004. Groundwater as an archive of climatic and environmental change: Europe to Africa. In: Battarbee R.W., Gasse F., and Stickley C.E. (eds), Past Climate Variability through Europe and Africa. Springer, Dordrecht, pp. 279–306.

Edwards T.W.D. 1993. Interpreting past climate from stable isotopes in continental organic matter. In: Swart P.K., Lohmann K.C., McKenzie J. and Savin S. (eds), Climate Change in Continental Isotopic Records. Geophys. Monogr. 78, Am. Geophys. Union, Washington, pp. 333–341.

Ehhalt D.H. 1974. Vertical profiles of HTO, HDO and H_2O in the troposphere. NCAR Tech. Note, NCAR-TN/STR-100.

Eichinger L., Rauert W., Stichler W., Bertleff B. and Egger R. 1984. Comparative study of different aquifer types in Central Europe, using environmental isotopes. In: Isotope Hydrology 1983. IAEA, Vienna, pp. 271–289.

Emblanch C., Zuppi G.M., Mundy J., Blavoux B. and Batiot C. 2003. Carbon-13 of TDIC to quantify the role of the unsaturated zone: the example of the Vaucluse karst systems (Southeastern France). J. Hydrol. 279: 262–274.

Emrich K., Ehhalt D.H. and Vogel J.C. 1970. Carbon isotope fractionation during the precipitation of calcium carbonate. Earth Planet. Sc. Lett. 8: 363–371.

EPICA community members. 2004. Eight glacial cycles from an Antarctic ice core. Nature 429: 623–628.

Epstein S. and Mayeda T.K. 1953. Variation of ^{18}O content of waters from natural sources. Geochim. Cosmochim. Ac. 4: 213–224.

Epstein S., Yapp C.J. and Hall J.H. 1976. The determination of the D/H ratio of non-exchangeable hydrogen in cellulose extracted from aquatic land plants. Earth Planet. Sc. Lett. 30: 241–251.

Farah E.A., Mustafa E.M.A. and Kumai H. 2000. Sources of groundwater recharge at the confluence of the Niles, Sudan. Environ. Geol. 39: 667–672.

Flora O. and Longinelli A. 1989. Stable isotope hydrology of a classical Karst area, Trieste, Italy. In: Isotope Techniques in the Study of the Hydrology of Fractured and Fissured Rocks. IAEA, Vienna, pp. 203–213.

Fontes J.C. and Zuppi G.M. 1976. Isotopes and water chemistry in sulphide-bearing springs of Central Italy. In: Interpretation of Environmental Isotope and Hydrogeochemical data in Ground Water Hydrology. IAEA, Vienna, pp. 143–158.

Fontes J.C. and Garnier J.M. 1979. Determination of the initial ^{14}C activity of the total dissolved carbon: A review of the existing models and a new approach. Water Resour. Res. 15: 399–413.

Fontes J.C., Pouchan P., Saliege J.F. and Zuppi G.M. 1980. Environmental isotope study of groundwater systems in the Republic of Djibouti. In: Arid Zone Hydrology. IAEA, Vienna, pp. 237–262.

Fontes J.C., Andrews J.N., Edmunds W.M., Guerre A. and Travi Y. 1991. Palaeorecharge by the Niger River (Mali) deduced from groundwater geochemistry. Water Resour. Res. 27: 199–214.

Fricke H.C. and O'Neil J.R. 1999. The correlation between $^{18}O/^{16}O$ ratios of meteoric water and surface air temperature: its use in investigating terrestrial climate change over geologic time. Earth Planet. Sc. Lett. 170: 181–196.

Friedman I. 1953. Deuterium content of natural waters and other substances. Geochim. Cosmochim. Ac. 4: 89–103.

Friedman, I. and O'Neil J.R. 1977. Compilation of stable isotope fractionation factors of geochemical interest. In: Fleischer M. (ed.) Data of Geochemistry, U.S. Geological Survey Professional Paper 440-KK, 6th ed., USGS, Reston VA.

Fritz P. 1981. River waters. In: Gat J.R. and Gonfiantini R (eds), Stable Isotope Hydrology: Deuterium and Oxygen-18 in the Water Cycle. Technical Reports Series No. 210. IAEA, Vienna, pp. 177–201.

Fritz P., Reardon E.J., Barker J., Brown R.M., Cherry J.A., Killey R.W.D. and McNaughton D. 1978. The carbon isotope geochemistry of a small groundwater system in northeastern Ontario. Wat. Resour. Res. 14: 1059–1067.

Fritz P., Mozeto A.A. and Reardon E.J. 1985. Practical considerations on carbon isotope studies on soil carbon dioxide. Chem. Geol. (Isot. Geosci.) 58: 89–95.

Gat J.R. 1981. Lakes. In: Gat J.R. and Gonfiantini R. (eds), Stable Isotope Hydrology — Deuterium and Oxygen-18 in the Water Cycle. Tech. Rep. Series No. 210, IAEA, Vienna, pp. 203–221.

Gat J.R. 1995. Stable isotopes of fresh and saline lakes. In: Lerman A., Imboden D. and Gat J.R. (eds), Physics and Chemistry of Lakes. Springer-Verlag, New York, pp. 139–166.

Gat J.R. 1996. Oxygen and hydrogen isotopes in the hydrological cycle. Annu. Rev. Earth Pl. Sc. 24: 225–262.

Gat J.R. 2000. Atmospheric water balance – the isotopic perspective. Hydrol. Process. 14: 1357–1369.

Gat J.R. and Tzur Y. 1967. Modification of the isotopic composition of rainwater by processes which occur before groundwater recharge. In: Isotopes in Hydrology. IAEA, Vienna, pp. 49–60.

Gat J.R. and Levy Y. 1978. Isotope hydrology of inland sabkhas in the Bardawil area, Sinai. Limnol. Oceanogr. 23: 841–850.

Gat J.R. and Bowser C.J. 1991. Heavy isotope enrichment in coupled evaporative systems. In: Taylor H.P., O'Neil J.R. and Kaplan I.R. (eds), Stable Isotope Geochemistry: A Tribute to Samuel Epstein. Spec. Publ. No. 3, The Geochem. Society, San Antonio, Texas, pp. 159–168.

Gat J.R. and Matsui E. 1991. Atmospheric water balance in the Amazon basin: An isotopic evapotranspiration model. J. Geophys. Res. 96(D7): 13,179–13,188.

Gedzelmann S.D. and Lawrence J.R. 1990. The isotopic composition of precipitation from two extratropical cyclones. Mon. Weather Rev. 118: 495–509.

Gehrels J.C., Peeters J.E.M., de Vries J.J. and Dekkers M. 1998. The mechanism of soil water movement as inferred from ^{18}O stable isotope studies. Hydrolog. Sci. J. 43: 579–594.

Geirnaert W. and Laeven M.P. 1992. Composition and history of ground water in the western Nile Delta. J. Hydrol. 138: 169–189.

Genty D., Baker A., Massulat M., Proctor C., Gilmour M., Poris-Branchu E. and Hamelin B. 2001. Dead carbon in stalagmites: carbonate bedrock paleodissolution vs ageing of soil organic matter. Implications for 13C variations in speleothems. Geochim. Cosmochim. Ac. 65: 3443–3457.

Genty D., Plagnes V., Causse C., Cattani O., Stievenard M., Falourd S., Blamart D., Ouahdi R. and Van-Exter S. 2002. Fossil water in large stalagmite voids as a tool for paleoprecipitation stable isotope composition reconstitution and paleotemperature calculation. Chem. Geol. 184: 83–95.

Gibson J.J. 2002. A new conceptual model for predicting isotope enrichment of lakes in seasonal climates. International Geosphere Biosphere Programme IGBP PAGES News 10: 10–11.

Gibson J.J. and Edwards T.W.D. 2002. Regional surface water balance and evaporation–transpiration partitioning from a stable isotope survey of lakes in northern Canada. Global Biogeochem. Cy. 16: 1026. doi:10.1029/2001GB001839.

Gibson J.J., Prepas E.E. and McEachern P. 2002. Quantitative comparison of lake throughflow, residency, and catchment runoff using stable isotopes: modelling and results from a survey of boreal lakes. J. Hydrol. 262: 128–144.

Gibson J.J., Edwards T.W.D., Birks S.J., St. Amour N.A., Buhay W., McEachern P., Wolfe B.B. and Peters D.L. 2005. Progress in Isotope Tracer Hydrology in Canada. Hydrol. Process. 19: 303-327.

Giggenbach W. 1971. Isotopic composition of waters of the Broadlands geothermal field. New Zeal. J. Sci. 14: 959–970.

Giggenbach W., Gonfiantini R. and Panichi C. 1983. Geothermal systems. In: Guidebook on Nuclear Techniques in Hydrology. Tech. Rep. Series No. 91, pp. 359–380.

Gleick P. 2000. The World's Water 2000-2001. Island Press, Washington D.C., 315 pp.

Goller R., Wilcke W., Leng M.J., Tobschall H.J., Wagner K., Valarezo C. and Zech W. 2005. Tracing water paths through small catchments under a tropical montane rain forest in south Ecuador by an oxygen isotope approach. J. Hydrol. 308: 67–80.

Golyshev S.I., Padalko N.L. and Pechenkin S.A. 1981. Fractionation of stable oxygen and carbon stable isotopes in carbonate systems. Geochem. Intl. 18: 85–99.

Gonfiantini R. 1986. Environmental isotopes in Lake Studies. In: Fritz P. and Fontes J.C. (eds), Handbook of Environmental Isotope Geochemistry: Vol 2, The Terrestrial Environment, B. Elsevier, Amsterdam, pp. 113–168.

Gonfiantini R. and Zuppi G.M. 2003. Carbon isotopic exchange rate of DIC in karst groundwater. Chem. Geol, 197: 319–336.

Gonfiantini R., Conrad G., Fontes J.C., Sauzay G. and Payne B.R. 1974. Etude isotopique de la nappe du Continental Intercalaire et de ses relations avec les autres nappes du Sahara septenrional. In: Isotope Techniques in Groundwater Hydrology, Vol. I. IAEA, Vienna, pp. 227–241.

Halas S. and Krouse H.R. 1982. Isotopic abundances of water of crystallization of gypsum from the Miocene evaporite formation, Carpathian Foredeep, Poland. Geochim. Cosmochim. Ac. 46: 293–296.

Halas S., Szaran J. and Niezgoda H. 1997. Experimental determination of carbon isotope equilibrium fractionation between dissolved carbonate and carbon dioxide. Geochim. Cosmochim. Ac. 61: 2691–2695.

Hedges R.W.E., Stevens R.E. and Koch P.L. This volume. Isotopes in bones and teeth. In: Leng M.J. (ed.), Isotopes in Palaeoenvironmental Research. Springer, Dordrecht, The Netherlands.

Hendy G. 1971. The isotopic geochemistry of speleothems. Geochim. Cosmochim. Ac. 35: 801–824.

Herczeg A.L. 1988. Early diagenesis of organic matter in lake sediments: a stable carbon isotope study of pore waters. Chem. Geol. (Isot. Geosci.) 72: 199–209.

Herrmann A. and Stichler W. 1980. Groundwater-runoff relationships. Catena 7: 251–264.

Hoefs J. 1973. Stable Isotope Geochemistry. Springer-Verlag, Berlin, 140 pp.

Hoffmann G., Werner M. and Heimann M. 1998. Water isotope module of the ECHAM atmospheric general circulation model: A study on timescales from days to several years. J. Geophys. Res. 103(D14): 16,871–16,896.

Holdsworth G., Fogarasi S. and Krouse H.R. 1991. Variation of the stable isotopes of water with altitude in the Saint Elias Mountains of Canada. J. Geophys. Res. 96(D4): 7483–7494.

Holmes J.A. 1996. Trace-element and stable-isotope geochemistry of non-marine ostracod shells in Quaternary paleoenvironmental reconstruction. J. Paleolimnol. 15: 223–235.

Horita J. 1990. Stable isotope paleoclimatology of brine inclusions in halite: modeling and application to Searles Lake, California. Geochim. Cosmochim. Ac. 54: 2059–2073.

Horita J. and Wesolowski D. 1994. Liquid-vapour fractionation of oxygen and hydrogen isotopes of water from the freezing to the critical temperature. Geochim. Cosmochim. Ac. 47: 2314–2326.

Hunt R.J., Bullen T.D., Krabbenhoft D.P. and Kendall C. 1998. Using stable isotopes of water and strontium to investigate the hydrology of a natural and a constructed wetland. Ground Water 36: 434–443.

Iacumin P. and Longinelli A. 2002. Relationship between $\delta^{18}O$ values for skeletal apatite from reindeer and foxes and yearly mean $\delta^{18}O$ values of environmental water. Earth Planet. Sc. Lett. 201: 213–219.

IAEA (International Atomic Energy Agency). 1992. Statistical treatment of environmental isotope data in precipitation. 2nd edition. Technical Report Series No. 331. IAEA, Vienna. 784 pp.

IAEA (International Atomic Energy Agency). 2002. GNIP Maps and Animations. IAEA, Vienna, http://isohis.iaea.org.

Inoue H. and Sugimura Y. 1985. Carbon isotopic fractionation during the CO_2 exchange process between air and seawater under equilibrium and kinetic conditions. Geochim. Cosmochim. Ac. 49: 2453–2460.

IPCC (Intergovernmental Panel on Climate Change). 2003. Workshop on Climate Sensitivity, Paris, 2003.

Ito E. 2001. Application of stable isotope techniques to inorganic and biogenic carbonates. In: Last W.M. and Smol J.P. (eds), Tracking Environmental Change Using Lake Sediments, Vol. 2: Physical and Chemical Techniques. Kluwer Academic Publishers, Dordrecht, pp. 351–371.

Jacob H. and Sonntag C. 1991. An 8-year record of the seasonal variations of ^{2}H and ^{18}O in atmospheric water vapour and precipitation at Heidelberg, Germany. Tellus 43B: 291–300.

Jimenez-Lopez C. and Romanek C.S. 2004. Precipitation kinetics and carbon isotope partitioning of inorganic siderite at 25°C and 1 atm. Geochim. Cosmochim. Ac. 68: 557–571.

Jimenez-Lopez C., Caballero E., Huertas F.J. and Romanek C.S. 2001. Chemical, mineralogical and isotope behaviour, and phase transformation during the precipitation of calcium carbonate minerals from intermediate ionic solution at 25°C. Geochim. Cosmochim. Ac. 65: 3219–3231.

Johnsen S.J., Dahl-Jensen D., Dansgaard W. and Gundestrup N. 1995. Greenland palaeotemperatures derived from GRIP borehole temperature and ice core isotope profiles. Tellus 47B: 624–629.

Johnsen S.J., Dahl-Jensen D., Gundestrup N., Steffensen J.P., Clausen H.B., Miller H., Masson-Delmotte V., Sveinbjornsdottir A.E. and White J. 2001. Oxygen isotope and palaeotemperature records from six Greenland ice-core stations: Camp Century, Dye-3, GRIP, GISP2, Renland and NorthGRIP. J. Quaternary Sci. 16: 299–307.

Johnson D.G., Jucks K.W., Traub W.A. and Chance K.V. 2001. Isotopic composition of stratospheric water vapour: Measurements and photochemistry. J. Geophys. Res. 106D: 12,211–12,217.

Jouzel J. 1999. Calibrating the isotopic paleothermometer. Science 286: 910–911.

Jouzel J. and Merlivat L. 1984. Deuterium and oxygen-18 in precipitation, modelling of the isotope effect during snow formation. J. Geophys. Res. 89D: 11,749–11,757.

Jouzel J., Koster R.D., Suozzo R.J., Russell G.L., White J.W.C. and Broecker W.S. 1991. Simulations of the HDO and H$_2$^{18}O atmospheric cycles using the NASA GISS General Circulation Model: Sensitivity experiments for present-day conditions. J. Geophys. Res. 96D: 7495–7507.

Jouzel J., Hoffmann G., Koster R.D. and Masson V. 2000. Water isotopes in precipitation: data/model comparison for present-day and past climates. Quaternary Sci. Rev. 19: 363–379.

Jouzel J., Koster R.D., Suozzo R.J. and Russell G.L. 1994. Stable water isotope behaviour during the last glacial maximum: A general circulation model analysis. J. Geophys. Res. 99D: 25,791–25,801.

Jouzel J., Vimeux F., Caillon N., Delaygue G., Hoffmann G., Masson-Delmonte V. and Parrenin F. 2003. Magnitude of isotope/temperature scaling for interpretation of central Antarctic ice cores. J. Geophys. Res. 108D: 4361. doi:10.1029/2002JD002677.

Kaufman A., Bar-Matthews M., Ayalon A. and Carmi I. 2003. The vadose flow above Soreq Cave, Israel: a tritium study of the cave waters. J. Hydrol. 273: 155–163.

Keeling C.D. 1979. The Suess effect: 13-Carbon – 14-Carbon interrelations. Environ. International 2: 229–300.

Keeling C.D., Mook W.G. and Tans P.P. 1979. Recent trends in the ^{13}C/^{12}C ratio of atmospheric carbon dioxide. Nature 277: 121–123.

Keeling C.D., Bacastow R.B. and Tans P.P. 1980. Predicted shift in the ^{13}C/^{12}C ratio of atmospheric carbon dioxide. Geophys. Res. Lett. 7: 505–508.

Keeling C.D., Whorf T.P., Wahlen M. and van der Plichtt J. 1995. Interannual extremes in the rate of rise and atmosphere carbon dioxide since 1980. Nature 375: 666–670.

Keith D.W. 2000. Stratosphere-troposphere exchange, Inferences from the isotopic composition of water vapour. J. Geophys. Res. 105D: 15,167–15,173.

Keith M.L. and Weber J.N. 1964. Isotopic composition and environmental classification of selected limestones and fossils. Geochim. Cosmochim. Ac. 28: 1787–1816.

Kendall C. and Coplen T.B. 2001. Distribution of oxygen-18 and deuterium in river waters across the United States. Hydrol. Process. 15: 1363–1393.

Knauth L.P. and Roberts S.K. 1991. The hydrogen and oxygen isotope history of the Silurian-Permian hydrosphere as determined by direct measurement of fossil water. In: Taylor H.P.,

O'Neil J.R. and Kaplan I.R. (eds), Stable Isotope Geochemistry: A Tribute to Samuel Epstein. Spec. Publ. No. 3, The Geochem. Society, San Antonio, Texas, pp. 91–104.

Knauth L.P., Brilli M. and Klonowski S. 2003. Isotope geochemistry of caliche developed on basalt. Geochim. Cosmochim. Ac. 67: 185–195.

Kortelainen N.M. and Karhu J.A. 2004. Regional and seasonal trends in the oxygen and hydrogen isotope ratios of Finnish groundwaters: a key for mean annual precipitation. J. Hydrol. 285: 143–157.

Krabbenhoft D.P., Bowser C.J. and Kendall C. 1994. Use of O-18 and deuterium to assess the hydrology of groundwater-lake systems. Advances in Chemistry Series 237: 67–90.

Krishnamurthy R.V. and Bhattacharya S.K 1991. Stable oxygen and hydrogen isotope ratio in shallow ground waters from Northern India and a study of the role of evapotranspiration in the Indian monsoon. In: Taylor H.P., O'Neil J.R. and Kaplan I.R. (eds), Stable Isotope Geochemistry: A Tribute to Samuel Epstein. Spec. Publ. No. 3, The Geochem. Society, San Antonio, Texas, pp. 1–7.

Krouse H.R. and McKay J.R. 1971. Application of the $H_2^{18}O/H_2^{16}O$ abundances to the problem of lateral mixing in the Liard/Mackenzie river system, Canada. Can. J. Earth Sci. 8: 1107–1115.

Kurita N., Yoshida N., Inoue, G. and Chayanova E.A. 2004. Modern isotope climatology of Russia: A first assessment. J. Geophys. Res., 109D: D03102. doi:10.1029/2003JD003404.

Lambs L. and Berthelot M. 2002. Monitoring water from the underground to the tree: first results with a new sap extractor on a riparian woodland. Plant Soil 242: 197–207.

Landon M.K., Delin G.N., Komor S.C. and Regan C.P. 1999. Comparison of the stable-isotopic composition of soil water collected from suction lysimeters, wick samplers, and cores in a sandy unsaturated zone. J. Hydrol. 224: 45–54.

Lange J., Greenbaum N., Husary S., Timmer J., Leibundgut C. and Schick A.P. 2003. Tracers for runoff generation studies in a Mediterranean region: comparison of different scales. IAHS Publ. 278: 117–123.

Laudon H., Seibert J., Kohler S. and Bishop K. 2004. Hydrological flow paths during snowmelt: Congruence between hydrometric measurements and oxygen-18 in meltwater, soil water, and runoff. Water Resour. Res. 40: W03102. doi:10.1029/2003WR002455.

Lawler H.A. 1987. Sampling for isotopic responses in surface waters. Earth Surf. Processes 12: 551–559.

Leng M.J., Lamb A.L., Heaton T.H.E., Marshall J.D., Wolfe B.B., Jones M.D., Holmes J.A. and Arrowsmith C. This volume. Isotopes in lake sediments. In: Leng M.J. (ed.), Isotopes in Palaeoenvironmental Research. Springer, Dordrecht, The Netherlands.

Lewis S. 1979. Environmental isotope balance of Lake Kinneret as a tool in evaporation rate estimates. In: Isotopes in Lake Studies. IAEA, Vienna, pp. 33-65.

Li H.C. and Ku T.L. 1997. $\delta^{13}C$-$\delta^{18}O$ covariance as a paleohydrological indicator for closed-basin lakes. Palaeogeogr. Palaeoecol. 133: 69–80.

Liu B., Phillips F., Hoines S., Campbell R. and Sharma P. 1995. Water movement in desert soil traced by hydrogen and oxygen isotopes, chloride, and chlorine-36, southern Arizona. J. Hydrol. 168: 91–110.

Love A.J., Herczeg A.L., Sampson L, Cresswell R.G. and Fifield L.K. 2000. Sources of chloride and implications for ^{36}Cl dating of old groundwater, southwestern Great Artesian Basin, Australia. Water Resour. Res. 36: 1561–1574.

Majoube M. 1971. Fractionnement en oxygene-18 et en deuterium entre l'eau et sa vapour. J. Chem. Phys. 197: 1423–1436.

Manzano M., Custodio E., Loosli H.H., Cabrera M.C., Riera X. and Custodio J. 2001. Palaeowater in coastal aquifers of Spain. In: Edmunds W.M. and Milne C.J. (eds), Palaeowaters in Coastal Europe. Geol. Soc. Lond. Spec. Publ. No. 189, pp. 107–138.

Marechal J.C. and Etcheverry D. 2003. The use of ^3H and ^{18}O tracers to characterize water inflows in Alpine tunnels. Appl. Geochem. 18: 339–351.

Mathieu R. and Bariac T. 1996. An isotopic study (^2H and ^{18}O) on water movements in clayey soils under a semiarid climate. Water Resour. Res. 32: 779–789.

Matsui E., Salati F., Friedman I. and Brinkman W.L.F. 1976. Isotopic hydrology in the Amazonia, 2, Relative discharges of the Negro and Solimoes rivers through [18]O concentrations. Water Resour. Res. 12: 781–785.

McCarroll D. and Loader N. This volume. Isotopes in tree rings. In: Leng M.J. (ed.), Isotopes in Palaeoenvironmental Research. Springer, Dordrecht, The Netherlands.

McDermott F., Schwarcz H. and Rowe P.J. This volume. Isotopes in speleothems. In: Leng M.J. (ed.), Isotopes in Palaeoenvironmental Research. Springer, Dordrecht, The Netherlands.

McDonnell J.J., Stewart M.K. and Owens I.F. 1991. Effect of catchment-scale subsurface mixing on stream isotopic response. Water Resour. Res. 27: 3065–3073.

McMahon P.B. and Chapelle F.H. 1991. Geochemistry of dissolved inorganic carbon in a Coastal Plain aquifer. 2. Modeling carbon sources, sinks, and $\delta^{13}C$ evolution. J. Hydrol. 127: 109–135.

Menot-Combes G., Burns S.J. and Leuenberger M. 2002. Variations of $^{18}O/^{16}O$ in plants from temperate peat bogs (Switzerland): implications for paleoclimatic studies. Earth Planet. Sc. Lett. 202: 419–434.

Merlivat L. and Jouzel J. 1979. Global climatic interpretation of the deuterium–oxygen-18 relationship for precipitation. J. Geophys. Res. 84: 5029–5033.

Mickler P.J., Banner J.L., Stern L., Asmerom Y., Edwards R.L. and Ito E. 2004. Stable isotope variations in modern tropical speleothems: Evaluating equilibrium vs. kinetic isotope effects. Geochim. Cosmochim. Ac. 68: 4381–4393.

Midwood A.J., Boutton T.W., Watts S. and Archer S. 1993. Natural abundance of ^{2}H and ^{18}O in soil moisture, rainfall and plants in a subtropical thorn woodland ecosystem: implications for plant water use. In: Isotope Techniques in the Study of Past and Current Environmental Changes in the Hydrosphere and Atmosphere. IAEA, Vienna, pp. 419–431.

Mook W.G. 1970. Stable carbon and oxygen isotopes of natural waters in the Netherlands. In: Isotope Hydrology 1970. IAEA, Vienna, pp. 163–189.

Mook W.G., Bommerson J.C. and Staverman W.H. 1974. Carbon isotope fractionation between dissolved bicarbonate and gaseous carbon dioxide. Earth Planet. Sc. Lett. 22: 169–176.

Mudd G.M. 2000. Mound Springs of the Great Artesian Basin in South Australia: a case study from Olympic Dam. Environ. Geol. 39: 463–476.

Nativ R., Gunay G., Hotzl H., Reichert B., Solomon D.K. and Tezcan L. 1999. Separation of groundwater-flow components in a karstified aquifer using environmental tracers. Appl. Geochem. 14: 1001–1014.

Navada S.V. and Rao S.M. 1991. Study of Ganga River–groundwater interaction using environmental [18]O. Isotopenpraxis 27: 380–384.

Nier A.O. and Gulbransen E.A. 1939. Variations in the relative abundance of the carbon isotopes. J. Am. Chem. Soc. 61: 697–698.

Niggemann S., Mangini A., Richter D.K. and Wurth G. 2003. A paleoclimate record of the last 17,600 years in stalagmites from the B7 cave, Sauerland, Germany. Quaternary Sci. Rev. 22: 555–567.

Nkotagu H.H. 1996. Application of environmental isotopes to groundwater recharge studies in a semi-arid fractured crystalline basement area of Dodoma, Tanzania. J. Afr. Earth Sci. 22: 443–457.

Ofterdinger U.S., Balderer W., Loew S. and Renard P. 2004. Environmental isotopes as indicators for ground water recharge to fractured granite. Ground Water 42: 868–879.

Oremland R.S., Miller L.G. and Whiticar M.J. 1987. Sources and fluxes of natural gases from Mono Lake, California. Geochim. Cosmochim. Ac. 51: 2915–2929.

Papadimitriou S., Kennedy H., Kottner G., Dieckmann G.S. and Thomas D.N. 2004. Experimental evidence for carbonate precipitation and CO_2 degassing during sea ice formation. Geochim. Cosmochim. Ac. 68: 1749–1761.

Parkhurst D.L. and Appelo C.A.J. 1999. User's guide to PHREEQC (Version 2) - A computer program for speciation, batch-reaction, one-dimensional transport, and inverse geochemical calculations: U.S. Geol. Surv. Water-Resources Investigations Rep. 99-4259, 310 pp. Code available at: http://wwwbrr.cr.usgs.gov/projects/GWC_coupled/phreeqci/index.html.

62 DARLING, BATH, GIBSON AND ROZANSKI

Payne B.R., Quijano L. and Latorre C. 1979. Environmental isotopes in a study of the origin of salinity of groundwater in the Mexicali Valley, Mexico. J. Hydrol. 41: 201–215.
Pazdur A., Pazdur M.F., Starkel L. and Szulc J. 1988. Stable isotopes of Holocene calcareous tufa in Southern Poland as paleoclimatic indicators. Quaternary Res. 30: 177–189.
Pearson F.J. and Hanshaw B.B. 1970. Sources of dissolved carbonate species in groundwater and their effects on carbon-14 dating. In: Isotope Hydrology 1970. IAEA, Vienna, pp. 271–286.
Perrin J., Jeannin P.Y. and Zwahlen F. 2003. Epikarst storage in a karst aquifer: a conceptual model based on isotopic data, Milandre test site, Switzerland. J. Hydrol. 279: 106–124.
Pichon A., Travi Y. and Marc V. 1996. Chemical and isotopic variations in throughfall in a Mediterranean context. Geophys. Res. Lett. 23: 531–534.
Plummer L.N. 1977. Defining reactions and mass transfer in part of the Floridan aquifer. Water Resour. Res. 13: 801–812.
Plummer L.N., McConnell J.B., Busenberg E., Drenkard S., Schlosser P. and Michel R.L. 1998. Flow of river water into a karstic limestone aquifer—1. Tracing the young fraction in groundwater mixtures in the Upper Floridan aquifer near Valdosta, Georgia. Appl. Geochem. 13: 995–1015.
Polya D.A., Foxford K.A., Stuart F., Boyce A. and Fallick A.E. 2000. Evolution and paragenetic context of low δD hydrothermal fluids from the Panasqueira W-Sn deposit, Portugal: new evidence from microthermometric, stable isotope, noble gas and halogen analyses of primary fluid inclusions. Geochim. Cosmochim. Ac. 64: 3357–3371.
Quade J. 2004. Isotopic records from ground-water and cave speleothem calcite in North America. Dev. Quatern. Sci. 1: 205–220.
Rank D., Rajner V., Nussbaumer W., Papesch W., Dreher J. and Reitinger J. 1984. Study of the interrelationships between groundwater and lake water at Neusiedlersee, Austria. In: Isotope Hydrology 1983. IAEA, Vienna, pp. 67-81.
Rank D., Adler A., Araguas-Araguas L., Froehlich K., Rozanski K. and Stichler W. 1998. Hydrological parameters and climatic signals derived from long term tritium and stable isotope time series of the river Danube. In: Isotope Techniques in the Study of Environmental Change. IAEA, Vienna, pp. 191–207.
Rao S.M., Jain S.K., Navada S.V., Nair A.R. and Shivana K. 1987. Isotopic studies on sea water intrusion and interpretations between water bodies: some field examples. In: Isotope Techniques in Water Resources Development. IAEA, Vienna, pp. 403-425.
Reardon E.J. and Fritz P. 1978. Computer modelling of groundwater ^{13}C and ^{14}C isotope compositions. J. Hydrol. 36: 201–224.
Ricken W., Steuber T., Freitag H., Hirschfeld M. and Niedenzu B. 2003. Recent and historical discharge of a large European river system — oxygen isotopic composition of river water and skeletal aragonite of Unionidae in the Rhine. Palaeogeogr. Palaeoecol. 193: 73–86.
Rindsberger M., Jaffe S., Rahmain S. and Gat J.R. 1990. Patterns of the isotopic composition of precipitation in time and space: data from the Israeli storm water collection program. Tellus 42B: 263–271.
Robertson I., Waterhouse J.S., Barker A.C., Carter A.H.C. and Switsur V.R. 2001. Oxygen isotope ratios of oak in east England: implications for reconstructing the isotopic composition of precipitation. Earth Planet. Sc. Lett. 191: 21–31.
Roedder E. 1984. Fluid inclusions. Min. Soc. America. Reviews in Mineralogy 12, 644 pp.
Romanek C.S., Grossman E.L. and Morse J.L. 1992. Carbon isotopic fractionation in synthetic aragonite and calcite: Effects of temperature and precipitation rate. Geochim. Cosmochim. Ac. 56: 419–430.
Romanek C.S., Zhang C.L., Li Y., Horita J., Vali H., Cole D.R. and Phelps T.J. 2003. Carbon and hydrogen isotope fractionations associated with dissimilatory iron-reducing bacteria. Chem. Geol. 195: 5–16.
Rozanski K. and Sonntag C. 1982. Vertical distribution of deuterium in atmospheric water vapour. Tellus 34: 135–141.

Rozanski K. and Araguas-Araguas L. 1995. Spatial and temporal variability of stable isotope composition of precipitation over the South American continent. Bull. Inst. Fr. Etudes Andines 23: 379–390.

Rozanski K., Sonntag C. and Münnich K.O. 1982. Factors controlling stable isotope composition of European precipitation. Tellus 34: 142–150.

Rozanski K., Araguas-Araguas L. and Gonfiantini R. 1992. Relation between long-term trends of oxygen-18 isotope composition of precipitation and climate. Science 258: 981–985.

Rozanski K., Araguas-Araguas L. and Gonfiantini R. 1993. Isotopic patterns in modern global precipitation. Climate Change in Continental Isotopic Records. Am. Geophys. Union, Geophys. Monogr. 78: 1–36.

Salati E., Dall'Olio A., Matsui E. and Gat J.R. 1979. Recycling of water in the Amazon basin: an isotopic study. Water Resour. Res. 15: 1250–1258.

Salomons W. and Mook W.G. 1986. Isotope geochemistry of carbonates in the weathering zone. In: Fritz P. and Fontes J.C. (eds), Handbook of Environmental Isotope Geochemistry, Vol. 1, The Terrestrial Environment, A. Elsevier, Amsterdam, pp. 239–270.

Saxena R.K. 1986. Estimation of canopy reservoir capacity and oxygen-18 fractionation in throughfall in a pine forest. Nordic Hydrol. 17: 251–260.

Schoch-Fischer H., Rozanski K., Jacob H., Sonntag C., Jouzel J., Oestlund G. and Geyh M.A. 1984. Hydrometeorological factors controlling the time variation of D, ^{18}O and ^{3}H in atmospheric water vapour and precipitation in the northern westwind belt. In: Isotope Hydrology 1983, IAEA, Vienna, pp. 3–31.

Schoell M. 1980. The hydrogen and carbon isotopic composition of methane from natural gases of various origins. Geochim. Cosmochim. Ac. 44: 649–661.

Schoell M. 1984. Stable isotopes in petroleum research. In Brooks J. and Wete D. (eds), Advances in Petroleum Geochemistry, Vol. 1. Academic Press, London, pp. 215–245.

Schuerch M., Kozel R., Schotterer U. and Tripet J.P. 2003. Observation of isotopes in the water cycle - the Swiss National Network (NISOT). Environ. Geol. 45: 1–11.

Schwalb A. 2003. Lacustrine ostracodes as stable isotope recorders of late-glacial and Holocene environmental dynamics and climate. J. Paleolimnol. 29: 265–351.

Schwarcz H.P. 1986. Geochronology and isotope geochemistry of speleothems. In: Fritz P. and Fontes J.C. (eds), Handbook of Environmental Isotope Geochemistry: Vol 2, The Terrestrial Environment, B. Elsevier, Amsterdam, pp. 271–303.

Serefiddin F., Schwarcz H.P., Ford D.C. and Baldwin S. 2004. Late Pleistocene paleoclimate in the Black Hills of South Dakota from isotope records in speleothems. Palaeogeogr. Palaeoecol. 203: 1–17.

Shackleton N.J. 1987. Oxygen isotopes, ice volume and sea level. Quaternary Sci. Rev. 6: 183–190.

Sharma M.L. and Hughes M.W. 1985. Groundwater recharge estimation using chloride, deuterium and oxygen-18 profiles in the deep coastal sands of Western Australia. J. Hydrol. 81: 93–109.

Siegenthaler U. and Oeschger H. 1980. Correlation of O-18 in precipitation with temperature and altitude. Nature 285: 314–317.

Simpkins W.W. and Parkin T.B. 1993. Hydrogeology and redox geochemistry of CH_4 in a late Wisconsinan till and loess sequence in central Iowa. Water Resour. Res. 29: 3643–3657.

Simpson H.J. and Herczeg A.L. 1991. Stable isotopes as indicators of evaporation in the River Murray, Australia. Water Resour. Res. 27: 1925–1935.

Skidmore M., Sharp M. and Tranter M. 2004. Kinetic isotopic fractionation during carbonate dissolution in laboratory experiments: Implications for detection of microbial CO_2 signature using δ^{13}C-DIC. Geochim. Cosmochim. Ac. 68: 4309–4317.

Sklash M.G. and Farvolden R.N. 1979. The role of groundwater in storm runoff. J. Hydrol. 43: 45–65.

Smith D.B., Otlet R.L., Downing R.A., Monkhouse R.A. and Pearson F.J. 1975. Stable carbon
    and oxygen isotope ratios of groundwaters from the Chalk and Lincolnshire Limestone. Nature
    257: 783–784.
Smith D.M., Jackson N.A., Roberts J.M. and Ong C.K. 1999. Reverse flow of sap in tree roots and
    downward siphoning of water by Grevillae robusta. Funct. Ecol. 13: 256–264.
Smith R.B. 1992. Deuterium in North Atlantic storm tops. J. Atm. Sci. 49: 2041–2057.
Sofer Z. 1978. Isotopic composition of hydration-water in gypsum. Geochim. Cosmochim. Ac.
    42: 1141–1149.
Sonntag C., Christmann D. and Munnich K.O. 1985. Laboratory and field experiments on
    infiltration and evaporation of soil water by means of deuterium and oxygen-18. In: Stable and
    radioactive isotopes in the study of the unsaturated soil zone, IAEA, Vienna, TECDOC-357,
    145-159.
Sonntag C., Rozanski K., Munnich K.O. and Jacob H. 1983. Variations of deuterium and oxygen-18
    in continental precipitation and groundwater and their causes. In: Street-Perrott F.A., Beran
    M.A. and Ratcliffe R.G. (eds), Variations in the Global Water Budget. D. Reidel Publishing
    Company, pp. 107–124.
Soulsby C., Malcolm R., Helliwell R.C., Ferrier R.C. and Jenkins A. 2000. Isotope hydrology of
    the Allt a'Mharcaidh catchment, Cairngorm mountains, Scotland: implications for hydrological
    pathways and water residence times. Hydrol. Process. 14: 747–762.
Stewart M.K. and Downes C.J. 1982. Isotope hydrology of Waikoropupu Springs, New Zealand.
    In: Perry E.C. and Montgomery C.W. (eds), Isotope Studies in Hydrologic Processes, Northern
    Illinois University Press, pp. 15–23.
Stiller M., Rounick J.S. and Shasha S. 1985. Extreme carbon-isotope enrichments in evaporating
    brines. Nature 316: 434–435.
Stuiver M. 1970. Oxygen and carbon isotope ratios of fresh-water carbonates as climatic indicators:
    J. Geophys. Res. 75: 5247–5257.
Sturchio N.C., Sultan M., Arehart G.B., Sano Y., Kamar Y.A. and Sayed M. 1996. Composition
    and origin of thermal waters in the Gulf of Suez area, Egypt. Appl. Geochem. 11: 471-479.
Stute M. and Schlosser P. 2000. Atmospheric noble gases. In: Cook P.G. and Herczeg A. (eds),
    Environmental Tracers in Subsurface Hydrology. Kluwer Academic Publishers, pp. 349–379.
Sultan M., Sturchio N., Hassan F.A., Hamdan M.A.R., Mahmood A.M., El Alfy Z. and Stein T.
    1997. Precipitation source inferred from stable isotopic composition of Pleistocene ground-
    water and carbonate deposits in the Western Desert of Egypt. Quaternary Res. 48: 29–37.
Szaran J. 1997. Achievement of carbon isotope equilibrium in the system HCO_3^- (solution) –
    CO_2(gas). Chem. Geol. 142: 79–86.
Talbot M.R. 1990. A review of the palaeohydrological interpretation of carbon and oxygen
    isotopic ratios in primary lacustrine carbonates. Chem. Geol. (Isot. Geosci.) 80: 261–279.
Talbot M.R. and Kelts K. 1990. Palaeolimnological signatures from carbon and oxygen isotopic
    ratios in carbonates from organic-rich lacustrine sediments. In: Katz B.J. (ed.), Lacustrine
    Exploration: Case Studies and Modern Analogues. Memoir No. 50, Am. Assoc. Petrol. Geol.,
    Tulsa, OK, pp. 99–112.
Taylor C.B. 1972. The vertical variations of isotopic compositions of tropospheric water vapour
    over continental Europe and their relations to tropospheric structure. Inst. Nucl. Sci. Rep. INS-
    R-107, Lower Hutt, New Zealand, 44 pp.
Telmer K. and Veizer J. 1999. Carbon fluxes, pCO_2 and substrate weathering in a large northern
    river basin, Canada: carbon isotope perspectives. Chem. Geol. 159: 61–86.
Thode H.G., Shima M., Rees C.E. and Krishnamurty K.V. 1965. Carbon-13 isotope effects in
    systems containing carbon dioxide, bicarbonate, carbonate, and metal ions. Can. J. Chemistry
    43: 582–595.
Thompson L.G. 2001. Stable isotopes and their relationships to temperature as recorder in
    low-latitude ice cores. In: Gerhard L.C., Harrison W.E. and Hanson B.M. (eds), Geological
    Perspectives of Global Climate Change, Studies in Geology No. 47, Am. Assoc. Petrol. Geol.,
    Tulsa, Oklahoma, pp. 99-119.

Thorstenson D.C., Fisher D.W. and Croft M.G. 1979. The Geochemistry of the Fox Hills-Basal Hill Creek aquifer in southeastern North Dakota and northwestern South Dakota. Water Resour. Res. 15: 1479–1498.

Turi B. 1986. Stable isotope geochemistry of travertines. In: Fritz P. and Fontes J.C. (eds), Handbook of Environmental Isotope Geochemistry: Vol 2, The Terrestrial Environment, B. Elsevier, Amsterdam, pp. 207–238.

Turner J.V. 1982. Kinetic fractionation of carbon-13 during calcium carbonate precipitation. Geochim. Cosmochim. Ac. 46: 1183–1191.

Turner J.V. and Fritz P. 1982. Enriched ^{13}C composition of interstitial waters in sediments of a freshwater lake. Can. J. Earth Sci. 20: 616–621.

Tyler S.C. 1986. Stable carbon isotope ratios in atmospheric methane and some of its sources. J. Geophys. Res. 91D: 13,232–13,238.

Usdowski E. and Hoefs J. 1986. ^{13}C/^{12}C partitioning and kinetics of CO_2 absorption by hydroxide buffer solutions. Earth Planet. Sc. Lett. 8: 130–134.

van Ommen T.D. and Morgan V. 1997. Calibrating the ice core paleothermometer using seasonality. J. Geophys. Res. 102D: 9351–9357.

Verhagen B.T., Geyh M.A., Froehlich K. and Wirth K. 1991. Isotope hydrological methods for the quantitative evaluation of groundwater resources in arid and semi-arid areas: Development of a methodology. Rep. Federal Ministry for Economic Cooperation, Bonn, FRG, 164 pp.

Vogel J.C. and Ehhalt D. 1963. The use of the carbon isotopes in groundwater studies. In: Radioisotopes in Hydrology. IAEA, Vienna, pp. 383–395.

Vogel J.C., Grootes P.M. and Mook W.G. 1970. Isotope fractionation between gaseous and dissolved carbon dioxide. Z. Phys. 230: 255–258.

Vogt H.J. 1976. Isotopentrennung bei der Verdampfung von Wasser. PhD thesis, University of Heidelberg, Germany.

von Grafenstein U., Erlenkeuser H., Müller J. and Alefs J. 1996. A 200-year mid-European temperature record preserved in lake sediments: an extension of the $\delta^{18}O_P$–air temperature relation into the past. Geochim. Cosmochim. Ac. 60: 4025–4036.

von Grafenstein U., Erlenkeuser H., Brauer A., Jouzel J. and Johnsen S.J. 1999. A Mid-European decadal isotope-climate record from 15,500 to 500 years B.P. Science 284: 1654–1657.

Wang X.F. and Yakir D. 2000. Using stable isotopes of water in evapotranspiration studies. Hydrol. Process. 14: 1407–1421.

Wassenaar L.I., Aravena R., Fritz P. and Barker J.F. 1991. Controls on the transport and carbon isotopic composition of dissolved organic carbon in a shallow groundwater system, Central Ontario, Canada. Chem. Geol. (Isot. Geosci.) 87: 39–57.

Welhan J.A. and Fritz P. 1977. Evaporation pan isotopic behavior as an index of isotopic evaporation conditions. J. Hydrol. 41: 682–686.

Wendt I. 1968. Fractionation of carbon isotopes and its temperature dependence in the system CO_2-Gas–CO_2 in solution and HCO_3–CO_2 in solution. Earth Planet. Sc. Lett. 4: 64–68.

Wendt I. 1971. Carbon and oxygen isotope exchange between HCO_3 in saline solution and solid $CaCO_3$. Earth Planet. Sc. Lett. 12: 439–442.

Werner M., Mikolajewicz U., Hoffmann G. and Heimann M. 2000a. Possible changes of the delta O 18 in precipitation caused by a meltwater event in the North Atlantic. J. Geophys. Res., 105D: 10161–10167.

Werner M., Mikolajewicz U., Heimann M., and Hoffmann G. 2000b. Borehole versus isotope temperature on Greenland: Seasonality does matter. Geophys. Res. Lett. 27: 723–726.

White J.W.C., Cook E.R., Lawrence J.R. and Broecker W.S. 1985. The D/H ratio of sap in trees: implications for water sources and tree ring D/H ratios. Geochim. Cosmochim. Ac. 49, 237–246.

White T., Gonzalez L., Ludwigson G. and Poulsen C. 2001. Middle Cretaceous greenhouse hydrologic cycle of North America. Geology 29: 363–366.

Wigley T.M.L., Plummer L.N. and Pearson F.J. 1978. Mass transfer and carbon isotope evolution in natural water systems. Geochim. Cosmochim. Ac. 42: 1117–1139.

Williams P.W. and Fowler A. 2002. Relationship between oxygen isotopes in rainfall, cave percolation waters and speleothem calcite at Waitomo, New Zealand. New Zeal. J. Hydrol. 41: 53–70.

Winograd I.J., Coplen T.B., Landwehr J.M., Riggs A.C., Ludwig K.R., Szabo B.J., Kolesar P.T. and Revesz K.M. 1992. Continuous 500,000-year climate record from vein calcite in Devils Hole, Nevada. Science 258: 255–260.

Winston W.E. and Criss R.E. 2003. Oxygen isotope and geochemical variations in the Missouri River. Environ. Geol. 43: 546–556.

Winston W.E. and Criss R.E. 2004. Dynamic hydrologic and geochemical response in a perennial karst spring, Water Resour. Res. 40: W05106. doi:10.1029/2004WR003054.

Wolfe B.B., Edwards T.W.D., Beuning K.R.M. and Elgood R.J. 2001. Carbon and oxygen isotope analysis of lake sediment cellulose: methods and applications. In: Last W.M. and Smol J.P. (eds), Tracking Environmental Change Using Lake Sediments, Vol. 2, Physical and Geochemical Methods, Developments in Paleoenvironmental Research, Kluwer Academic Publishers, Dordrecht, pp. 373–400.

Woodward D.F., Farag A., Hubert W.A., Goldstein J.N. and Meyer J.S. 2000. Effects of geothermal effluents on rainbow trout and brown trout in the Firehole River, Yellowstone National Park, Wyoming. Rep. Columbia Environ. Res. Center, USGS, 124 pp.

Yeh H.W. and Wang W.M. 2001. Factors affecting the isotopic composition of organic matter. (1) Carbon isotopic composition of terrestrial plant materials. Proc. Nat. Sci. Council. ROC(B), 25: 137–147. http://nr.stic.gov.tw/ejournal/ProceedingB/v25n3/137-147.pdf.

Yonge C.J., Ford D.C., Gray J. and Schwarcz H.P. 1985. Stable isotope studies of cave seepage water. Chem. Geol. 58: 97–105.

Yoshimura K., Liu Z., Cao J., Yuan D., Inokura Y and Noto M. 2004. Deep source CO_2 in natural waters and its role in extensive tufa deposition in the Huanglong Ravines, Sichuan, China. Chem. Geol. 205: 141–153.

Yurtsever Y. and Gat J.R. 1981. Atmospheric waters. In: Gat J.R. and Gonfiantini R. (eds), Stable Isotope Hydrology: Deuterium and Oxygen-18 in the Water Cycle. Tech. Rep. Series No. 210. IAEA, Vienna, pp. 103–142.

Zhang C.L., Fouks B.W., Bonheyo G.T., Peacock A.D., White D.C., Huang Y. and Romanek C.S. 2004. Lipid biomarkers and carbon-isotopes of modern travertine deposits (Yellowstone national Park, USA): Implications for biogeochemical dynamics in hot-spring systems. Geochim. Cosmochim. Ac. 68: 3157–3169.

Zhang C.L., Horita J., Cole D.R., Zhou J., Lovley D.R. and Phelps T.J. 2001. Temperature-dependent oxygen and carbon isotope fractionations of biogenic siderite. Geochim. Cosmochim. Ac. 65: 2257–2271.

Zhang J., Quay P.D. and Wilbur D.O. 1995. Carbon isotope fractionation during gas-water exchange and dissolution of CO_2. Geochim. Cosmochim. Ac. 59: 107–114.

Zimmerman U., Ehhalt D., Münnich K.O. 1967. Soil-water movement and evapotranspiration: changes in the isotopic composition of the water. In: Isotopes in Hydrology. IAEA, Vienna, pp. 567–584.

2. ISOTOPES IN TREE RINGS

DANNY McCARROLL (d.mccarroll@swansea.ac.uk)
NEIL J. LOADER (n.j.loader@swansea.ac.uk)
Department of Geography
University of Wales
Swansea SA2 8PP, UK

Key words: Oxygen isotopes, carbon isotopes, hydrogen isotopes, tree rings, dendrochronology, dendroclimatology.

Introduction

Of all the natural archives of palaeoenvironmental information, tree rings hold perhaps the greatest potential for reconstructing past climates at annual or even sub-annual resolution. Carefully constructed tree ring series are perfectly dated and the stable isotopes of carbon, hydrogen and oxygen within the wood record temporal changes in the isotope composition of the source materials used by the tree during wood formation and in their fractionation by the tree in response to its external environment. Suitable trees are widespread, and by piecing together the records from living, dead and sub-fossil wood it is possible to produce long and continuous chronologies (Eronen et al. 2002; Grudd et al. 2002; Hantemirov and Shiyatov 2002). Where each ring from each tree can be analysed individually, it is also possible to quantify the variability of the isotope ratios, so that statistically-defined confidence limits can be placed around mean values and translated into confidence intervals around quantitative palaeoclimate estimates.

Tree ring archives

As trees grow, cell layers are added incrementally around the stem, formed initially of cellulose, but strengthened by lignin. Whereas some of the other constituents of wood, including resins, may move around within the tree, mixing and dampening any temporal signal, the cellulose and lignin framework is effectively fixed and does not change until the wood decomposes (Suberkropp and Klug 1976; Spiker and Hatcher 1987; Benner et al. 1987; Schleser et al. 1999a).

In *extra*-tropical environments, where there are marked seasonal changes in tree growth, and in particular where there is a dormant period when the leaves are no longer actively photosynthesising and no wood cells are formed, trees form annual rings, recognizable because of large differences in the size and shape of the wood cells that form before and after the dormant season. In such cases a discreet layer of wood can be attributed to each growth season, providing an annually-resolved physical archive that extends for the life of the tree. Since there are many tree species that live for several

67

M J. Leng (ed.), 2005. *Isotopes in Palaeoenvironmental Research.* Springer. Printed in The Netherlands.

hundreds of years, and a few that live for millennia, tree rings have the potential to provide long and continuous archives, across a wide variety of environments.

In practice, building long tree ring chronologies is a difficult and time-consuming process. Even where trees are very long-lived, as with the bristlecone pines (*Pinus longaeva*. Bailey) of California (Schulman 1958), a chronology cannot be constructed simply by counting backwards from the bark. It is not unusual, particularly in harsh environments, for individual trees to produce no wood cells in some years, at least around part of the trunk, so that the ring is effectively 'missing' or only partially formed. Similarly, a tree may stop growing and then start again within a single season, so that two rings form within one year, producing a 'false' ring. By measuring the width of each ring, and comparing the ring widths of many individual trees and radii, it is possible to recognise and exclude both 'missing'/locally absent and false rings and so construct a 'master chronology' that provides a continuous and unambiguous record of tree growth (Fritts 1976; Schweingruber 1988; Schweingruber and Briffa 1996).

By comparing the ring widths of an individual tree with the master chronology it is often possible to place a numerical age on each tree ring with a very high degree of certainty. This is the basis of 'dendrochronology' and the technique is used routinely in archaeology. Cross-dated trees also provided the absolute time-frame, and wood samples, that were used to calibrate the radiocarbon time scale (Suess 1970; Pilcher et al. 1984; Becker 1993; Stuiver and Reimer 1993). The physical and chemical characteristics of those rings may also provide an annually-resolved record of changing climate.

The most common way to extract a palaeoclimate signal from tree rings is to measure the width of the rings (Fritts 1976). Ring-widths represent the amount of wood formed in a particular year and thus are likely to reflect, to some extent, net photosynthesis during the growing season. Where growth is limited by some environmental factor, such as summer temperature or drought, variability in the ring widths is likely to reflect variability in that limiting factor, providing a palaeoclimate signal. The relationship between ring widths and climate during a 'calibration' period can be quantified, using regression-based techniques, and checked against similar data from a 'verification' period (Cook and Kairiukstis 1990). Alternatively, verification may be provided using 'bootstrap' approaches, based on multiple random resampling from the same data set (Wigley et al. 1984; Briffa and Jones 1990).

Although ring-width chronologies provide some of the best high-resolution palaeoclimate data sets available (Hughes and Graumlich 1996; Jones et al. 1998; Briffa 2000; Jones and Mann 2004), particularly for the last millennium, there are a number of limitations that restrict their potential. One of the most serious has become known as the 'segment length curse' and derives from the fact that, as trees grow, their ring widths tend to decline (Cook et al. 1995). This phenomenon is largely geometric because, once the tree has reached maturity, the photosynthates used to produce wood cells have to be spread around an ever-increasing surface area. It is easy to remove the decline statistically, using regression-based techniques (usually fitting a negative exponential curve), but this also removes any climate signal that occurs over a time span similar to the life-span of the tree (the segment length). The effect is that tree ring width chronologies faithfully record high-frequency, year-to-year changes in climate, but where statistical detrending has been employed it is much more difficult to retain the lower-frequency long-term changes in climate that are usually of most interest. Alternative data treatments (e.g., Regional Curve Standardisation) that attempt to retain a greater proportion of the low-frequency information are available, but require very large data sets (Cook et al. 1995; Briffa et al. 1996; Esper et al. 2002).

Another significant problem with using ring widths to reconstruct climate is that the correlation between ring widths and climate may not be very strong. Trees use their resources in many ways, such as producing leaves, shoots, roots, flowers and seeds, and adding girth is unlikely to be the most pressing necessity. Variations in ring width are best seen as a general reflection of changes in net photosynthesis rather than as a direct response to a particular climate parameter. The effect is that individual trees in a stand are likely to respond in different ways to the climate of a given year, so that there is considerable variability in the relative widths of the rings formed. The ring widths of trees at different sites may also differ because local edaphic and microclimate conditions temper their response to the regional climate that is of interest (Schweingruber and Briffa 1996). This problem of between-tree and between-site variability can be overcome to some extent by using a large number of trees, so that these variations (noise) are effectively cancelled out and the common signal is enhanced. The strength of the common signal can be calculated using analysis of variance and equivalent correlation-based techniques. The larger the variability, the more trees are required to produce a sufficiently strong common signal.

Sampling a lot of trees does not entirely solve the problem, however. Where the calibration with climate is based on a large sample, it may be statistically highly significant whilst actually explaining a small percentage of the variance (low correlation coefficient). This causes problems when the regression equation is then used to reconstruct the climate of the past, because the usual approach (reverse calibration) produces a bias towards the mean, the magnitude of which depends upon the strength of the correlation (Birks 1995). Climate reconstructions based on calibration data sets with low correlation are likely to be biased towards the mean and thus to underestimate both the magnitude and the frequency of climate extremes (von Storch et al. 2004).

By using stable isotopes, it may be possible to exploit the great advantages of tree ring chronologies whilst avoiding some of the problems associated with reconstructing past climate using ring width measurements. In particular, it may not be necessary to statistically de-trend isotope series, so that palaeoclimate information is retained at all temporal frequencies and there is no 'segment length' effect. The strong mechanistic link between environmental variables and isotope fractionation may also lessen the reliance on statistical inference in environmental reconstruction.

Isotope fractionation in trees

The main constituents of wood are carbon, oxygen and hydrogen and variability in the isotope composition of each of these could potentially provide a palaeoenvironmental signal (McCarroll and Loader 2004). The water isotopes enter the tree through the roots and move up the xylem to the leaves, where evaporation leads to preferential loss of the lighter isotopes. The source of carbon is atmospheric CO_2, which enters the leaves through the stomata. Photosynthesis in the leaves combines the three elements to produce sugars, some of which are transported down the phloem and used to build wood cells around the stem, forming the tree rings.

The first source of isotope variation in tree rings is differences in the isotope composition of the source water and of atmospheric carbon dioxide. The isotope composition of precipitation is determined by a number of factors (Craig 1961; Dansgaard 1964; Gat 1980; Clark and Fritz 1997; Darling et al. this volume). The isotope composition and temperature of the evaporating water body (usually the ocean), and the humidity of the air, determines the initial isotope composition of the water vapour. When the water vapour condenses to produce precipitation there is a temperature-dependent fractionation of both the oxygen and hydrogen isotopes. However, with each precipitation event the isotope characteristics of the cloud alter, so that the isotope ratios of the precipitation change as the cloud moves away from its source area. The isotope ratios of precipitation are not, therefore, a simple proxy for air temperature and must be interpreted with caution. Any changes in the source of air masses carrying precipitation are likely to have a major influence that may not necessarily be independent of temperature at the site where precipitation occurs (Hammarlund et al. 2002; Rosqvist et al. 2004; Leng et al. this volume).

Trees take their water from the soil, not directly from precipitation, so fractionation and mixing can occur before the water is incorporated (Darling 2004). The critical factors here are the seasonality of the precipitation in relation to the growing season and also the depth of rooting of the tree. Deep-rooted trees, for example, may have access to groundwater with a different isotope composition to soil water, and soil water with different residence times may differ substantially due to differences in the source of precipitation at different times of the year (Dawson 1993; Dawson and Pate 1996; Darling 2004). For example, Robertson et al. (2001) have shown that oak trees in one of the drier areas of eastern England utilise water that fell as precipitation during the winter rather than during the growing season. In theory, evaporation from the soil surface should lead to enrichment of the heavier water isotopes in soil water, but this effect is probably limited to a thin layer very close to the surface and therefore not normally utilised by tree roots. There is no fractionation when water is taken-up by roots (Wershaw et al. 1966), so xylem water, which moves up the stem and into the leaves, has the same isotope ratios as the source.

Within the leaf there is fractionation of carbon and the water isotopes prior to photosynthesis, mediated in each case by the action of the stomata. Since water molecules are smaller than those of carbon dioxide, when the stomata are open to allow gases into the leaf they inevitably lose water vapour by evaporation (transpiration). When water evaporates, molecules that comprise the lighter isotopes of hydrogen and oxygen are preferentially lost, leading to an isotope enrichment of the leaf water that remains.

The degree of ^{18}O enrichment of leaf water above xylem water at the sites of evaporation ($\Delta^{18}O_e$) is given by:

$$\Delta^{18}O_e = \varepsilon^* + \varepsilon_k + (\Delta^{18}O_v - \varepsilon_k)e_a/e_i \qquad (1)$$

Where: ε^* is the proportional depression of water vapour pressure by the heavier water molecules, ε_k is the fractionation as water diffuses through the stomata and leaf boundary layer, $\Delta^{18}O_v$ is the oxygen isotope composition of water vapour in the atmosphere (relative to source water), and e_a and e_i are the ambient and intercellular vapour pressures (Craig and Gordon 1965; Dongmann et al. 1974; Barbour et al. 2001; 2002). The same model applies to the fractionation of hydrogen, though the fractionation factors are different. Temperature has a small influence on ε^*, but leaf enrichment depends mainly on the difference between the isotope compositions of the xylem water and the ambient moisture vapour, and on the difference in vapour pressures inside and outside of the leaf. At constant temperature, and where xylem water and atmospheric vapour have the same isotope signature, the degree of enrichment due to evaporation is linearly dependent on $1-e_a/e_i$ (Barbour et al. 2001). This model accounts for the isotope enrichment that occurs at the sites of evaporation, but it overestimates the degree of enrichment of leaf water, because as the enrichment diffuses backwards, it is opposed by the convection of (lighter) xylem water (a Péclet effect: Barbour and Farquhar (2000); Barbour et al. (2001)).

As carbon dioxide diffuses through the stomata into the intercellular spaces there is a fractionation against ^{13}C. This 'fractionation due to diffusion' remains at about $-4.4‰$, irrespective of changes in temperature and vapour pressure and is even insensitive to changes in the size of the stomatal opening. The difference in diffusion of $^{13}CO_2$ and $^{12}CO_2$ is simply due to the fact that as the molecules bounce around, the lighter ones bounce furthest, and so are more likely to pass through the stomatal opening. Only when the stomatal apertures are very small (≤ 0.1 μm) does fractionation increase markedly, because interactions with the guard cells, as well as the other gas molecules, start to become important. This is only likely to occur in species, such as citrus trees, with a high frequency of very small stomata (Farquhar and Lloyd 1993).

Photosynthesis combines leaf water and intercellular CO_2 to produce sugars, and this step also involves fractionation of carbon and the water isotopes (Farquhar et al. 1982; Francey and Farquhar 1982; Ehleringer et al. 1993). The fractionation of carbon is about $-27‰$, so that the ratios in leaf sugars are depleted relative to the intercellular CO_2. However, if photosynthesis uses CO_2 faster than it can be replenished by stomatal conductance, then the intercellular concentration of CO_2 falls and that CO_2 becomes enriched in ^{13}C. Although the fractionation remains near constant at about $-27‰$, the ratio of ^{13}C in leaf sugars will increase as the ratio in the source (intercellular CO_2) increases. The stable carbon isotope ratios of leaf sugars are therefore controlled principally by changes in the isotope ratios of the CO_2 within the leaf which, if the ratios in ambient air do not change, is in turn controlled by the internal concentration of CO_2. Low internal concentrations lead to higher concentrations of ^{13}C and therefore higher (less negative) $\delta^{13}C$ values in leaf sugars. In fact the isotope ratios of atmospheric CO_2 have changed since industrialisation, due to release of isotopically light CO_2 from fossil fuel combustion and cement manufacture (Keeling et al. 1979; Mook et al. 1983; Pearman and Hyson 1986; Switsur and Waterhouse 1998) (see Data Analysis). The interplay between these factors in relation to carbon isotope fractionation in trees may be described in terms of the following equation (Farqhar et al. 1982)

$$\delta^{13}C_{plant} = \delta^{13}C_{air} - a - (b-a) \cdot (c_i/c_a) \qquad (2)$$

Where: $\delta^{13}C_{plant}$ is the stable carbon isotopic ratio of the carbon fixed by the plant, $\delta^{13}C_{air}$ is the stable carbon isotopic ratio of the source carbon dioxide (the atmospheric pool), a is the fractionation caused by diffusion of CO_2 from the atmosphere into the intercellular spaces of the leaves; b is the biochemical fractionation; c_i/c_a the ratio of CO_2 concentration in the intercellular spaces of the leaves (c_i) to that of the atmosphere (c_a).

The water isotopes are also fractionated during photosynthesis. Exchange of carbonyl oxygen in organic molecules with water leads to enrichment of about 27‰ (Sternberg et al. 1986), but there is a strong discrimination against deuterium (estimated to be between −100‰ and −171‰; Yakir (1992)) so that, relative to leaf water, leaf sugars are enriched in ^{18}O but strongly depleted in deuterium.

Sugars produced in the leaf may move down the phloem of the stem and be used to produce wood cells, or they may be stored as starch and used later. Where sugars are used directly to produce wood, further fractionations of each of the three elements occurs. The main components of tree ring cells are cellulose and lignin, and these have different $\delta^{13}C$ values. In oak trees, for example, the offset is about 3‰ and cellulose is isotopically heavier than lignin (Loader et al. 2003; Robertson et al. 2004a). For the water isotopes, changes in the isotope composition at the site of wood synthesis are more complicated because of the potential for exchange of both oxygen and hydrogen molecules with xylem water (Hill et al. 1995; Sternberg et al. 1986). This is a complex process and it is difficult to predict the percentage of exchange that will occur during cellulose and lignin synthesis. As the magnitude of exchange increases, the isotope signature of the wood moves away from that of the leaf sugars and towards that of the source water. When wood forms there is a strong discrimination in favour of deuterium (estimated to be between +144‰ and +166‰ ; Yakir 1992), which almost balances the strong discrimination against deuterium during photosynthesis.

It should be clear that fractionation occurs at several steps in the path from source water or CO_2 to wood (Figure 1, Figure 2) and that there is no single environmental factor that completely controls the stable isotope ratios of carbon, oxygen or hydrogen (McCarroll and Loader 2004). Carbon isotopic ratios in tree rings reflect the concentration of CO_2 within the leaves during the growing season and thus the balance between the rate at which CO_2 enters, by stomatal conductance, and the rate at which it is removed by photosynthesis. The water isotopes reflect the isotope ratios in the source (precipitation or ground water), and the degree of evaporative enrichment in the leaf, the dominant control on which is vapour pressure (air humidity). The relative strength of the source and leaf enrichment signals is controlled by the degree of exchange with xylem water that occurs at the site of wood synthesis.

Sample selection and preparation

Species and site selection
The first principle in choosing sites and trees for isotope work should be that the trees can be properly cross-dated. Absolute chronologies cannot be reliably produced simply by counting backwards from the bark, even in mild temperate environments (Schweingruber 1988; Fritts 1976), and without an absolute time scale the great

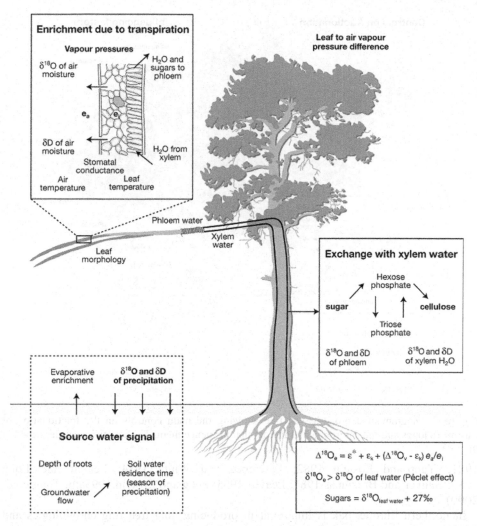

Enrichment due to transpiration

Vapour pressures

$\delta^{18}O$ of air moisture

H_2O and sugars to phloem

e_a e_i

δD of air moisture

H_2O from xylem

Stomatal conductance

Air temperature   Leaf temperature

Leaf to air vapour pressure difference

Phloem water

Xylem water

Leaf morphology

Exchange with xylem water

Hexose phosphate

sugar   cellulose

Triose phosphate

$\delta^{18}O$ and δD of phloem

$\delta^{18}O$ and δD of xylem H_2O

Evaporative enrichment

$\delta^{18}O$ and δD of precipitation

Source water signal

Depth of roots

Soil water residence time (season of precipitation)

Groundwater flow

$$\Delta^{18}O_e = \varepsilon^* + \varepsilon_k + (\Delta^{18}O_v - \varepsilon_k)\, e_a/e_i$$

$$\delta^{18}O_e > \delta^{18}O \text{ of leaf water (Péclet effect)}$$

$$\text{Sugars} = \delta^{18}O_{leaf\ water} + 27\text{‰}$$

Figure 1. Diagram of a needle-leaf tree showing the main controls on the fractionation of the water isotopes and the environmental factors that influence them. The equations are described in the text.

advantage of tree rings is lost. Reliable cross dating is particularly critical if wood samples from several trees are to be pooled prior to analysis, otherwise the signals from more than one year may be mixed resulting in a reduction in signal strength.

Many tree species have been used in isotope analysis, including the long-lived bristlecone pines of California (Tang et al. 1999; Feng and Epstein 1994) and *Fitzroya cupressoides* of Chile (Leavitt and Lara 1994), though in neither case was the isotope chronology very long or annually resolved. The thin rings of old, long-lived trees make isotope work difficult. Many other trees that have been used in the Americas including pine, juniper, fir, oak and spruce (Craig 1954; Epstein and Yapp 1976; Burk and Stuiver

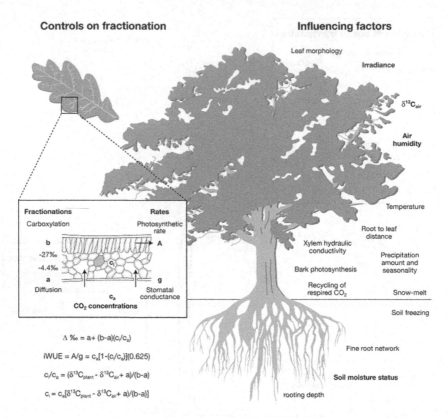

Figure 2. Diagram of a broad-leaved tree showing the main controls on the fractionation of carbon isotopes and the environmental factors that influence them. The equations are described in the text.

1981; Yapp and Epstein 1982; Lawrence and White 1984; Leavitt and Long 1985; Stuiver and Braziunas 1987; Leavitt 1993; Feng and Epstein 1995a;b; Tang et al. 2000).

In northern Europe, oak is important in producing long tree ring chronologies and there have been several studies of the carbon isotopic ratios in oak tree rings (Farmer and Baxter 1974; Freyer and Belacy 1983; Robertson et al. 1997a;b) and a few that included the water isotopes (Libby and Pandolfi 1974; Libby et al. 1976; Switsur et al. 1994; 1996; Hemming et al. 1998; Robertson et al. 2001). At the northern European timberline the important archive species are Scots pine (*Pinus sylvestris*), spruce (*Picea* spp.) and larch (*Larix* spp.), but isotope work here has so far focussed mainly on Scots pine (Sonninen and Jungner 1995; McCarroll and Pawellek 2001). This is one of the most widespread of species, extending naturally from the northern European tree line to the Mediterranean and it has also been planted in many areas of the world. There have been isotope studies on this species in a range of environments including Eurasia (Waterhouse et al. 2000; Saurer et al. 2002), the United Kingdom (Dubois 1984; Loader and Switsur 1996; Hemming et al. 1998), and the French Alps (Gagen et al. 2004). Also in Europe there have been isotope studies using fir trees (*Abies alba*) in Switzerland,

France and southern Germany (Lipp and Trimborn 1991; Bert et al. 1997; Anderson et al. 1998; 2002; Edwards et al. 2000), and of beech (*Fagus sylvatica*) in Switzerland (Saurer and Siegenthaler 1989; Saurer et al. 1995) and France (Dupouey et al. 1993; Duquesnay et al. 1998).

Beyond the Americas and Europe, where the best-equipped isotope laboratories are located, tree ring isotope studies are sparse. Some work has been done in India (Ramesh et al. 1985; 1986), in China (Liu et al. 1996; Feng et al. 1999), Taiwan (Sheu et al. 1996) and on the Tibetan plateau (Zimmermann et al. 1997), as well as in Japan (Libby et al. 1976; Kitagawa and Matsumoto 1995; Okada et al. 1995). Krishnamurthy and Epstein (1985) conducted early investigations into the climatic significance of a hydrogen isotope record from Kenya with some encouraging results. February and Stock (1999) tried to extract an isotope signal from the long-lived *Widdringtonia cedarbergensis* of South Africa, but this proved more difficult. More recently Tarhule and Leavitt (2004) have conducted similar pionnering work with other African species. In recent years there has been increasing interest in attempting to use stable isotopes in wood as a way of extracting palaeoclimate signals for the tropics, even where trees do not form annual rings (Robertson et al. 2004b; Verheyden et al. 2004; Evans and Schrag 2004).

The selection of suitable species and field sites for studying tree ring stable isotopes should be guided principally by the aims of the study. Where tree ring widths are used to reconstruct past climates, the advice is often to seek 'sensitive sites' where tree growth is likely to be strongly limited by the climate parameter that is of interest. At latitudinal and altitudinal tree lines this is often summer temperature, and in dry environments precipitation. However, in choosing sites for isotope studies it should be remembered that stable isotopes, unlike ring widths, are not a proxy for net annual growth, so that a factor that is strongly growth limiting need not necessarily be the dominant factor influencing isotope fractionation. Species and site selection, therefore, needs to be guided by knowledge of isotope theory and of the ecophysiology of the species involved.

Carbon isotopes, for example, reflect the internal concentration of CO_2 and therefore the balance between the rate at which CO_2 enters and is removed from the stomatal chambers (fixed). The dominant controls are therefore stomatal conductance, which responds to moisture stress and air humidity, and photosynthetic rate, which responds to temperature and sunlight. If the aim of the study is to reconstruct past changes in moisture stress, then trees and sites should be chosen so that stomatal conductance is likely to limit strongly internal CO_2 concentrations. Sites with well-drained soils and trees with shallow roots, without access to groundwater, are likely to give the best record. On the other hand, if the aim is to reconstruct changes in summer temperature and sunshine, it would be better to choose sites where the trees are unlikely to have been water-stressed, so that moist soils and deep-rooted trees are preferable. Several studies have shown marked differences in the stable carbon isotopic ratios, and the palaeoclimate signal that they contain, at relatively moist and dry sites (Saurer et al. 1995; 1997; McCarroll and Pawellek 2001).

For the water isotopes too, the choice of site and species should be guided by isotope theory. The stable water isotope ratios (oxygen and hydrogen) in tree rings essentially conflate two quite separate signals: the isotope ratios of source water and the enrichment due to evaporation in the leaf. If the aim is to produce an annually resolved record that reflects changes in the isotope ratios of precipitation, then it is critical to ensure that the species and site are chosen so that recent precipitation dominates the source water accessed by the trees. In some cases this may be the precipitation that falls during the growing season, but where precipitation and soil moisture recharge is strongly seasonal the trees may use winter precipitation during subsequent summer growth (Robertson et al. 2001). Where the aim is to use water isotope ratios to reconstruct changes in summer humidity, then it would be better to choose sites where the source water signal is muted, for example by using deep rooted trees that have access to well mixed groundwater.

There is one very important exception to the wisdom of choosing sites and trees that are predicted, on theoretical grounds, to be sensitive to a particular climate parameter. Where the aim of the study is to use living trees to provide a calibration with climate, that can then be applied to much longer chronologies, deliberately sampling sensitive trees and sites is likely to result in a biased and unrealistic calibration. Long chronologies are typically constructed using a range of samples including standing dead wood, building timbers, river-rafted logs and sub-fossil material from lakes and mires. It is rarely possible to determine the local environmental conditions under which such trees grew. To interpret realistically the signals retained in such chronologies, the calibration needs to be based on a sample of trees that is characteristic of the range of environments that the trees might have lived in, not just the most sensitive sites. Trees that are sampled in this way are likely to produce more variable isotope results, and may be more difficult to interpret, but they will provide a much more realistic guide to interpreting the long chronologies.

Sampling techniques

Most of the early studies, and many recent studies, of tree ring isotopes sampled material from disks of wood from trees that had been felled. However, modern on-line techniques now allow isotope measurements to be conducted on much smaller samples and it is often possible to obtain sufficient material from increment cores, which can be taken without harming the tree. Standard increment cores used for dendrochronology are quick and relatively easy to use but provide a wooden rod with a diameter of only 4 mm, so where the rings are thin it may be necessary to combine samples from more than one core. The larger increment borers are usually 12 mm in external diameter, and can be very difficult to insert and even more difficult to extract, but they provide a rod with a diameter of *ca.*10 mm, and thus more than six times as much wood per ring. Dendrochonologists often oil the borers to ease insertion, but for isotope work this must be avoided to prevent contamination. Teflon-coated borers work adequately without lubrication.

Field safety
Felling trees is an exceptionally dangerous procedure and best left to experienced foresters. However, even taking increment cores involves some risk. One of the greatest dangers is vibration from the coring causing dead branches to fall, which is particularly common when sampling standing dead wood. Standing dead trees with many remaining

branches are sometimes known as widow-makers! A helmet and eye protection is a sensible precaution when carrying out any work on trees. Attempting to extract a jammed corer can also be very dangerous. When coring wood that is soft, or partly rotten, it is not uncommon for the corer to twist without biting into the wood and so to be extremely difficult to extract. A common 'trick' is to attach a rope to the corer and tie to it a tree or suitable fixed point in line with the long-axis of the jammed borer. As the corer spins the rope twists, shortens and pulls the corer out. The forces involved, however, are enormous and there is potential for very serious injury, either from a flying or broken corer or a snapped rope. Commercially available core extractors are, unfortunately, difficult to obtain. Fallen dead wood can sometimes be sampled with an increment corer or with a bow saw. Chain saws should never be used by untrained personnel and without full protective clothing and equipment.

Living trees are always cored using hand-powered borers, but building timbers, particularly old oak beams, can be far too hard for this and electrical power tools are necessary. This is problematic for isotope work because it is difficult to avoid charring the wood, which may cause fractionation. A full risk assessment when sampling ancient buildings and structures is essential. Recovering sub-fossil logs from arctic lakes is not for the faint hearted and best left to the experts.

Laboratory methods

The first task in the laboratory is to measure and cross date the tree rings to establish the chronology. Where disks or cores are provided by an experienced dendrochonologist they will be marked to show each decade and century, but for isotope work the marks should be pin-holes rather than pencil or ink marks, which would contaminate the samples. If measurements are to be conducted in-house, the samples need to be either cut or sanded smooth to show the rings, and this inevitably involves some loss of sample. Rings widths can be measured manually on a specially adapted long-travel microscope stage or using scanning techniques. Various commercial and free-ware programmes are available for cross-dating, but this step requires some experience and access to a master chronology for the area. If in doubt it is probably better to collaborate with a dendrochronologist than to waste time and money on running samples that are not precisely dated, because the results are unlikely to be publishable.

Having dated the tree rings, the next stage is to decide what material to analyse. Decisions to make include whether to take blocks of wood that cover several rings or to work at annual resolution, whether to take whole rings or split them into early wood and late wood, and then what chemical fraction of the wood to analyse. Samples can be analysed separately for each tree or pooled prior to analysis. Most early studies, that required large samples, used blocks of whole wood spanning several annual rings, whereas most studies now try to achieve annual resolution.

For calibration studies, comparing isotope results with instrumental climate records, annual resolution is essential. However, when wood cells start to form in the spring, some of the photosynthates used may have been fixed in the previous or earlier growing seasons and have been remobilized from storage, so those earlywood cells do not contain stable isotopic signatures from the year in which the ring formed. This is certainly true for some oak trees, because earlywood cells start to form before the new leaves have opened, but the extent to which it applies to other species, including pines, is uncertain. To be sure that the material sampled represents the climate of the year that the tree ring formed, therefore, many studies have used only the latewood, which forms during the summer. The amount of latewood present and the width relative to earlywood varies greatly between species, so it is not always feasible to separate it.

Separation of latewood can be achieved using a scalpel or razor blade under magnification, and a skilled and experienced technician can work with rings that are less than 0.5 mm in width. In some species the procedure is greatly aided by the difference in texture of earlywood, which comprises large open cells and latewood, where the cells are smaller and the wood much harder. Working in from the bark, as a soft earlywood layer is scraped away, it ends abruptly at a ring boundary where the last of the annual latewood cells are small and the wood much harder. In Scots pine, the difference at the ring boundary is usually so great that it is possible to scrape the surface very clean of earlywood without cutting into it. The latewood can then be scraped away, usually forming thin tight curls rather than the flatter shavings of earlywood. The boundary between the latewood and the earlywood of the same year is more diffuse, but the consequences of incorporating the last of the earlywood cells into a latewood sample are probably not great.

A critical decision in any study is whether to run the samples from each tree separately or to pool samples from several trees, thereby reducing the number of isotope analyses and therefore the cost. Pooling samples was often the only realistic option in early studies because off-line preparation and analysis was so labour intensive. Leavitt and Long (1984) suggested that pooling samples from four trees is sufficient to provide an accurate absolute $\delta^{13}C$ value. McCarroll and Pawellek (1998), however, working on young Scots Pine trees in Finland, found that the between-tree variability was substantial and argued that larger samples would be required.

The central problem with pooling samples is that it masks the between-tree variability, which is the key to determining the degree of statistical confidence that can be placed in estimates of palaeoclimate. This ability to place statistically-defined confidence intervals around reconstructions of past climate is one of the great advantages of using tree rings. The magnitude of the uncertainty that surrounds estimates of natural climate variability over the last millennium, for example, is critical to our understanding of the impact of anthropogenic 20th Century warming (Mann et al. 1999; Esper et al. 2004; von Storch et al. 2004). Tree ring isotope studies could contribute significantly to that debate, but only if the samples from individual trees are analysed separately. The other problem with pooling samples prior to analysis is that it is impossible to trace any errors in the chronology due to mistakes in cutting very thin rings. Isotope values cross-date very strongly, so cutting errors are relatively easy to identify by comparing the chronologies from several trees, but if the samples are cut and pooled then any errors will lead to a mixing of wood from adjacent years until the chronologies fall into sequence again. Pooling samples also masks any changes in the strength of the common signal over time, which might indicate important changes in the response of trees to their environment.

The decision about whether to pool samples prior to analysis needs to be guided by the aims of the project and by constraints of cost. However, with modern on-line techniques for mass-spectrometry, much of the cost and effort of producing an isotope chronology goes into collecting, cutting and treating the samples, so the savings of pooling are not as large as they once were. Wherever it is feasible, analysing each tree sample individually is to be recommended.

Once the sample has been dated and cut the decision must be made as to which component of the wholewood should be analysed. Isotope analysis can be conducted on whole wood, resin-extracted wood, cellulose, alpha cellulose or lignin. Early work used whole wood, but it was later argued that since the relative abundance of the various components can vary, it is preferable to isolate a single component, the most common (and abundant) being cellulose, often purified further to alpha cellulose. However, several studies have now shown that although the isotope ratios of cellulose and lignin are offset, they nevertheless remain parallel and so carry very similar signals (Loader et al. 2003; Robertson et al. 2004a). With improved techniques for on-line combustion and mass spectrometry, the chemical separation of cellulose has become the rate-limiting step in many laboratories, and some have argued that it would be just as effective to work with whole wood. Even though there may be variations in the relative proportions of cellulose and lignin between years, the influence on isotope ratios is likely to be well within the natural variability between trees growing at the same site.

Resin extraction
Resin represents a mobile fraction of the tree ring and the tree's defence mechanism against disease and damage. The mobility of the resin component is exemplified in the field where trees that have been sampled several months previously, heal the 'wound' caused by the increment borer during sampling through deposition of resistant resin. The isotope composition of resin differs significantly from that of the cellulose, and since the distribution of resin around the tree is variable with time it is likely to comprise carbon fixed at times different from those used to form the cellulose and lignin it encrusts. As a consequence of this potential 'blurring' of the environmental signal associated with mobile resinous components they may need to be removed from the wholewood prior to isotope analysis.

Traditionally the extraction of resin has taken place prior to the purification of cellulose from wholewood using continuous reflux of an azeotropic mixture of benzene and ethanol (in the ratio 2:1) in a Soxhlet apparatus for approximately 6 hours. Safety concerns regarding the carcinogenic nature of benzene initiated the use of a safer toluene:ethanol (2:1) extracting mixture.

Whilst removal of resins is not required for hardwoods such as oak, this procedure required for coniferous softwoods and some tropical species significantly increased the time required for sample preparation. Recent studies using infra-red spectroscopy (Rinne et al. in Press.) have demonstrated that this resin extraction stage may not be required when processing softwood samples to cellulose. This discovery removes a significant rate-limiting step in sample preparation and should permit a greater throughput of samples in the future. This advance only applies however, where alpha-cellulose is being prepared for analysis. If the desire is to measure the isotope ratio of homogenised wood samples then resin extraction is still necessary. Removal of resins, using a Soxhlet apparatus, is a much simpler and quicker process than isolation of cellulose, and since this procedure is routinely applied to wood laths used for X-ray densitometry, resin-extracted wood from such prepared laths may be suitable for isotope analysis of conifers and other resinous trees.

Purification of cellulose
A range of methods are available for the purification of cellulose from wholewood samples (Green (1963) and references therein; Leavitt and Danzer 1993; Loader et al. 1997; MacFarlane et al. 1999; Brendel 2000). Perhaps the most widely used at present are based upon the oxidation of lignin using an acidified sodium chlorite solution followed by removal of hemicelluloses in sodium hydroxide and thorough washing. Different laboratories have developed their own variants relating to the concentration of the solution used, the extraction time required and the temperature for the reactions. An alternative approach to the acidified sodium chlorite methods is that of an acidic (acetic-nitric) digestion (Brendel et al. 2000). The advantage of this method is that it is relatively quick. Additional processing with sodium hydroxide is required to yield alpha-cellulose.

During the lignin oxidation process the sodium chlorite solution reacts to form highly toxic and potentially explosive chlorine dioxide gas. Exposure should therefore be minimised and extractions should always be carried out in a well ventilated area and within a fumecupboard. To minimise exposure further and to enable more efficient processing of larger numbers of small wood samples batch methods have been developed which can now be used to routinely prepare in excess of 100 samples per day. This advance was achieved primarily by reducing the quantities of reagents required and the level of intervention/exposure required at the different processing stages. The application of specially made extraction tubes or glassfibre pouches to this technique not only facilitated the processing of larger batches of samples, but also reduced sample losses when working with particularly small samples (Figure 3).

When preparing cellulose from tree rings and associated plant material the time required for the oxidation of lignin will vary between species. To facilitate complete extraction the initial starting material should always be cut finely enough (*ca.* 40 μm) (Switsur and Waterhouse 1998) or homogenised (Borella et al. 1998) to permit penetration of the reagents. Where possible the bleaching times required for each species under study should be tested either isotopically or using infra-red spectroscopy. At the end of the process the samples should be white and fibrous. Practically, this is sometimes not the case and the cellulose is seen to discolour when sodium hydroxide is added, it is likely that this reflects that the sample material was either initially too thick or too great in amount to enable complete bleaching in the time provided and consequently residual lignin may be present. Under these circumstances the discoloured samples, once thoroughly washed, may be subjected to a final (1 hour) bleaching step which appears to remove any remaining discolouration from the flocculent cellulose product which is then rinsed and dried as normal (a *second* sodium hydroxide 'wash' is normally not required in such situations). Where material has been initially pooled or woody structure remains further ultra-sonic homogenisation ensures high levels of sample homogeneity. The yield of holocellulose from these methods is typically between 40-50% and following purification of alpha-cellulose between 30-40%. The exact proportions reflect the characteristics of the starting material and the nature of the

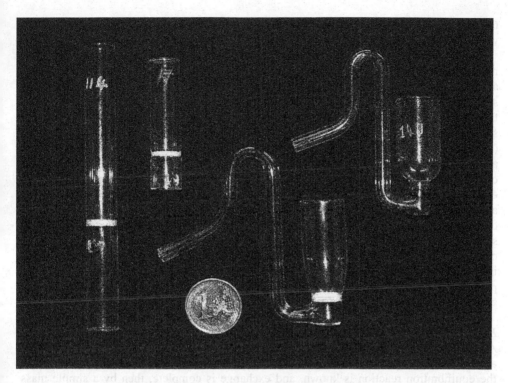

Figure 3. Variants of glass extraction vessels (pipe and cylinder form) routinely used for batch processing of wholewood to cellulose/cellulose nitrate. Note the etching on the tubes and the use of inorganic staining of the glass sinter filter (top right) to assist the recovery of the final product.

bleaching. The magnitude of these losses should be considered when planning analytical strategies, especially where sample material is limited.

Nitration
Hydrogen isotope analysis of raw alpha-cellulose is problematic owing to the presence of hydroxyl-bound hydrogen in the cellulose molecule. Since this hydrogen (*ca.* 30%) is free to exchange with atmospheric moisture or water during processing it is necessary to remove it from the cellulose structure prior to mass spectrometry to ensure that the isotope ratio measured reflects that of the water used by the plant during cellulose synthesis. To achieve this chemically, the sample is nitrated, a process which replaces the hydroxyl hydrogen with nitro groups. A mixture of fuming nitric acid with phosophorous pentaoxide or acetic anhydride is used (Bennett and Timmell 1955; Green 1963; Ramesh 1984; Robertson et al. 1995). This procedure is potentially hazardous and should be carried out in a fume cupboard with suitable protection.

Nitration of the cellulose involves the samples being added to the nitrating solution and stirred gently for over 12 hours. After this time the samples are filtered and thoroughly washed and dried. As this procedure may not completely nitrate the cellulose a pure sample of cellulose nitrate can be obtained by subsequently dissolving the 'raw' cellulose nitrate in dry acetone and then reprecipitating it in cold distilled water.

This procedure typically requires large amounts (*ca.*1 g) of cellulose as a starting material and adds considerable time and cost to the preparation of samples for isotope analysis. Understandably, this is reflected in the relatively limited number of tree ring studies exploring hydrogen isotope variability. However, as with many of the methods transferable to tree ring research this procedure can be scaled down to handle smaller samples and combined to run as a batch. Using these methods, as many as 50 samples (as small as 5 mg) can be nitrated with little difficulty, each batch requiring approximately 2 days processing time. To ensure that the nitration process has been completed, the degree of nitration of the cellulose nitrate can be measured using an elemental analyser. Where the degree of nitration falls below theoretical levels additional nitration time may be required. Once dry, the cellulose nitrate is prone to degradation and cannot be stored indefinitely. To slow this process samples may be stored in a freezer. The resulting cellulose nitrate may be analysed for its hydrogen or carbon isotopic ratios, however it is generally thought not to be suitable for oxygen isotope determination owing to the large amount of oxygen added during the nitration process. Cellulose nitrate is also explosive, and this should be considered when archiving large amounts of this material.

Equilibration
An alternative approach to nitration is the equilibration of the exchangeable hydrogen atoms with hydrogen of a known isotope composition. If the ratio of the water used in the equilibration reaction is known, and exchange is complete, then by a simple mass balance it is possible to determine the isotope composition of the non-exchangeable hydrogen in cellulose. This procedure has the benefit that the cellulose end product is not chemically modified and hence remains available for carbon, hydrogen and oxygen isotope determinations (Krishnamurthy and Machavaram 1998; Schimmelmann 1993).

The rate of equilibration is temperature dependent, with more rapid equilibration occurring with steam than with liquid water. Homogenous, finely divided samples also equilibrate more easily and reliably. Once equilibrated the sample is normally quick frozen to prevent further exchange during drying and the water removed under vacuum. New advances in rapid on-line equilibration may potentially revolutionise the analysis of hydrogen isotopes in tree rings. Sample preparation can now take as little as 15 minutes to perform and isotope determinations can be made on samples as small as 0.2 mg on-line.

Wholewood and lignin analysis
There is significant debate relating to the requirement for preparation of (α-) cellulose from tree ring samples for isotope analysis over the benefits of greater sample throughput attainable through analysis of resin extracted wholewood (McCarroll and Loader 2004; Barbour et al. 2002; Loader et al. 2003; Schleser et al. 1999b; Borella et al. 1999). Several studies have shown significant similarity in response preserved in the carbon isotopes of cellulose and wholewood fractions, which would tend to support the current questioning of the need to extract cellulose in dendroclimatology. However, certain circumstances will still remain where the analysis of cellulose is either essential

or desirable, including its use to 'verify' the wholewood signal, or to test for diagenetic modification in sub-fossil timbers. More importantly, where modelling or insight into specific plant processes is sought, a greater level of clarity and confidence can only be attained through the analysis of a single plant component.

Few studies have considered the palaeoclimatic significance of tree ring lignin (Loader et al. 2003; Robertson et al. 2004a). Of these, an environmental signal has been identified in the carbon isotopic ratios which appears to be similar to that of both the cellulose or wholewood. To what extent this relationship may be extrapolated for oxygen or hydrogen isotopes is yet to be established. Lignin analysis may be useful in situations where preservation conditions have favoured preservation of the lignin over that of the cellulose and hemicellulose components.

Lignin extraction is relatively simple to perform, requiring a concentrated sulphuric acid digestion of the milled or powdered wood (TAPPI 1988). A fumecupboard is essential for this work and care should be taken when working with concentrated acids.

Chemical hazards
Chemical hazards represent perhaps the major risk in sample purification and analysis for which risk data should be supplied at point of sale. Where this is not the case additional information relating to chemical risk, storage, transport and emergency precautions/first aid can normally be obtained free of charge from the manufacturer or through web-sites such as the University of Oxford, Physical Chemistry Department (http://physchem.ox.ac.uk/MSDS/).

Mass spectrometry

To measure the stable isotopic ratios of carbon, hydrogen or oxygen in wood or a wood component, it is first necessary to convert each sample into a gas species suitable for isotope ratio mass spectrometry. Traditionally this preparation and purification was carried for each sample individually using a diverse range of lengthy chemical procedures 'off-line' prior to mass spectrometric analysis (Figure 4). However, the development, refinement and wider application of "on-line" (continuous flow) methods for stable isotope analysis which integrate elemental analysers with isotope ratio mass spectrometers (Preston and Owens 1985) have revolutionised and advanced greatly the scope and potential for progress in the field of isotope dendroclimatology. For carbon isotopic analysis CO_2 is produced by combustion, and this can be accomplished either off-line, in sealed glass tubes, or on-line using a furnace and elemental analyser interfaced with an isotope ratio mass spectrometer. The off-line method (Buchanan and Corcoran 1959; Sofer 1980; Boutton et al. 1985; Boutton 1991) involves placing the sample in a Pyrex tube with an excess of copper oxide. The tube is evacuated and sealed (with a flame) and then placed in a furnace at 450°C for 18 hours. The copper oxide releases oxygen, ensuring complete combustion to form CO_2 and water. The sealed tubes (A) are cracked in a vacuum line and drawn through a water-trap (B) where the water is removed by freezing. The CO_2 is subsequently frozen-down in a second cold trap (C) prior to thawing and analysis by mass spectrometry (D). Cellulose standard materials can be treated in precisely the same way as the samples, so that precision can

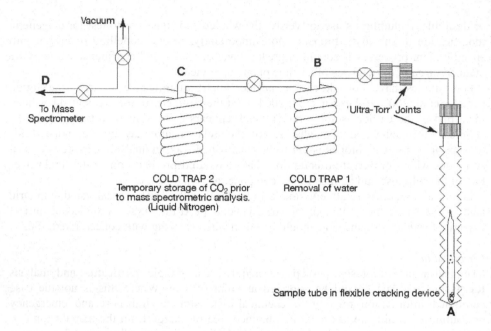

Figure 4. Schematic diagram of the vacuum line for purification of CO_2 samples for mass spectrometry following off-line combustion.

be monitored. This procedure is laborious but reliable, regularly exhibiting precision of about 0.1‰.

On-line carbon isotope analysis

The on-line method for carbon isotope analysis involves packing small samples into tin capsules and dropping them through a furnace tube heated to 1000-1100°C. Combustion of tin releases heat, so the samples are flash-combusted and the products carried on a stream of helium. A pulse of oxygen is admitted with the sample to assist combustion and the resulting gases passed through a silica column packed with chemical reagents. Whilst the precise combinations and composition of these reagents vary with manufacturers' recommendations these typically comprise chromium oxide and copper oxide which serve as further oxygen donors to ensure that the sample is completely combusted and that the carbon is fully oxidised to CO_2. In some experimental set-ups this initial stage also includes a packing of silver wool to remove any halogens and sulphur which may be present in the samples. Following the initial combustion stage the sample gases comprise CO_2, water vapour and oxides of nitrogen. Since some of these oxides of nitrogen could potentially interfere with the precise measurement of the carbon isotopic ratios of the CO_2 admitted to the source, the nitrogen oxides are reduced to nitrogen gas over hot copper warmed in a second furnace (Figure 5 A,B). Following removal of any water vapour via a chemical trap (magnesium perchlorate) or Nafion drier the gas mixture is resolved to CO_2 and N_2 using a gas chromatography column packed with a stationary phase such as Porapak $Q^®$. With run times of approximately 8 minutes per sample and long batch runs possible using autosamplers this robust

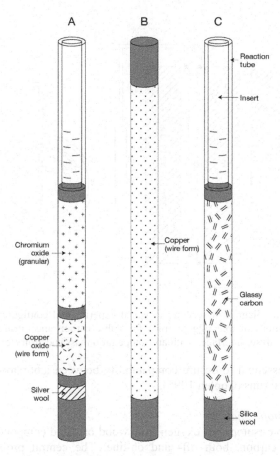

Figure 5. Typical reaction tube packings for sample preparation and on-line analysis of tree ring samples. Tubes A and B are for carbon and nitrogen isotope analysis, Tube C for hydrogen and oxygen isotope analysis. Tube inserts are typically used to retrieve ash and debris from Tubes A and C between batch runs. Alternative de-ashing procedures (e.g., vacuum extraction) are also available.

technique is rapid and attains levels of precision comparable to, or better than those of more traditional off-line methods (Figure 6). This method is less suitable for large samples (>1 mg C), but samples containing as little as 50 μg of carbon can be routinely measured with confidence. For tree ring work small samples are not necessarily an advantage, because the variability within tree rings will not be averaged if only a few fibres are analysed. With analysis of such small samples now possible homogeneity becomes an important consideration although the preparation of samples using ultra-sonic homogenisation methods reduces related problems significantly (see below). As with all continuous flow methods, factors such as sample size effects, memory and drift should be fully quantified and minimised/removed in order to obtain meaningful isotope data. The use of well characterised (externally calibrated) international and laboratory

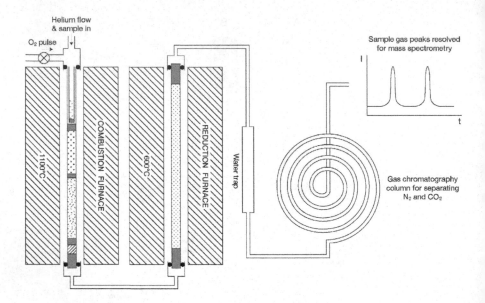

Figure 6. A schematic diagram describing a typical experimental configuration for the on-line preparation (combustion) of tree ring samples for stable carbon and nitrogen isotope analysis using an elemental analyser interfaced with an isotope ratio mass spectrometer.

standards is also essential to ensure comparability between laboratories, machines and individual analytical runs (Coplen 1995).

Off-line oxygen isotope analysis
Measuring the stable isotopes of oxygen from wood or wood components is much more difficult than for carbon, both off- and on-line. The central problem is achieving complete conversion of the sample oxygen to a single product (carbon dioxide), without further addition of oxygen as required for carbon isotope analysis. Many approaches have been developed (Bunton et al. 1956; Aggett et al. 1965; Taylor and Chen 1970; Ferhi et al. 1983; Wong and Klein 1986; Wong et al. 1987; Mullane et al. 1988; Farquhar et al. 1997; Loader and Buhay 1999), but the two most common off-line methods involve pyrolysis using either mercuric chloride or nickel tubes.

The mercury$_{(II)}$ chloride approach (Rittenberg and Ponticorvo 1956; Dunbar and Wilson 1983; Field et al. 1994; Sauer and Sternberg 1994) involves pyrolysis of dry cellulose under vacuum with mercury$_{(II)}$ chloride, producing both carbon dioxide and carbon monoxide, with mercury and hydrochloric acid (Figure 7). The sample tubes are cracked in a vacuum line (B) and the gases purified prior to analysis. In a lengthy process, hydrochloric acid is removed from the products using an ion exchange resin or quinoline and CO converted to CO_2 using an electrical discharge (A), Further cryogenic purification and yield determination (C) is required prior to transfer of the carbon dioxide to a sealable sample tube (D) for off-line mass spectrometry. The method yields reliable results, with a precision of about 0.2‰, but it is laborious and involves exposure to very toxic substances. Several laboratories use this approach, but given the

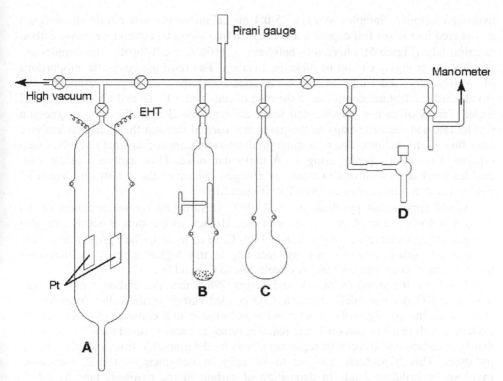

Figure 7. Schematic diagram of the vacuum line for purification of samples pyrolised using the mercury$_{(II)}$ chloride method for off-line oxygen isotope analysis (after Field et al. 1994).

dangers involved and the availability of safer, more rapid alternatives, it is generally not recommended.

Nickel-tube pyrolysis (Thompson and Gray 1977; Brenninkmeijer and Mook 1981; Gray et al. 1984; Motz et al. 1997) involves placing the dry samples in evacuated pure nickel tubes (or 'bombs'). The pyrolysis products are hydrogen and carbon monoxide, but at >950°C the hydrogen is able to diffuse through the nickel tube, preventing it reacting with the oxygen bound to the CO. As the tube cools, most of the CO converts to CO_2 and any remaining CO is converted using an electrical discharge. This approach is as laborious but less hazardous than using mercuric chloride and the level of analytical precision is about the same. It is more expensive, since the high purity nickel tubes have a limited durability.

On-line oxygen isotope analysis
On-line techniques for oxygen isotopic analysis are still under development (Santrock and Hayes 1987; Begley and Scrimgeour 1996; 1997; Werner et al. 1996; Farquhar et al. 1997; Koziet 1997, Saurer et al. 1998a; b; Loader and Buhay 1999). There is some variability in the procedures being used, but most are based upon high temperature pyrolysis to produce carbon monoxide as the target gas. Experimentally, the system for pyrolysis of tree ring samples (cellulose/cellulose nitrate) is similar for both oxygen and

hydrogen isotopes. Samples of ca. 0.25-0.4 mg cellulose/cellulose nitrate are weighed and placed into silver foil capsules and dropped into a quartz, ceramic or glassy carbon reaction tube (Figure 5C) heated to between *ca.* 1100°C and 1400°C. The samples are pyrolised over glassy carbon or nickelised carbon. The resulting gases, the proportions of which vary according to pyrolysis conditions, the 'water-gas reactions' and Boudouard equilibrium comprise, in the case of cellulose; CO, H_2 and traces of CO_2 and water. Traces of carbon dioxide and water are removed chemically using magnesium perchlorate and ascarite traps as the gases are carried through the elemental analyser on a flow of dry helium. The remaining products are chromatographically resolved into separate CO and H_2 peaks using a 5Å molecular sieve. This stationary phase also enables good separation of CO from any nitrogen present in the gas stream, which is vital when determining masses 28 and 30 (Figure 8).

It would appear that pyrolysis at *ca.* 1100°C is sufficient for measurement of the oxygen isotopic ratio of tree ring cellulose. Higher temperature systems are also commercially available which pyrolise at 1400°C. In addition to their greater versatility for use with other sample types, by operating at this higher temperature they may provide a more complete conversion of cellulose to CO and H_2.

It has been suggested (Santrock and Hayes 1987) that the carbon bound to the oxygen in CO derives solely from the C-O bonded carbon in the cellulose molecule and that during pyrolysis this C-O bond is not broken and consequently not free to exchange with residual carbon in the reaction furnace. Theory would therefore suggest that both carbon and oxygen isotopic ratios may be determinable from a single sample pyrolysis. This hypothesis has yet to be fully or convincingly tested. Complete pyrolysis of cellulose leads to deposition of carbon in the pyrolysis tube as a by-product of the reaction and if reliable dual isotope measurement is to be achieved, it is essential that the pyrolysis must be both complete and free from extraneous sources of oxygen (e.g., moisture) entering with the sample, which could effectively scavenge this deposited carbon. Initial results based upon pyrolysis of cellulose at *ca.*1100°C (Loader et al. in press) suggest that whilst the ^{13}C signal is *generally* preserved, the precision and reliability of the resulting data are not as reliable as those obtained with either on- or off-line combustion methods. This variability may reflect the lower temperature of pyrolysis used during these tests and consequently more encouraging results may be obtained from cellulose pyrolised more completely to CO at higher temperatures (*ca.*1400°C).

Hydrogen isotopes
Hydrogen is by far the most difficult of the three elements to analyse, either on or off-line. The main problem is that a proportion of the hydrogen atoms (*ca.* 30%) found in cellulose molecules are labile and can exchange, with atmospheric moisture for example, and so in order to attain a record of hydrogen isotopes that reflects the hydrogen used by the tree during cellulose synthesis, these hydroxyl hydrogens need to be replaced, usually by nitration, or equilibrated with water of known isotope composition. In off-line analysis the samples are combusted in sealed quartz tubes (A) with an excess of copper oxide to yield CO_2 and water, which are separated cryogenically (B) (Figure 9). The water is transferred to a tube containing 'Indiana zinc' (a specially produced zinc product with catalytic impurities produced by the University

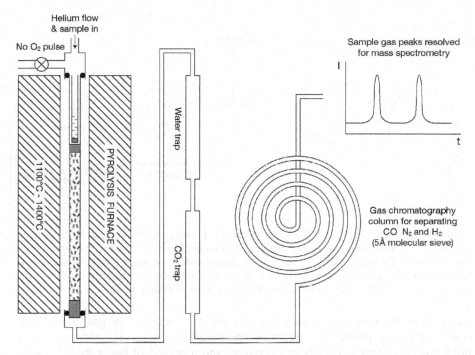

Figure 8. A schematic diagram describing a typical experimental configuration for the on-line preparation (pyrolysis) of tree ring samples for stable oxygen and hydrogen isotope analysis using an elemental analyser interfaced with an isotope ratio mass spectrometer.

of Indiana) (C) which reacts when heated off-line to form quantitatively zinc oxide and hydrogen gas, which is then analysed as the target gas for the mass spectrometer (Stump and Frazer 1973; Northfeldt et al. 1981; Coleman et al. 1982; Heaton and Chenery 1990; Schimmelmann 1991, Schimmelmann and DeNiro, 1993; Robertson et al. 1995).

An alternative approach is to place a silica sleeve around the nickel bombs used for pyrolysis of oxygen and thus trap the hydrogen that diffuses out of them (Gray et al. 1984; Motz et al. 1997). Although both of these approaches work reasonably well, they are very slow and labour intensive, so the number of studies that have applied the technique to tree rings is still small, and most of these are based on a single tree or a few trees and on blocks of wood rather than individual rings (Libby and Pandolfi 1974; Schiegl 1974; Epstein and Yapp 1976; Libby et al. 1976; Yapp and Epstein 1982; Dubois 1984; Gray and Se 1984; Lawrence and White 1984; Krishnamurthy and Epstein 1985; Epstein and Krishnamurthy 1990; Becker et al. 1991; Lipp and Trimborn 1991; Lipp et al. 1991; Switsur et al. 1994, 1996; Buhay and Edwards 1995; Feng and Epstein 1995b; Edwards et al. 2000; Pendall 2000; Tang et al. 2000).

On-line, continuous flow approaches to hydrogen isotope analysis are very similar to those used for oxygen (see above), but there are additional difficulties, particularly the mass spectrometry (Prosser and Scrimgeour 1995; Tobias et al. 1995; Tobias and Brenna 1996; Brenna et al. 1998; Kelly et al. 1998). Dry samples are wrapped in silver foil and dropped into a pyrolysis oven at > 1100°C. Any traces of water and CO_2 are

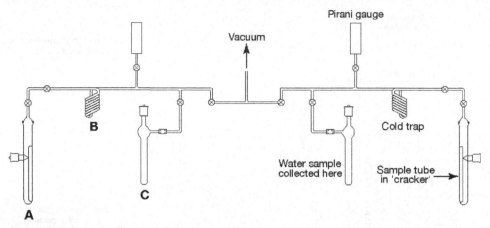

Figure 9. Schematic diagram of the vacuum line for purification of combusted cellulose nitrate samples for off-line hydrogen isotope analysis. The vacuum line presented is 'mirrored' to enable greater sample throughput and is heated with electrical heating tape to assist in the transfer of water within the vacuum system. Samples of water are cryogenically transferred to the removable sample vessels (C) for off-line reduction to hydrogen immediately prior to mass spectrometric analysis.

trapped and the remaining gasses (H_2, CO and N_2) separated by gas chromatography (GC-EA) before moving to the mass spectrometer (Figure 8). Although the main problems now seem to have been overcome, and precision of better than 2.5‰ can be achieved, few laboratories are yet running tree ring samples routinely.

Laser-ablation

One of the most exciting developments likely to impact isotope dendroclimatology in the coming years is the advent of laser ablation isotope ratio mass spectrometry. Laser ablation systems have been used for sampling in elemental studies for a number of years, however their application for stable isotope analysis of tree rings has so far been limited.

Through its capability to sample very small amounts of wood/cellulose (<10 µg) at very high resolution, laser ablation offers the potential for effectively 'non-destructive' sampling to leave the archive intact for X-ray densitometry, ringwidth and elemental analysis. The key to applying these methods to organic samples such as tree rings lies in the development of a suitable interface to link the laser and mass spectrometer. Commercial systems are currently in development, however at present the only published works report a small number of technical notes and evaluation studies obtained from laboratory prototype systems (Schulze et al. 2004; Wieser and Brand 1999).

Early applications of laser ablation systems were based upon infra-red (CO_2) lasers which used the laser energy to combust the sample. The resulting CO_2 was collected cryogenically or analysed on-line. When these systems were applied to tree ring samples, however, it was found that the localised heating of the laser beam caused charring and a halo effect around the point of sampling, which also led to low precision and likely fractionation of the combustion products. The advent of UV laser ablation led to a reassessment of the approach for tree ring sampling. UV lasers do not heat the sample, but instead photoablate the target thereby converting it into a powder without charring or associated effects. The fine particulates are then carried upon a stream of helium into a combustion interface where they are converted to CO_2 over copper oxide or chromium oxide. Traces of water are removed through a Nafion drier or cryogenic trap and the gas peak resolved by gas chromatography. Additional cryogenic focussing may also be employed where necessary to further concentrate the signal and reduce peak 'tailing'.

One of the most complete published assessments of the technique demonstrates the capability for analysing $\delta^{13}C$ from tree ring material both as wholewood and cellulose (Schulze et al. 2004). Whilst this paper exhibits the immense potential of the method, particularly in the field of high-resolution isotope analyses, it also highlights the current state-of-the-art of this technique, and the need for future development and refinement if this technological advance is to transfer to the development of long tree ring isotope time series.

Future access to this technology for palaeoclimatology involves significant challenges. In particular, analytical precision is currently lower than can be obtained with on- or off-line methods and there are limitations of sample size. Core sections must also be sub-divided into 2 cm lengths prior to analysis, which represents a significant disadvantage and conflicts with the concept of high sample throughput and non-destructive sampling. High levels of user intervention are currently required and frequent obstruction of gas-flow lines by particulates are also factors which need consideration as they limit capacity for unattended analyses (batch running), as is currently possible with EA-IRMS. Further to these practical obstacles, new analytical and data handling protocols will also need to be developed which address the selection and nature of the sample matrix (i.e., wholewood, cellulose (extracted in core form), or resin extracted wood). With such a precise analytical tool capable of sampling isotope variability at cellular levels, standardisation of sampling protocols will also require consideration in order to cope with intra-annual variability and the retrieval of a robust, reproducible and representative signal from the tree ring. These challenges are not insurmountable, and the great potential benefits of laser ablation techniques within stable isotope research will ensure that they are addressed within the next decade.

High Resolution stable isotope analyses

It has been known for a number of years that tree rings exhibit significant intra-annual variability from the early- to latewood fractions of each annual growth band, however it is only recently that a new generation of very high resolution intra-annual isotope analyses have provided greater insight into the nature of these trends within the growing season. These will inform future sampling regimes, assist in the mechanistic modelling of tree-physiology, and in the isolation of robust intra-annual climatic information. They also offer potential for the extraction of environmental signals from tropical trees which exhibit little or no physical evidence of periodic growth.

Early studies of intra-annual isotope variability were based upon relatively coarse sub-division of wide tree rings (Wilson and Grinsted 1977; Leavitt and Long 1982;

Loader et al. 1995; Switsur et al. 1995; Ogle and McCormack 1994; Robertson et al. 1996; Walcroft et al. 1997). It was then hypothesised that analysis of the isotope composition of the cellulose, lignin or nitrated cellulose extracted from individual increments would provide a signal based upon the growth period of the tree and potentially a record of climatic variability experienced by the tree during the growing season.

Samples were extracted from each tree ring, typically using a scalpel and dissecting microscope, or manual microtome. However a new generation of computer-controlled microtomes combined with digital photographic capabilities enable sampling of a block of wood at significantly enhanced resolutions (<10 μm) with the capability to regularly monitor and update the rate of removal of material (Schleser et al. 1999b; Helle and Schleser 2004). Ultimately, the sampling resolution adopted reflects the funds available as well as the minimum sample size acceptable for preprocessing and analysis by the EA-IRMS.

The resulting thin sections are typically removed using tweezers and placed into pre-labelled micro-centrifuge tubes. Where water or steam has been used to lubricate or soften the wood surface for cutting, samples should be dried prior to weighing or isotope analysis. Excess use of water can also cause the wood surface to swell, which can lead to uneven thickness in sample removal. However, regular visual analysis, of digital images where available, can provide an archive of the progress of cutting and assist in linking individual slices to their original position in the sample.

Whilst the earlier 'lower resolution' studies of within-ring isotope variability proved useful in identifying the nature of isotope trends in tree rings, the elucidation of plant physiological processes and the development of sampling strategy, the results produced through this new generation of high-resolution studies have provided further new information. High resolution analyses of trees which exhibit well defined annual rings have been presented by Schleser et al. (1999b) and Helle and Schleser (2004; 2003). Using an increment thickness of only 10 μm they were able to identify a series of rhythmic trends in the tree ring series from broad leaved and needle leaf trees. Helle and Schleser (2004) suggest that metabolic processes during growth dormancy and remobilised stored photosynthates are the driving forces underlying these cycles. It is now suspected that the initial climatic interpretation of the earlier studies is an over simplification of the tree system and it is likely that, instead of the tree ring acting as a passive recorder of environmental forcing, the climatic signal preserved intra-annually, which represents the source of the inter-annual record, is imposed onto a rhythmic biological trend. High resolution analysis of tree rings represents a powerful tool for improving our understanding of how trees function that is vital for robust palaeoclimatic interpretation of longer annually-resolved timeseries. However, further monitoring, analyses and reinterpretation are undoubtedly required over the coming years in order to test these hypotheses conclusively and to make further advances that will elucidate more fully the nature of signal preservation.

Tropical trees, which do not physically exhibit annual ring structure, also exhibit rhythmic trends not dissimilar to those observed in temperate species (Leavitt and Long 1991; Helle and Schleser 2004). Evans et al. (2004) demonstrated a link between the oxygen isotope composition of trees in Costa Rica and El Niño/Southern Oscillation (ENSO) variability. The signal in the trees most likely reflecting the changing source waters associated with this large scale phenomenon. Verheyden et al. (2004) have applied microtome sampling to mangroves from Africa and also report clear rhythmic trends which reflect plant response to environmental forcing and a possible ENSO-related deviation. Such studies represent a significant step forward in our understanding of the factors influencing the tree ring isotope record. More importantly, perhaps, the rhythmic tendencies now repeated in a number of tropical species offer potential for extracting terrestrial palaeoclimate information from the tropics, something which is rarely possible using physical tree ring proxies. Radiocarbon dating of tropical forest trees (Robertson et al. 2004b; Chambers 1998) suggest significant potential for developing such records which extend beyond the last 500, possibly 1,000 years. However, whilst this potential is great the nature of these fluctuations must first be characterised. They may reflect annual or seasonally-controlled periods of growth and dormancy, perhaps driven by external factors, which may prove useful for the development of absolute chronologies and cross-dating. Alternatively, they may reflect asynchronous changes in canopy cover, disturbance or fruit development which may vary quasi-randomly between individuals across a site such that extraction and interpretation of a climate signal may prove more challenging. Answering these questions will require further high resolution studies to be carried out with high levels of replication and firm (absolute) chronological control.

Data analysis

As with all proxy palaeoclimate data, stable isotope values from tree rings need to be interpreted with care. The aim must be to retain as much of the desired signal as possible whilst minimising the influence of any other factors.

Atmospheric $\delta^{13}C$ trend

The burning of fossil fuels and associated activities of industrialisation have released isotopically light carbon into the atmosphere, so that the $\delta^{13}C$ value of atmospheric CO_2 has declined by about 1.5‰ since industrialisation. Since fractionation has an additive effect, this 'atmospheric decline' will influence the $\delta^{13}C$ values in tree rings and must be removed prior to comparison with climate data.

The simplest way to remove the atmospheric decline is to statistically de-trend the data. There is a wide range of techniques available, many of which are routinely applied to tree ring width series. The simplest is to fit a straight line or a curve through the data and de-trend by subtraction or division from those values. However, such a solution based upon statistical de-trending is most unsatisfactory. De-trending certainly removes the atmospheric decline, but it also removes any other trends that occur over the same time scale. The high frequency, year-to-year, differences in isotope ratios would remain but the lower frequency changes, which are often of most interest, would be removed. This is particularly undesirable for the industrial period since there have been many changes during this time that might have affected the response of trees to their environment, not least the increase in atmospheric CO_2 concentrations. De-trending would also remove any climate change signal that parallels the increase in atmospheric CO_2 concentrations, so any 'global warming' signal, for example, would be removed.

There are two ways to remove the 'atmospheric decline' without statistically de-trending. Tree ring $\delta^{13}C$ values can be expressed in terms of discrimination against ^{13}C, or the changes in atmospheric $\delta^{13}C$ can be removed mathematically and the $\delta^{13}C$ values of the tree rings expressed relative to a pre-industrial standard value. Both approaches require an annually resolved record of the changing $\delta^{13}C$ values of atmospheric CO_2.

Ice cores provide a record of past changes in atmospheric $\delta^{13}C$ values and these are supplemented in recent years by air samples, but the data are incomplete and there is considerable scatter (Keeling et al. 1979; Mook et al. 1983; Pearman and Hyson 1986). The decline in $\delta^{13}C$ is non-linear, so a range of techniques can be used to place a line through the data and thus estimate annual values. Although several authors have used such data to remove the atmospheric decline from tree ring $\delta^{13}C$ series, precisely how the annual $\delta^{13}C$ values were calculated is rarely clear. This makes it difficult to compare directly the results of different studies. McCarroll and Loader (2004) have suggested that a common data set for atmospheric $\delta^{13}C$ values be used, based on extrapolation of a high-precision record obtained from Antarctic ice cores (Francey et al. 1999). Although there are other data sets available that might include measurements closer to the areas where tree rings are sampled, McCarroll and Loader (2004) argue that these data are sufficiently precise to provide a standard approach. The data can be summarised very simply by an annual decline of 0.0044‰ between 1850 and 1961 and a steeper annual decline of 0.0281‰ between 1962 and 1980. The values for years after 1980 are based on a linear extrapolation of the ice core data presented by Francey et al. (1999), but they accord well with those obtained from firn (compacted snow) in the same area.

By using annual values for atmospheric $\delta^{13}C$, tree ring $\delta^{13}C$ values can be translated into values that represent discrimination against ^{13}C (Δ) using the equation:

$$\Delta = (\delta^{13}C_a - \delta^{13}C_p)/(1 - \delta^{13}C_p/1000) \qquad (3)$$

Where $\delta^{13}C_a$ is the carbon isotopic ratio of the air (the source) and $\delta^{13}C_p$ is the carbon isotopic ratio of the product (leaf sugars). A correction factor also needs to be applied to account for the difference in $\delta^{13}C$ of leaf sugars and the wood component that is being measured.

Discrimination values are commonly used in the plant physiology literature, particularly where the aim is to examine any physiological response of trees to increased concentrations of atmospheric CO_2 (Berninger et al. 2000). However, this procedure is cumbersome when applied to tree ring series intended for palaeoclimate reconstruction, since it requires all values to be changed, even though the industrial period may be a small part of the complete record.

An alternative, and we would argue preferable solution, is to remove the effect of the atmospheric decline in $\delta^{13}C$ mathematically, effectively correcting modern tree ring $\delta^{13}C$ values to a pre-industrial standard value. Since fractionation is an additive process, this is achieved by simply adding the difference between the atmospheric value and the standard value. McCarroll and Loader (2004) have suggested that a reasonable standard value for the pre-industrial $\delta^{13}C$ of the atmosphere is −6.4‰, which is close to the values around AD1850 and also before the Little Ice Age (Francey et al. 1999). The

$\delta^{13}C$ of the atmosphere was not constant in pre-industrial times, but the use of a standard is convenient. The correction for each year since 1850 is provided in Table 1.

It should be stressed that the aim of this correction for changes in the $\delta^{13}C$ of the atmosphere is not to de-trend the data. On the contrary, any trend that remains after the correction may provide valuable evidence for other changes, either environmental or physiological, that have occurred over the industrial period. It is essential therefore, that the same procedure is applied to all tree ring series, whether or not they show a decline. Fractionation acts to change the $\delta^{13}C$ of the source gas, and if the $\delta^{13}C$ values of the source changes then this **must** be carried through to the $\delta^{13}C$ values of the product. Although some tree ring series have been reported not to show an atmospheric decline (e.g., Stuiver (1978); Freyer (1979a); Tans and Mook (1980); Francey (1981); Robertson et al. (1997a); Anderson et al. (1998); Duquesnay et al. (1998)), they must, nevertheless, have responded in the same way as all other trees. The absence of a decline in $\delta^{13}C$ actually represents a reduction in discrimination against ^{13}C over that period. Most tree ring series do show an atmospheric decline (e.g., Freyer (1979b); Freyer and Belacy (1983); Leavitt and Long (1988); Epstein and Krishnamurthy (1990); Leavitt and Lara (1994); Feng and Epstein (1995a); February and Stock (1999); Treydte et al. (2001)).

It has been argued that correction for the atmospheric decline in $\delta^{13}C$ should include a factor that accounts for changes in the response of trees to increasing atmospheric concentrations of CO_2 (Treydte et al. 2001). Alpine spruce trees, for example, showed a change of 0.007‰ in $\delta^{13}C$ per unit (ppmv) increase in atmospheric CO_2 (Treyde et al. 2001). There is a rapidly growing literature on the physiological response of trees to changing atmospheric CO_2 concentrations (Feng and Epstein 1995a; Marshall and Monserud 1996; Picon et al. 1996; Kürschner 1996; Feng 1998; 1999; Krishnamurthy and Machavaram 2000; Luomala et al. 2003), so it is perhaps premature to apply a standard correction, but reasonable values for atmospheric CO_2 concentrations (Robertson et al. 2001) are, nevertheless, provided by McCarroll and Loader (2004).

Juvenile effect

Several studies have shown an increase in the $\delta^{13}C$ values of young trees and this has become known as the 'juvenile effect'. Whilst not believed to reflect age directly, one explanation is that it is caused by recycling of respired air close to the forest floor (Schleser and Jayasekera 1985). This is likely to occur in forests and would certainly have the effect of lowering the $\delta^{13}C$ value in young trees because, as with the 'atmospheric decline', any change in the source gas must be carried through into the products of photosynthesis. However, the juvenile effect has been reported even in trees where it is difficult to envisage significant recycling of respired air; on exposed mountain slopes, for example, where trees are well spaced (Gagen et al. 2004). Although recycling of respired air is likely to be important in many cases, it is unlikely to be the full explanation for the juvenile effect (McCarroll and Loader 2004).

Another possible explanation is that the increase in $\delta^{13}C$ values reflects changes in hydraulic conductivity as trees age and increase in height (Ryan and Yoder 1997; McDowell et al. 2002). Leaf water potential declines with height and this would cause a

Table 1. Annual correction factors used to express stable carbon isotopic ratios relative to a standard value of –6.4‰ AD1850-2004. Correction values (‰) for each year should be added to the corresponding tree ring $\delta^{13}C$ value.

Year	Add	Year	Add	Year	Add	Year	Add
1850	0.01	1890	0.19	1930	0.37	1970	0.75
1851	0.02	1891	0.19	1931	0.37	1971	0.77
1852	0.02	1892	0.20	1932	0.38	1972	0.80
1853	0.03	1893	0.20	1933	0.38	1973	0.83
1854	0.03	1894	0.21	1934	0.38	1974	0.86
1855	0.03	1895	0.21	1935	0.39	1975	0.89
1856	0.04	1896	0.22	1936	0.39	1976	0.92
1857	0.04	1897	0.22	1937	0.40	1977	0.94
1858	0.05	1898	0.22	1938	0.40	1978	0.97
1859	0.05	1899	0.23	1939	0.41	1979	1.00
1860	0.06	1900	0.23	1940	0.41	1980	1.03
1861	0.06	1901	0.24	1941	0.42	1981	1.06
1862	0.07	1902	0.24	1942	0.42	1982	1.08
1863	0.07	1903	0.25	1943	0.42	1983	1.11
1864	0.07	1904	0.25	1944	0.43	1984	1.14
1865	0.08	1905	0.26	1945	0.43	1985	1.17
1866	0.08	1906	0.26	1946	0.44	1986	1.20
1867	0.09	1907	0.26	1947	0.44	1987	1.22
1868	0.09	1908	0.27	1948	0.45	1988	1.25
1869	0.10	1909	0.27	1949	0.45	1989	1.28
1870	0.10	1910	0.28	1950	0.46	1990	1.31
1871	0.11	1911	0.28	1951	0.46	1991	1.34
1872	0.11	1912	0.29	1952	0.46	1992	1.37
1873	0.11	1913	0.29	1953	0.47	1993	1.39
1874	0.12	1914	0.30	1954	0.47	1994	1.42
1875	0.12	1915	0.30	1955	0.48	1995	1.45
1876	0.13	1916	0.30	1956	0.48	1996	1.48
1877	0.13	1917	0.31	1957	0.49	1997	1.51
1878	0.14	1918	0.31	1958	0.49	1998	1.53
1879	0.14	1919	0.32	1959	0.50	1999	1.56
1880	0.15	1920	0.32	1960	0.50	2000	1.59
1881	0.15	1921	0.33	1961	0.50	2001	1.62
1882	0.15	1922	0.33	1962	0.52	2002	1.65
1883	0.16	1923	0.34	1963	0.55	2003	1.67
1884	0.16	1924	0.34	1964	0.58	2004	1.69
1885	0.17	1925	0.34	1965	0.61	-	-
1886	0.17	1926	0.35	1966	0.63	-	-
1887	0.18	1927	0.35	1967	0.66	-	-
1888	0.18	1928	0.36	1968	0.69	-	-
1889	0.18	1929	0.36	1969	0.72	-	-

reduction in stomatal conductance and hence increase the $\delta^{13}C$ values of tree rings as the tree grows. The explanation is complicated by the fact that in some trees the sapwood to leaf area ratio increases with age, so that more water is potentially available.

The combination of declining leaf water potential but increasing sapwood to leaf area ratios has now been used to explain the juvenile effect in both deciduous and conifer trees (Schäfer et al. 2000; Monserud and Marshall 2001; McDowell et al. 2002).

The juvenile effect in the $\delta^{13}C$ value of tree rings is a serious limitation for palaeoenvironmental reconstruction because it reflects changes in tree development or height, rather than an environmental signal. It must be removed before tree ring $\delta^{13}C$ series can be used. If this is achieved by statistical de-trending then any environmental signal of the same length as the juvenile effect will also be removed. Alternatively, the juvenile years can be removed from the series, so that no de-trending is required. The need to statistically de-trend is the curse of palaeoenvironmental reconstructions based on tree ring width and to a lesser extent density (Cook et al. 1995), so if it can be avoided in isotope studies it will be a great advantage. Discarding isotope data that are costly to produce is of course undesirable, so in many studies the length of the juvenile effect has been estimated or assumed and those early rings have not been sampled. The common procedure of pooling samples prior to analysis limits the amount of information that is available on the length of the juvenile effect for different species and environments, but it is clearly very variable. Where the early years are to be ignored, it would seem prudent to run a few individual trees to check the length of the juvenile effect (Figure 10).

If changes in hydraulic conductivity are at least partly responsible for the juvenile effect seen in the $\delta^{13}C$ values of tree rings, then a similar though more muted effect would be expected in the water isotopes. This is because fractionation of water isotopes in the leaf is controlled to some extent by stomatal conductance (Equation 1). Very few stable water isotope data are available for individual trees, so any juvenile effect in the water isotopes remains to be demonstrated conclusively, however, there are indications that it may be present in both hydrogen (Gray and Se 1984; Switsur et al. 1996) and oxygen (Treydte, Pers. Comm).

Variability, sample size and precision

Until recently, the cost and effort of producing tree ring isotope measurements using off-line methods was very high. This led to an emphasis on the precision of each individual measurement. Effort was focused on the preparation of very pure samples, usually of cellulose or alpha cellulose, and the wood of several trees was often pooled to reduce the number of analyses required. High analytical precision is of course desirable, but when tree ring stable isotopic measurements are used for palaeoenvironmental reconstruction, the analytical precision is only one of several sources of variability that must be taken into account. Where the aim is to reconstruct the climate of a region, for example, what is required is a representative average value for the isotope ratio of the trees in that region for each year, and the difference between trees is the largest source of variability. A good analogy is tree height. To obtain a representative average height for all of the trees in a forest, it is much better to measure a lot of trees to the nearest metre than to measure just a few trees to the nearest millimetre. Where the variability between trees is large, as with stable isotopic measurements, adding more trees to the sample is much more efficient than increasing the analytical precision.

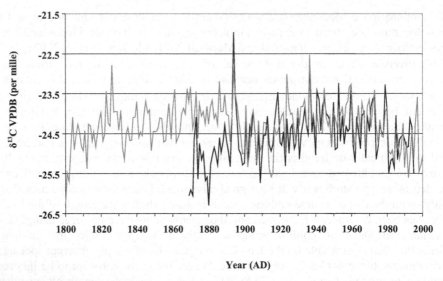

Year (AD)

Figure 10. Comparison of the stable carbon isotopic ratios from a single tree (thin black line) compared with the mean ratios obtained from five much older trees (grey line). Note that there is a juvenile effect lasting, in this instance, for about 23 years before the results fall into line. These data are from the latewood cellulose of *Pinus sylvestris* trees in northern Finland.

The number of trees required to provide a representative mean value is a critical question, but the answer depends on the aims of the study. For stable carbon isotopes, for example, Leavitt and Long (1984) have suggested that pooling four cores from four trees at a site will yield an accurate absolute $\delta^{13}C$ value. This conclusion was based on results from eight piñon pine (*Pinus edulis*) trees at a site in Arizona. For each increase in sample size the results were compared with the overall average result obtained when all eight trees were included, using the Pearson's correlation coefficient and the sum of the square of the differences. In both cases, however, the variables are not independent, and as the sample size increases the mean must, perforce, approach the overall mean of the whole sample. Their results do not define the confidence that can be placed in mean values based on a sample of four (or indeed eight) trees (McCarroll and Pawellek 1998).

Two aspects of the variability of isotope measurements must be considered when attempting to define an adequate sample size; the degree to which trees vary in parallel through time (signal strength), and the absolute difference in isotope values between trees, and hence the precision of mean values.

Signal strength
The degree to which trees vary in parallel from year to year is a reflection of the strength of the common signal, which can be quantified using analysis of variance or correlation-based techniques. Analysis of variance, as commonly applied (Fritts 1976), requires that all series be of the same length, thus limiting the analysis to the length of the shortest series of measurements, so the equivalent 'Mean Correlation Technique' (Wigley et al. 1984; Briffa and Jones 1989) is often preferable. Briffa and Jones (1989) suggest that at least five trees and a minimum of 30 years should be used.

In the 'mean correlation technique', the signal is defined as the mean correlation calculated between all possible pairs of trees ($\bar{r}_{bt}$). This is equivalent to the fractional variance component in Analysis of Variance. The variance that is not common between trees (the 'noise') is cancelled in direct proportion to the number of trees (t) that are combined to produce a common signal. Since the values represent a (presumably unbiased) sample, the degree to which a particular sample chronology portrays the hypothetically perfect chronology can be quantified using the chronology signal expressed as a fraction of the total chronology variance. This has been termed the Expressed Population Signal, or EPS (Wigley et al. 1984; Briffa and Jones 1989):

$$\text{EPS (t)} = (t.\bar{r}_{bt})/(t.\bar{r}_{bt} + (1-\bar{r}_{bt})) \qquad (4)$$

Although there is no formal definition of the minimum acceptable EPS, Wigley et al. (1984) suggest 0.85 as a reasonable value. The EPS is very sensitive to the number of trees (t) in a chronology (Briffa and Jones 1989). The EPS values can also be converted to a measure of chronology error equivalent to the Standard Error calculated from the component variances in an Analysis of Variance approach, where:

$$SE^2 = (1 - \bar{r}_{bt})/(t) \qquad (5)$$

The mean correlation technique can be used to measure the strength of the common signal within a single site, between sites and when sites are combined to produce regional chronologies. This approach was used by McCarroll and Pawellek (1998) to quantify the signal strength in stable carbon isotopic ratios obtained from 25 *Pinus sylvestris* trees at 5 sites from the Arctic Circle to the northern limit of pine in Finland. Expressed Population Signals (EPS) from samples of five trees were very high (0.89 to 0.95) and combining 15 trees from the three northernmost sites yielded an EPS of 0.97. These results should be treated with caution, since many of the trees used were young, and neither the juvenile effect nor atmospheric decline were removed prior to the analysis. However, McCarroll and Pawellek (2001) presented a larger data set from the same sites and found that correcting for the atmospheric decline and de-trending to remove any juvenile effects, increased rather than decreased the common signal strength (Table 2). These results suggest that the common $\delta^{13}C$ signal in pine trees growing close to the northern European tree line is very strong, and that sites can be combined across a very large area to produce a representative regional $\delta^{13}C$ signal. The common signal in $\delta^{13}C$ in this area is much stronger than in tree ring widths and the former are much less sensitive to local site conditions.

A strong common signal and relative insensitivity to local site conditions is even more clearly displayed by Gagen et al. (2004), who measured $\delta^{13}C$ of latewood cellulose from two species of pine growing at different altitudes in the French Alps. The $\delta^{13}C$ results were so similar that the sites and species could be combined to produce a single chronology, despite large differences in ring widths and density. At the high latitude Finnish sites, carbon isotope fractionation is strongly controlled by photosynthetic rate, whereas at the high altitude Alpine sites it is strongly controlled by stomatal conductance. At less extreme sites, where no single environmental factor strongly

Table 2. Number of trees (n), mean correlation (r) and expressed population signal (EPS) obtained for $\delta^{13}C$ of latewood cellulose (*Pinus sylvestris*) from three sites in northern Finland for the period 1961 to 1995 and from the three sites combined. The same data sets were used raw, after mathematically removing the atmospheric decline in $\delta^{13}C$ (corrected), and after linear de-trending to remove any juvenile effects (McCarroll and Pawellek 2001).

	n	Raw		Corrected		De-trended	
		r	EPS	r	EPS	r	EPS
Utsjoki	7	0.77	0.96	0.76	0.96	0.82	0.97
Kaamanen	5	0.80	0.95	0.84	0.96	0.85	0.97
Laanila	8	0.68	0.95	0.67	0.94	0.73	0.96
Combined	**20**	**0.63**	**0.97**	**0.63**	**0.97**	**0.67**	**0.98**

controls carbon isotope fractionation, the common signal is likely to be much weaker. The paucity of results from individual (as opposed to pooled) tree ring samples precludes a similar assessment of the strength of the common signal in the water isotopes.

Absolute differences and the precision of the mean
Measures of signal strength, such as the EPS, measure the degree to which signals vary in parallel, but ignore differences in the absolute values obtained from different trees or sites. This is irrelevant when they are used to describe tree ring widths or density, because these are routinely de-trended and indexed so that they share a common mean. One of the greatest potential benefits of tree ring isotopes is that it may be possible to forego any statistical de-trending, so that the original values are retained and no potential low-frequency palaeoclimate information is lost. In this case the absolute differences between trees becomes critical, for it determines the precision of the mean values upon which palaeoclimate reconstructions will be based. If the precision can be defined statistically, then it should also be possible to place statistically-defined confidence limits around the quantitative estimates of palaeoclimate.

The variability of absolute values can be expressed in a number of different ways using, for example, the population standard deviation, the sample standard deviation, the standard error or, commonly, twice the standard error. However, the statistical significance of each of these varies as the sample size changes, so it is much better to present mean values with statistically-defined confidence limits that take account of the sample size.

Statistically-defined confidence limits are based on the assumption that the data represent an unbiased sample from a population and use the variability of the results to predict that the true population mean will lie within a certain range of the sample mean. The level of confidence required is set in advance, and represents the 'chances of being wrong'. A confidence level of 95% ($p < 0.05$) represents a one in twenty chance of the true population mean lying beyond the confidence limits that are defined.

The 95% confidence limits are often approximated using the equation:

$$1.96 \, (\sigma/n^{\frac{1}{2}}) \tag{6}$$

where σ is the standard deviation and n the sample size. This is the equation used in some spreadsheets (such as EXCEL). However, this is only correct for large samples. For the small samples typical of isotope studies the correct formula is:

$$t \, (\sigma/n^{\frac{1}{2}}) \tag{7}$$

where t is taken from the student t distribution (1.96 is the asymptotic value). For samples of 30 or more the difference is very small and can be ignored, but for very small samples the difference is substantial. For sample sizes of 4 and 8, for example, the t values at $p < 0.05$ are 3.18 and 2.36 respectively. Given a standard deviation of 0.72, the correct values for samples of 8 and 4 are ± 0.60 and ± 1.14, as opposed to ± 0.50 and ± 0.71 if the approximate equation is used. For convenience, the appropriate t values for some small sample sizes are provided in Table 3.

The ability to assign statistically defined confidence intervals to mean absolute isotope values is one of the great advantages of tree ring stable isotope studies. The great sensitivity of confidence limits to sample size, however, again underlines the importance of increasing the number of trees that are sampled. Pooling samples prior to analysis is a not a solution because confidence limits cannot be calculated without quantifying the variability of the individual results.

Number of trees
It is not feasible to define the number of trees required to provide a representative sample for all situations. The number required depends on the variability of the results and on the precision required by the study. If the aim is to focus only on high frequency, inter-annual variability, then the absolute differences between trees are irrelevant and the number of trees required can be defined on the basis of the Expressed Population Signal (≥ 0.85) or an equivalent. Where a single environmental factor strongly controls fractionation then a small number of trees (4 or 5) is likely to suffice (McCarroll and Pawellek 2001; Gagen et al. 2004). If the aim is to retain low-frequency, long-term changes then it is necessary to use the absolute isotope values, which requires that the precision of the mean be quantified using confidence intervals. The confidence level used and the acceptable width of the confidence band will depend on the aims of the study. Where the results are to be used to reconstruct climate, then the uncertainty in the mean value for each year must be combined with the uncertainties in the calibration with the climate parameter to produce realistic confidence limits around palaeoclimate estimates. High precision of the palaeoclimate estimates will depend much more on the number of trees than on the level of analytical precision achieved in the laboratory.

Environmental signals

Contrary to early suggestions (e.g., Libby et al. 1976), none of the stable isotopes in tree rings provides a simple palaeothermometer. Nor can any other single palaeoclimate

Table 3. Values from the Student *t* distribution used for calculating confidence limits for small samples. Sample size is n, and values are given for three confidence levels.

n	10%	5%	1%
4	2.35	3.18	5.84
5	2.13	2.78	4.60
6	2.01	2.57	4.03
8	1.89	2.36	3.50
10	1.83	2.26	3.25
12	1.80	2.20	3.11
15	1.76	2.15	2.98
20	1.73	2.09	2.86
25	1.71	2.06	2.80
30	1.70	2.04	2.75

signal be apportioned to a particular isotope ratio in all cases. Extracting an environmental signal from stable isotopes in tree rings requires an understanding of both isotope theory and of the ecophysiology of the trees sampled. Correlation and regression techniques are a useful tool for quantifying the relationships, but they are no substitute for mechanistic understanding. Correlation does not necessarily imply causation.

Stable carbon isotopic ratios in tree rings mainly reflect the fractionation of atmospheric CO_2 in the leaf. Fractionation occurs as CO_2 enters the stomata (–4.4‰) and again as it is taken up by the photosynthetic enzyme (–27‰), but in each case the degree of fractionation is almost constant, and does not vary significantly in response to the environment. What does change, however, is the relative rate at which CO_2 enters and is removed, and therefore the internal concentration of CO_2. The $\delta^{13}C$ values of leaf sugars are therefore a reflection of the internal concentration of CO_2 and consequently of the balance between stomatal conductance and photosynthetic rate. If some of the sugars are used quickly to form wood, then the signal should be imparted into the cellulose and lignin framework of the rings. If the sugars are stored as starch and later remobilized then the signal may be much more complicated (Helle and Schleser 2004), which is why it is an advantage to isolate the summer or latewood, rather than the early wood, at least in broadleaf trees.

Since the dominant controls on $\delta^{13}C$ are stomatal conductance and photosynthetic rate, the environmental signal for a particular tree species at a particular site will depend on the relative strength of these two controls. At dry sites, where the dominant control is stomatal conductance, the likely environmental signals are air humidity and antecedent precipitation. This is the case in the western Alps for example (Gagen et al. 2004). Where moisture supply is plentiful, the dominant control may be photosynthetic rate, which may be controlled either by temperature or, more likely, the amount of sunlight (McCarroll and Loader 2004). This appears to be the case for pine trees growing close to the northern European timberline in Finland (McCarroll and Pawellek 2001). Where trees are growing under mild conditions, and the internal concentration of CO_2 is not controlled strongly by either stomatal conductance or photosynthetic rate, the carbon isotopic ratios in the tree rings may not carry a strong environmental signal and will, at least in isolation, be very difficult to interpret.

There is currently a great deal of interest in attempting to reconstruct variations in the isotope ratios of meteoric water, largely as a way of testing Global Circulation Models (e.g., the International Atomic Energy Agency ISOMAP Programme). It must be stressed that the water isotopes in tree rings do not provide a direct record of changes in the isotope ratios of meteoric water. The water isotope ratios in tree rings conflate two quite separate environmental signals. The first is the isotope ratio of the water taken up by the roots, the second is an evaporative enrichment signal produced in the leaf. The relative strength of these signals is difficult to predict, not least because of varying potential for exchange of enriched phloem water with xylem (source) water at the site of wood synthesis. Water isotopes in tree rings from different sites have nevertheless been interpreted as recording the isotope signature of source water and/or growing season humidity (Buhay and Edwards 1995; Pendall 2000; Saurer et al. 2002; Robertson et al. 2001; Anderson et al. 2002) and parallel progress made in the mechanistic modelling of multiple (water) isotopes in source water determination and plant physiology (e.g., Roden et al. (2000); Edwards and Fritz (1986)).

The most common environmental parameter to be linked statistically to the stable isotope ratios in tree rings, and to be reconstructed, is growing season temperature. However, although the correlation with temperature is often strong, that must not be taken to mean that temperature is the main control on fractionation of any of the isotopes. Temperature certainly has an effect, but it is also correlated with most of the other factors that influence fractionation, such as humidity, precipitation and sunshine. If the link with temperature is indirect, then temperature reconstructions will rely on the assumption that the link between temperature and the other parameters has remained constant over time. In fact there is good reason to suspect that this may not have been the case, particularly during the last century. Anthropogenic changes in the opacity of the atmosphere, for example, might result in a decline in effective sunshine (photon flux) without an accompanying change in temperature. More locally, changes in the density of a forest, as it grows following disturbance or an advance of the tree line for example, might be associated with changes in local air humidity, access to light and in competition for moisture resources. All of these could impart a trend in isotope ratios that is quite independent of changes in regional air temperature.

This is not to suggest that stable isotopes in tree rings cannot be used to reconstruct past changes in temperature. On the contrary, reconstructing low-frequency, long term changes in temperature is one of the most exciting challenges facing isotope dendroclimatology. However, ignoring a problem does not make it disappear, and it is better to acknowledge the difficulties and plan sampling strategies to deal with them. The answer probably lies in combining several different palaeoclimate proxies, from tree rings and other archives, to tease apart the desired signal from the many potential sources of confusion and noise.

Multi-proxy dendroclimatology

Palaeoenvironmental research using ice cores has, in recent years, revolutionised our understanding of the variability of past climate, and particularly of the potential rates of climate change. The richness of the information derived from ice cores reflects the many different techniques that have been used to extract information. The ice is a physical archive, and this allows a multi-proxy approach. Trees also represent physical archives, but they are rarely treated as such. More commonly the ring widths, or rarely relative densities, are measured and these data are archived for use more widely. The archives of tree ring data are a tremendous resource, but by treating trees as a physical (rather than numerical) archive, and applying a multi-proxy approach, we can greatly increase the amount of palaeoenvironmental information that can be extracted. As well as supplying a wider range of palaeoenvironmental information, a multi-proxy approach should allow us to test some of the assumptions we are forced to make in reconstructing the past.

A multi-proxy approach to dendroclimatology, including the measurement of stable isotopes, may provide the most powerful means available to reconstruct the climate of the Holocene. Where several proxies are used to reconstruct a single environmental variable, for example, some of the noise associated with each of the measures may be cancelled, enhancing the palaeoclimate signal (McCarroll et al. 2003). This approach should work even where the climate parameter being reconstructed is not the strongest direct control on the proxy being measured. The important assumption is that the sources of error (noise) for each of the proxies are not the same. This approach is likely to work particularly well for reconstructing summer temperature.

The combination of proxies, including multiple stable isotopic measurements, may also be useful for deriving palaeoenvironmental information that cannot be obtained from any single proxy. In northern Finland, for example, where tree growth is strongly controlled by summer temperature and trees are rarely moisture stressed, the combination of maximum density and stable carbon isotopic ratios allows dry summers to be identified (Figure 11).

Combinations of proxies may also allow us to test some of the assumptions that we are forced to make in palaeoenvironmental reconstruction. For example, if a modern training set is used to quantify the relationship between a given proxy and a climate parameter, then we need to assume that the relationship has not changed through time. One way to test this may be to monitor the relationship between more than one proxy. For example, if a modern training set shows a strong correlation between summer temperature and ring widths, a palaeotemperature reconstruction relies on the assumption that the relationship between ring widths and temperature has not changed through time. Times when rings were narrow are assumed to have been cool and times when ring widths were broad are assumed to have been warm.

However, temperature is not the only factor that can limit tree growth and there may have been periods of, for example, drought, frost, disease or insect attack that occurred in the past but are not present during the training period. The co-variance of different proxies during the training period should give a good indication of the way that they should co-vary in the past under similar constraints. If there are conditions in the past that are not present in the training period then they may be apparent as an unexpected

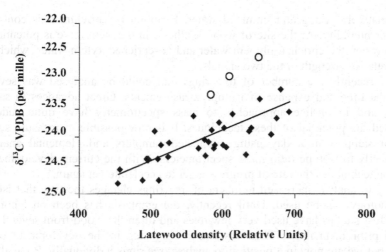

Figure 11. In northern Finland, both latewood density and stable carbon isotopic ratios (‰ VPDB) respond to the amount of energy received during the summer. However, soil moisture status influences the isotopic ratios (via stomatal conductance) but not wood density. When stable carbon isotopic ratios are regressed on latewood density, therefore, the residuals (open circles) represent dry summers. This multi-proxy approach to dendroclimatology can be used to reconstruct summer dryness, even in areas where water stress is not growth-limiting (after McCarroll et al. (2003)).

change in the co-variance of proxies. Either hot dry conditions or cool summers, for example, might result in reduced ring widths and densities, but these scenarios would have very different effects on the stable isotope ratios of both carbon and the water isotopes. The covariance of ring widths and isotope ratios should indicate which condition prevailed.

Summary

Trees provide one of the best natural archives of recent and Holocene climate and environmental change. They are widespread, sensitive to their environments, and they produce a near continuous physical archive of potential palaeoclimate information locked into the physical and chemical characteristics of their rings. Stable isotope measurements provide one of the most powerful means of realising that potential.

The stable isotopes of carbon record the balance between the rate at which CO_2 enters leaves via the stomata (stomatal conductance) and the rate at which it is removed (photosynthetic rate). Where one side of this balance dominates, then the environmental signal can be very strong. This is the case in dry environments, where stomatal conductance dominates and the main signals are air humidity and soil moisture content, and in moist environments where photosynthetic rate dominates and the signals are summer temperature and sunshine. Changes in the carbon isotopic composition of the atmosphere are also recorded.

The water isotopes conflate two separate signals, the isotope composition of the source water and isotope enrichment due to evaporation (transpiration) in the leaves. The former is only a proxy for the isotope composition of precipitation if the trees are accessing relatively young and unmixed water. Evaporative enrichment is controlled mainly by air humidity. At the site of wood synthesis in the stem there is potential for exchange between the enriched phloem water and un-enriched xylem water, which will change the relative strength of the two signals.

Until very recently the number of tree rings that could be analysed was severely limited by the time and expense of isotope measurements. Great advances in sample preparation, and in on-line approaches to mass spectrometry have quite suddenly revolutionised the potential of these techniques. It is now possible to measure several hundreds of samples in a day using automated samplers and elemental analysers attached directly to isotope ratio mass spectrometers. With the current development of laser ablation techniques the rate of progress is set to accelerate yet again.

The ability to analyse increased numbers of tree rings changes the way that isotope dendrochronology can be used. Until recently, the emphasis has been on laboratory precision. Most studies have used very few trees and often the wood from several trees was pooled prior to analysis. This was a necessary step in the development of the techniques, but we are now in a position to analyse tree rings individually. By analysing each tree independently, and combining the results from several trees, it is possible to quantify the variability, at different temporal frequencies, and thus to place statistically defined confidence limits around the mean values. If, as we suspect, the isotope results do not need to be statistically de-trended to remove age-related effects, then tree ring stable isotopes should retain palaeoclimate information at all temporal frequencies, including the low-frequency, long-term changes that are so difficult to capture using more traditional methods.

By treating trees as physical archives, and combining stable isotopes with other measurements in a multi-proxy approach, it should be possible to increase the precision of palaeoclimate estimates, extend the range of palaeoenvironmental signals that can be extracted and also test some of the assumptions inherent in palaeoenvironmental reconstruction.

References

Aggett J., Bunton C.A., Lewis T.A., Llewellyn D.R., O'Connor C. and Odell A.L. 1965. The isotopic analysis of oxygen in organic compounds and in co-ordination compounds containing organic ligands. Int. J. Appl. Radiat. Is. 16: 165–170.

Anderson W.T., Bernasconi S.M. and McKenzie J.A. 1998. Oxygen and carbon isotopic record of climatic variability in tree ring cellulose (*Picea abies*): an example from central Switzerland (1913–1995). J. Geophys. Res. 103/D24: 31625–31636.

Anderson W.T., Bernasconi S.M., McKenzie J.A., Saurer M. and Schweingruber F. 2002. Model evaluation for reconstructing the oxygen isotopic composition in precipitation from tree ring cellulose over the last century. Chem. Geol. 182: 121–137.

Barbour M.M., Andrews T.J. and Farquhar G.D. 2001. Correlations between oxygen isotope ratios of wood constituents of *Quercus* and *Pinus* samples from around the world. Aust. J. Plant Physiol. 28: 335–348.

Barbour M.M. and Farquhar G.D. 2000. Relative humidity– and ABA–induced variation in carbon and oxygen isotope ratios of cotton leaves. Plant Cell Environ. 23: 473–485.

Barbour M.M., Walcroft A.S. and Farquhar G.D. 2002. Seasonal variation in $\delta^{13}C$ and $\delta^{18}O$ of cellulose from growth rings of *Pinus radiata*. Plant Cell Environ. 25: 1483–1499.

Becker B. 1993. An 11,000-year German oak and pine dendrochronology for radiocarbon calibration. Radiocarbon 35: 201–213.

Becker B., Kromer B. and Trimborn P. 1991. A stable isotope tree ring timescale of the Late Glacial/Holocene boundary. Nature 353: 647–649.

Begley I.S. and Scrimgeour C.M. 1996. On-line reduction of H_2O for δ^2H and $\delta^{18}O$ measurement by continuous-flow isotope ratio mass spectrometry. Rapid Commun. Mass Sp. 10: 969–973.

Begley I.S. and Scrimgeour C.M. 1997. High precision δ^2H and $\delta^{18}O$ measurement for water and volatile organic compounds by continuous-flow pyrolysis isotope ratio mass spectrometry. Anal. Chem. 69: 1530–1535.

Benner R., Fogel M.L., Sprague E.K. and Hodson R.E. 1987. Depletion of ^{13}C in lignin and its implications for stable carbon isotope studies. Nature 329: 708–710.

Bennett C.F. and Timmell T.E. 1955. Preparation of cellulose trinitrate. Sven. Papperstidn. 58: 281–286.

Berninger F., Sonninen E., Aalto T. and Lloyd J. 2000. Modelling ^{13}C Discrimination in Tree Rings. Global Biogeochem. Cy. 14: 213–223.

Bert D., Leavitt S.W. and Dupouey J-L. 1997. Variations of wood $\delta^{13}C$ and water use efficiency of *Abies alba* during the last century. Ecology 78(5): 1588–1596.

Birks H.J.B. 1995. Quantitative paleoenvironmental reconstructions. In: Maddy D. and Brew J.S. (eds.), Statistical Modelling of Quaternary Science Data, Technical Guide 5, Quaternary Res. Association, Cambridge, pp. 161–254.

Borella S. and Leuenberger M. 1998. Reducing uncertainties in $\delta^{13}C$ analysis of tree rings: pooling, milling and cellulose extraction. J. Geophys. Res. 103: 19519–19526.

Borella S., Leuenberger M. and Saurer M. 1999. Analysis of $\delta^{18}O$ in tree rings: wood-cellulose comparison and method dependent sensitivity. J. Geophys. Res. 104: 19267–19273.

Boutton T.W. 1991. Stable carbon isotope ratios of natural materials. 1. Sample preparation and mass spectrometric analysis. In: Coleman D.C. and Fry B. (eds.), Carbon isotope techniques. Academic Press, New York. Pp. 155–171.

Boutton T.W., Wong W.W., Hachey D.L., Lee L.S., Cabera M.P. and Klein P.D. 1985. Comparison of quartz and Pyrex tubes for combustion of organic samples for stable carbon isotope analysis. Anal. Chem. 55: 1832–1833.

Brendel O., Iannetta P.P.M. and Stewart D. 2000. A rapid and simple method to isolate pure alpha-cellulose. Phytochem. Analysis 11: 7–10.

Brenna J.T., Tobias H.J. and Corso T.N. 1998. High precision deuterium and ^{13}C measurement by continuous flow IRMS: organic and position specific isotope analysis. In: Griffifths H. (ed.), Stable isotopes, pp.1-10. BIOS scientific publishers, Oxford.

Brenninkmeijer C.A.M. and Mook W.G. 1981. A batch process for direct conversion of organic oxygen and water to CO_2 for $^{18}O/^{16}O$ analysis. Int. J. Appl. Radiat. Is. 32: 137–141.

Briffa K.R. 2000. Annual climate variability in the Holocene: interpreting the message of ancient trees. Quaternary Sci. Rev. 19: 87–105.

Briffa K.R. and Jones P.D. 1989. Basic chronology statistics and assessment In: Cook E. and Kairiukstis L. (eds.), Methods of dendrochronology: applications in the environmental sciences. Kluwer Academic Publishers, Dordrecht. pp. 137-152.

Briffa K.R, Jones P.D, Schweingruber F.H., Karlen W. and Shiyatov G. 1996. Tree ring variables as proxy climate indicators: problems with low-frequency signals. In: Jones P.D, Bradley R.S. and Jouzel J. (eds.). Climatic Variations and Forcing Mechanisms of the last 2000 years. Springer-Verlag. Berlin. Heidelberg, pp. 9–41.

Buchanan D.L. and Corcoran B.J. 1959. Sealed tube combustion for the determination of carbon-14 and total carbon. Anal. Chem. 31: 1635–1638.

Buhay W.M. and Edwards T.W.D. 1995. Climate in southwestern Ontario, Canada, between AD 1610 and 1885 inferred from oxygen and hydrogen isotopic measurements of wood cellulose from trees in different hydrologic settings. Quaternary Res. 44: 438–446.

Bunton C.A., Lewis T.A. and Lewellyn D.R. 1956. Oxygen–18 tracer studies in the formation of reactions of organic peracid. Journal of the Chemical Society, Pt.1: 1226–1230.

Burk R.L. and Stuiver M. 1981.Oxygen isotope ratios in trees reflect mean annual temperature and humidity. Science 211: 1417–1419.

Clark I. and Fritz P. 1997. Environmental Isotopes in Hydrogeology, Lewis Publishers, Boca Raton, New York, 328 pp.

Chambers J.Q. 1998. Ancient trees in Amazonia. Nature 391: 135–136

Coleman M.L., Shepherd T.J., Durham J.J., Rouse J.E. and Moore G.R. 1982. Reduction of water with zinc for hydrogen isotope analysis. Anal. Chem 66: 2989–2991.

Cook E.R., Briffa K.R., Meko D.M., Graybill D.A. and Funkhouser G. 1995. The 'segment length curse' in long tree ring chronology development for palaeoclimatic studies. Holocene 5: 229–237.

Cook E.R. and Kairiukstis L.A. 1990. Methods of dendrochronology. Dordrecht, Boston, London. Kluwer Academic Publishers, 394 pp.

Coplen T.B. 1995. Discontinuance of SMOW and PDB. Nature 373: 285.

Craig H. 1954. Carbon-13 variations in Sequoia rings and the atmosphere. Science 119: 141–144.

Craig H. 1961. Isotopic variations in meteoric waters. Science 133: 1702–1703.

Craig H. and Gordon L.I. 1965. Deuterium and oxygen-18 variations in the ocean and marine atmospheres. In: Tongiorgi E. (ed.), Proceedings of a Conference on stable Isotopes in Oceanographic Studies and Palaeotemperatures. Lischi and Figli, Pisa, Italy. pp. 9–130.

Dansgaard W. 1964. Stable isotopes in precipitation. Tellus 16: 436–468.

Darling W.G. 2004. Hydrological factors in the interpretation of stable isotopic proxy data present and past: a European perspective. Quaternary Sci. Rev. 23: 743–770.

Darling W.G., Bath A.H., Gibson J.J. and Rozanski K. This volume. Isotopes in Water. In: Leng M.J. (ed.), Isotopes in Palaeoenvironmental Research. Springer, Dordrecht, The Netherlands.

Dawson T.E. 1993. Water sources of plants as determined from xylem-water isotopic composition: perspectives on plant competition, distribution and water relations. In: Ehleringer J.R., Happ A.E. and Farquhar G.D. (eds.), Stable isotopes and plant carbon-water relations. Academic Press, New York. pp. 465–496.

Dawson T.E. and Pate J.S. 1996. Seasonal water uptake and movement in root systems of Australian phraeatophytic plants of dimorphic root morphology: a stable isotope investigation. Oecologia 107: 13–20.

Dongmann G., Nürnberg H.W., Förstel H. and Wagener K. 1974. On the enrichment of $H_2^{18}O$ in leaves of transpiring plants. Radiat. Environ. Biophys. 11: 41–52.

Dubois A.D. 1984. On the climatic interpretation of the hydrogen isotope ratios in recent and fossil wood. Bulletin de la Société Belge de Géologie 93: 267–270.

Dunbar J. and Wilson A.T. 1983. Re-evaluation of the $HgCl_2$ pyrolysis technique for oxygen isotope analysis. Int. J. Appl. Radiat. Is. 6: 932–934.

Dupouey J-L., Leavitt S.W., Choisnel E. and Jourdain S. 1993. Modeling carbon isotope fractionation in tree rings based upon effective evapotranspiration and soil-water status. Plant Cell Environ. 16: 939–947.

Duquesnay A., Bréda N., Stievenard M. and Dupouey J.L. 1998. Changes of tree ring $\delta^{13}C$ and water-use efficiency of beech (*Fagus sylvatica* L.) in north-eastern France during the past century. Plant Cell Environ. 21: 565–572.

Edwards T.W.D. and Fritz P. 1986. Assessing meteoric water composition and relative humidity from ^{18}O and ^{2}H in wood cellulose: paleoclimatic implications for southern Ontario, Canada. Appl. Geochem. 1: 715–723.

Edwards T.W.D., Graf W., Trimborn P., Stichler W. and Payer H.D. 2000. $\delta^{13}C$ response surface resolves humidity and temperature signals in trees. Geochim. Cosmochim. Ac. 64: 161–167.

Ehleringer J.R., Hall A.E. and Farquhar G.D. (eds). 1993. Stable isotopes and plant carbon-water relations. Academic Press. New York. 555 pp.

Epstein S. and Krishnamurthy R.V. 1990. Environmental information in the isotopic record in trees. Phil. Trans. R. Soc. 330A: 427–439.

Epstein, S. and Yapp, C.J. 1976. Climatic implications of the D/H ratio of hydrogen in C-H groups in tree cellulose. Earth Planet Sc. Lett. 30: 252–261.

Eronen M., Zetterberg P., Briffa K.R., Lindholm M., Meriläinen J. and Timonen M. 2002. The supra-long Scots pine tree ring record for Finnish Lapland: Part 1, chronology construction and initial inferences. Holocene 12: 673–680.

Esper J., Cook E.R., Schweingruber F.H. 2002. Low-frequency signals in long tree ring chronologies for reconstructing past temperature variability. Science 295: 2250–2253.

Esper J., Frank D., Wilson R.J.S. 2004. Climate Reconstructions: Low frequency ambition and High-Frequency Ratification. EOS 85: 113-120.

Evans M.N. and Schrag D.P. 2004. A stable isotope-based approach to tropical dendroclimatology. Geochim Cosmochim. Ac. 68 (16): 3295–3305.

Farmer J.G. and Baxter M.S. 1974. Atmospheric carbon dioxide levels as indicated by the stable isotope record in wood. Nature 247: 273–275.

Farquhar G.D. and Lloyd J. 1993. Carbon and oxygen isotope effects in the exchange of carbon dioxide between terrestrial plants and the atmosphere. In: Ehleringer, J.R. Hall A.E. and Farquhar G.D. (eds.), Stable isotopes and plant carbon-water relations. Academic Press. New York. pp 47–70.

Farquhar G.D., Henry B.K. and Styles J.M. 1997. A rapid on-line technique for the determination of oxygen isotope composition of nitrogen containing organic compounds and water. Rapid Commun. Mass Spectrom. 11: 1554–1560.

Farquhar G.D. O'Leary M.H. and Berry J.A. 1982. On the relationship between carbon isotope discrimination and intercellular carbon dioxide concentration in leaves. Austr. J. Plant Physiol. 9: 121–137.

February E.C. and Stock W.D. 1999. Declining trends in the $^{13}C/^{12}C$ ratio of atmospheric carbon dioxide from tree rings of South African Widdringtonia cedarbergensis. Quaternary Res. 52: 229–236.

Feng X.H. 1998. Long-term c_i/c_a response of trees in western North America to atmospheric CO_2 concentration derived from carbon isotope chronologies. Oecologia 117: 19–25.

Feng X.H. 1999. Trends in intrinsic water-use efficiency of natural trees for the past 100–200 years: A response to atmospheric CO_2 concentration. Geochim. Cosmochim. Ac. 63: 1891–1903.

Feng X.H. and Epstein S. 1994. Climatic implications of an 8000 year hydrogen isotope time series from Bristlecone Pine trees. Science 265: 1079–1081.

Feng X. and Epstein S. 1995a. Carbon isotopes of trees from arid environments and implications for reconstructing atmospheric CO_2 concentration. Geochim. Cosmochim. Ac. 59: 2599–2608.

Feng X. and Epstein S. 1995b. Climatic temperature records in δD data from tree rings. Geochim. Cosmochim. Ac. 59: 3029–3037.

Feng X., Cui H., Tang K. and Conkey L.E. 1999. Tree ring delta-D as an indicator of Asian monsoon intensity. Quaternary Res. 51: 262–266.

Ferhi A., Baraic T., Jusserand C. and LeTolle R. 1983. An integrated method for the isotopic analysis of oxygen from organic compounds, air, water vapour and leaf water. Int. J. Appl. Radiat. Is. 34: 1451–1457.

Field E.M., Switsur V.R. and Waterhouse J.S. 1994. An improved technique for determinations of oxygen stable isotope ratios in wood cellulose. Int. J. Appl. Radiat. Is. 45: 177–181.

Francey R.J. 1981. Tasmanian tree rings belie suggested anthropogenic $^{13}C/^{12}C$ trends. Nature 290: 232–235.

Francey R.J., Allison C.E., Etheridge D.M., Trudinger C.M., Enting I.G., Leuenberger M., Langenfelds R.L., Michel E. and Steele L.P. 1999. A 1000-year high precision record of δ^{13}C in atmospheric CO_2. Tellus 51B: 170–193.

110   McCARROLL AND LOADER

Francey R.J. and Farquhar G.D. 1982. An explanation for the $^{12}C/^{13}C$ variations in tree rings. Nature 297: 28–31.

Freyer H.D. 1979a. On the ^{13}C record in tree rings. Part 2 registration of microenvironmental CO_2 and anomalous pollution effect. Tellus 31: 308–312.

Freyer H.D. 1979b. On the ^{13}C record in tree rings. Part 1 ^{13}C variations in northern hemispheric trees during the last 150 years. Tellus 31: 124–137.

Freyer H.D. and Belacy N. 1983. $^{12}C/^{13}C$ records in Northern Hemispheric trees during the past 500 years: anthropogenic impact and climatic superpositions. J. Geophys. Res. 88: 6844–6852.

Fritts H.C. 1976. Tree Rings and Climate. Academic Press, New York. 567 pp.

Gagen M., McCarroll D. and Edouard J-L. 2004. The effect of site conditions on pine tree ring width, density and $\delta^{13}C$ series in a dry sub-Alpine environment. Arct. Antarct. Alp. Res. 36: 166–171.

Gat J.R. 1980. The isotopes of hydrogen and oxygen in precipitation. In: Fritz P. and Fontes J. Ch. (eds.), Handbook of Environmental Isotope Geochemistry, Elsevier, Amsterdam, pp. 21–47.

Gray J. and Se J.S. 1984. Climatic implications of the natural variation of D/H ratios in tree ring cellulose. Earth Planet Sc. Lett. 70: 129–138.

Gray J., Yonge C. and Thompson P. 1984. Simultaneous determination of $^{18}O/^{16}O$ and D/H in compounds containing C, H and O. Int. J. Appl. Radiat. Is. 6: 525–530.

Green J.W. 1963. Wood Cellulose. In: Whistler R.L. (ed.), Methods in carbohydrate chemistry III. Academic Press, New York, 407 pp.

Grudd H., Briffa K.R., Karlén W., Bartholin T.S., Jones P.D. and Kromer B. 2002. A 7400–year tree ring chronology in northern Swedish Lapland: natural climatic variability expressed on annual to millenial timescales. Holocene 12: 657–665.

Hammarlund D., Barnekow L., Birks H.J.B., Buchardt B. and Edwards T.W.D. 2002. Holocene changes in atmospheric circulation recorded in the oxygen-isotope stratigraphy of lacustrine carbonates from northern Sweden. Holocene 12 : 339–351.

Hantemirov R.M. and Shiyatov S.G. 2002. A continuous multimillenial ring-width chronology from Yamal, northwestern Siberia. Holocene 12: 717–726.

Heaton T.H.E. and Chenery C.A. 1990. Use of zinc turnings in the reduction of water to hydrogen for isotopic analysis. Anal. Chem. 55: 995–998.

Helle G. and Schleser G.H. 2003. Seasonal variations of stable carbon isotopes from tree rings of Quercus petraea. In: Schleser G., Winiger M., Bräuning A., Gärtner H., Helle G., Jansma E., Neuwirth B. and Treydte K. (eds.), TRACE: Tree Rings in Archaeology, Climatology and Ecology 1, pp. 66–70.

Helle G. and Schleser G.H. 2004. Beyond CO2-fixation by Rubisco - an interpretation of C-13/C-12 variations in tree rings from novel intra-seasonal studies on broad-leaf trees Plant. Cell Environ. 27 (3): 367–380.

Hemming D.L., Switsur V.R., Waterhouse J.S., Heaton T.H.E. and Carter A.H.C. 1998. Climate and the stable carbon isotope composition of tree ring cellulose: an intercomparison of three tree species. Tellus 50B: 25–32.

Hill S.A., Waterhouse J.S., Field E.M., Switsur V.R. and ap Rees T. 1995. Rapid recycling of triose phosphates in oak stem tissue. Plant Cell Environ. 18: 931–936.

Hughes M.K. and Graumlich L.J. 1996. Multimillennial dendroclimatic records from the western United States. In: Climatic Variations and Forcing Mechanisms of the last 2000 Years, R.S. Bradley, P.D. Jones, and J. Jouzel. (eds.), NATO Advanced Studies Workshop Series. Springer-Verlag, pp. 109–124.

Jones P.D., Briffa K.R., Barnett T.P. and Tett S.F.B. 1998. High resolution palaeoclimatic records for the last millennium:integration, interpretation and comparison with general circulation model control run temperatures. Holocene 8: 455–471.

Jones P.D. and Mann M.E. 2004. Climate over past Millennia. Rev. Geophys. 42: RG2002 DOI:10.1029/2003RG000143.

Keeling C.D., Mook W.G. and Tans P.P. 1979. Recent trends in the ^{13}C-^{12}C ratio of atmospheric carbon- dioxide. Nature 277: 121–123.

Kelly S.D., Parker L.G., Sharman M. and Dennis M.J. 1998. On-line quantitative determination of ^{2}H/^{1}H isotope ratios in organic and water samples using an elemental analyser coupled to an isotope ratio mass spectrometer. J. Mass Spectrom. 33: 735–738.

Kitagawa H. and Matsumoto E. 1995. Climatic implications of δ^{13}C variations in a Japanese cedar (Cryptomeria japonica) during the last two millenia. Geophys. Res. Lett. 22: 2155–2158.

Koziet J. 1997. Isotope ratio mass spectrometric method for the on-line determination of oxygen-18 in organic matter. J. Mass Spectrom. 32: 103–108.

Krishnamurthy R.V. and Epstein S. 1985. Treering D/H ratio from Kenya, East Africa and its palaeoclimatic significance. Nature 317: 160–162.

Krishnamurthy R.V. and Machavaram M.V. 1998. Hydrogen isotope exchange in thermally stressed cellulose, Chem. Geol. 125: 85–96.

Krishnamurthy R.V. and Machavaram M.V. 2000. Is there a stable isotope evidence for CO_2 fertilization effect? P. Indian Acad. Sci. 109: 141–144.

Kürschner W.M. 1996. Leaf stomata as biosensors of palaeoatmospheric CO_2 levels. LPP Contribution Series 5: 152.

Lawrence J.R. and White J.W.C. 1984. Growing season precipitation from D/H ratios of Eastern White Pine. Nature 311: 558–560.

Leavitt S.W. 1993. Environmental information from ^{13}C/^{12}C ratios of wood. Geophysical Monographs 78: 325–331.

Leavitt S.W. and Danzer S.R. 1993. Method for batch processing of small wood samples to holocellulose for stable carbon isotope analysis. Anal. Chem 65: 87–89.

Leavitt S.W. and Lara A. 1994. South American tree rings show declining δ^{13}C trend. Tellus 46B: 152–157.

Leavitt S.W. and Long A. 1982. Evidence for ^{13}C/^{12}C fractionation between tree leaves and wood. Nature 298: 742–744.

Leavitt S.W. and Long A. 1985. An atmospheric ^{13}C/^{12}C reconstruction generated through removal of climate effects from tree ring ^{13}C/^{12}C measurements. Tellus 35B: 92–102.

Leavitt S.W. and Long A. 1984. Sampling strategy for stable carbon isotope analysis of tree rings in pine. Nature 311: 145–147.

Leavitt S.W. and Long A. 1988 Stable carbon isotope chronologies from trees in the southwestern United States. Glob. Biogeochem. Cy. 2(3): 189–198.

Leavitt S.W. and Long A. 1991. Seasonal stable-carbon isotope variability in tree rings: possible palaeoenvironmental signals. Chem. Geol. 87: 59–70.

Leng M.J., Lamb A.L., Heaton T.H.E., Marshall J.D., Wolfe B.B., Jones M.D, Holmes J.A. and Arrowsmith C. This volume. Isotopes in lake sediments. In: Leng M.J. (ed.), Isotopes in Palaeoenvironmental Research. Springer, Dordrecht, The Netherlands.

Libby L.M. and Pandolfi L.J. 1974. Temperature dependence of isotope ratios in tree rings. P. Natl. A. Sci. USA 71: 2482–2486.

Libby L.M., Pandolfi L.J., Payton P.H., Marshall J(III)., Becker B. and Giertz-Siebenlist V. 1976. Isotopic tree thermometers. Nature 261: 284–290.

Lipp J. and Trimborn P. 1991. Long-term records and basic principles of tree ring isotope data with emphasis on local environmental conditions. Paläoklimaforschung 6: 105–117.

Lipp J., Trimborn P., Fritz P., Moser H., Becker B. and Frenzel B. 1991. Stable isotopes in tree ring cellulose and climatic change. Tellus 43B: 322–330.

Liu Y., Wu X., Leavitt S.W. and Hughes M.K. 1996. Stable carbon isotope in tree rings from Huangling, China and climatic variation. Sci. China Ser. D(39) 2: 152–161.

Loader N.J. and Buhay W.M. 1999. Rapid catalytic oxidation of CO to CO_2 – on the development of a new approach to on-line oxygen isotope analysis of organic matter. Rapid Commun. Mass Sp. 13: 1828–1832.

Loader N.J., Helle G., Laumer W. and Schleser G.H. In press. Analysis of carbon, oxygen and nitrogen isotopes by continuous flow isotope ratio mass spectrometry. Jülbericht Research Report Series. ICG-V Forschungszentrum Jülich, Gmbh. Jülich, Germany.

Loader N.J., Robertson I., Barker A.C., Switsur V. R. and Waterhouse J.S. 1997. A modified method for the batch processing of small whole wood samples to α-cellulose. Chem. Geol. 136: 313–317.

Loader N.J., Robertson I. and McCarroll D. 2003. Comparison of stable carbon isotope ratios in the whole wood, cellulose and lignin of oak tree rings. Palaeogeogr. Palaeocl. 196: 395–407.

Loader N.J. and Switsur V.R. 1996. Reconstructing past environmental change using stable isotopes in tree rings. Botanical Journal of Scotland. 48: 65–78.

Loader N.J., Switsur V.R. and Field E.M. 1995. High-resolution stable isotope analysis of tree rings: implications of 'microdendroclimatology' for palaeoenvironmental research. Holocene 5: 457–460.

Luomala E.-M., Laitinen K., Kellomäki S. and Vapaavuori E. 2003. Variable photosynthetic acclimation in consecutive cohorts of Scots pine needles during 3 years of growth at elevated CO_2 and elevated temperature. Plant Cell Environ. 26: 645–660.

MacFarlane C., Warren C.R., White D.A. and Adams M.A. 1999. A rapid and simple method for processing wood to crude cellulose for analysis of stable carbon isotopes in tree rings. Tree Physiol. 19: 831–835.

Mann M.E., Bradley R.S. and Hughes M.K. 1999. Northern hemisphere temperatures during the past millennium: Inferences, uncertainties and limitations, Geophys. Res. Lett. 26: 759–762.

Marshall J.D. and Monserud R.A. 1996. Homeostatic gas-exchange parameters inferred from 13C/12C in tree rings of conifers during the Twentieth Century. Oecologia 105: 13–21.

McCarroll D., Jalkanen R., Hicks S., Tuovinen M., Pawellek F., Gagen M., Eckstein D., Schmitt U., Autio J. and Heikkinen O. 2003. Multi-proxy dendroclimatology: a pilot study in northern Finland. Holocene. 13 (6): 829–838.

McCarroll D. and Loader N.J. 2004. Stable isotopes in tree rings. Quat. Sci. Rev. 23: 771–801.McCarroll D. and Pawellek F. 1998. Stable carbon isotope ratios of latewood cellulose in Pinus sylvestris from northern Finland: variability and signal strength. Holocene 8: 693–702.

McCarroll D. and Pawellek F. 2001. Stable carbon isotope ratios of Pinus sylvestris from northern Finland and the potential for extracting a climate signal from long Fennoscandian chronologies. Holocene 11: 517–526.

McDowell N., Phillips N., Lunch C., Bond B.J. and Ryan M.G. 2002. An investigation of hydraulic limitation and compensation in large, old Douglas-fir trees. Tree Physiol. 22: 763–774.

Monserud R.A. and Marshall J.D. 2001. Time-series analysis of $\delta^{13}C$ from tree rings. I. Time trends and autocorrelation. Tree Physiol. 21: 1087–1102.

Mook W.G., Koopmans, M., Carter, A.F. and Keeling C.D. 1983. Seasonal, latitudinal, and secular variations in the abundance and isotopic-ratios of atmospheric carbon-dioxide. 1. Results from land stations. J. Geophys. Res. 88: 915–933.

Motz J.E., Edwards T.W.D. and Buhay W.M. 1997. Use of nickel-tube pyrolysis for hydrogen-isotope analysis of water and other compounds. Chem. Geol. 140: 145–149.

Mullane M.V., Waterhouse J.S. and Switsur V.R. 1988. On the development of a novel technique for the determination of stable oxygen isotope ratios in cellulose. Int. J. Appl. Radiat. Is. 10: 1029–1035.

Northfeldt D.W., DeNiro M.J. and Epstein S. 1981. Hydrogen and carbon isotopic ratios of cellulose nitrate and saponifiable lipid fractions prepared from annual growth rings of a Californian redwood. Geochim. Cosmochim. Ac. 45: 1895–1898.

Ogle N. and McCormac F.G. 1994. High resolution δ^{13}C measurements of oak show a previously unobserved spring depletion. Geophys. Res. Lett. 21: 12373–2375.

Okada N., Fujiwara T., Ohta S. and Matsumoto E. 1995. Stable carbon isotopes of Chamaecyparis obtusa grown at a high altitude region in Japan: within and among-tree variations. In: Ohta S., Fujii T., Okada N., Hughes M.K., Eckstein D. (eds.), Tree rings: from the past to the future. Proceedings of the International workshop on Asian and Pacific Dendrochronology. Forestry and Forest Products Research Institute Scientific Meeting Report 1: 165–169.

Pearman G.I. and Hyson P. 1986 Global transport and inter-reservoir exchange of carbon dioxide with particular reference to stable isotope distribution. J. Atmos. Chem. 4: 81–124.

Pendall E. 2000. Influence of precipitation seasonality on Piñon pine cellulose δD values. Global Change Biol. 6, 287–301.

Picon C., Guehl J.M. and Aussenac G. 1996. Growth dynamics, transpiration and water-use efficiency in Quercus robur plants submitted to elevated CO_2 and drought. Ann. Sci. Forest 53: 431–446.

Pilcher J.R., Baillie M.G.L., Schmidt B. and Becker B. 1984. A 7272-year tree ring chronology for western Europe. Nature. 312: 150–152.

Preston T. and Owens N.J.P. 1985. Preliminary [13]C Measurement using a gas chromatograph interfaced to an isotope ratio mass spectrometer. Biomed. Mass Spectrom. 12: 510–513.

Prosser S.J. and Scrimgeour C.M. 1995. High precision determination of ^{2}H/^{1}H in H_2 and H_2O by continuous flow isotope ratio mass spectrometry. Anal. Chem 34: 1992–1997.

Ramesh R. 1984 Stable isotope systematics in plant cellulose: implications to past climate Ph.D. thesis, Gujurat University, India.

Ramesh R., Bhattacharya S.K. and Gopalan K. 1985. Dendroclimatological implications of isotope coherence in trees from Kashmir Valley, India. Nature. 317: 802–804.

Ramesh R., Bhattacharya S.K. and Gopalan K. 1986. Climatic correlations in the stable isotope records of silver fir (Abies pindrow) tree from Kashmir, India. Earth Planet Sc. Lett. 79: 66–74.

Rinne K.T., Böttger T., Loader N.J., Robertson I., Waterhouse J.S. In press. On the purification of α-cellulose from resinous wood for stable isotope (H, C and O) analysis. Chem. Geol.

Rittenberg D. and Ponticorvo L. 1956. A method for the determination of the [18]O concentration of the oxygen of organic compounds. Int. J. Appl. Radiat. Is. 1: 208–214.

Robertson I., Field E.M., Heaton T.H.E., Pilcher J.R., Pollard M., Switsur V.R. and Waterhouse J.S. 1995. Isotope coherence in oak cellulose. Paläoklimaforschung 15: 129–140.

Robertson I., Froyd C.A., Walsh R.P.D., Newbery D.M., Woodborne S. and Ong R.C. 2004b. The dating of dipterocarp tree rings: establishing a record of carbon cycling and climatic change in the tropics. J. Quat. Sci. 19: 657–664.

Robertson I., Loader N.J., McCarroll D., Carter A.H.C., Cheng L. and Leavitt S .W. 2004a. δ ^{13}C of tree ring lignin as an indirect measure of climate change. Water Air Soil Poll. 4: 531–544.

Robertson I., Pollard A.M., Heaton T.H.E. and Pilcher J.R. 1996. Seasonal changes in the isotopic composition of oak cellulose. In: Dean J.S., Meko D.M. and Swetnam T.W. (eds.), Tree Rings, Environment and Humanity. Radiocarbon, University of Arizona, Arizona, pp. 617–628.

Robertson I., Rolfe J., Switsur V.R., Carter A.H.C., Hall M.A., Barker A.C. and Waterhouse J.S. 1997a. Signal strength and climate relationships in ^{13}C/^{12}C ratios of tree ring cellulose from oak in southwest Finland. Geophys. Res. Lett. 24: 1487–1490.

Robertson I., Switsur V.R., Carter A.H.C., Barker A.C., Waterhouse J.S., Briffa K.R. and Jones P.D. 1997b. Signal strength and climate relationships in ^{13}C/^{12}C ratios of tree ring cellulose from oak in east England. J. Geophys. Res. 102: 19507–19519.

Robertson I., Waterhouse J.S., Barker A.C., Carter A.H.C. and Switsur V.R. 2001. Oxygen isotope ratios of oak in east England: implications for reconstructing the isotopic composition of precipitation. Earth Planet Sc. Lett. 191: 21–31.

Roden J.S., Lin G. and Ehleringer J.R. 2000. A mechanistic model for interpretation of hydrogen and oxygen isotope ratios in tree ring cellulose. Geochim. Cosmochim. Ac. 64: 21–35.

Rosqvist G., Jonsson C., Yam R., Karlèn W. and Shemesh A. 2004. Diatom oxygen isotopes in pro-glacial lake sediments from northern Sweden: a 5000 year record of atmospheric circulation Quaternary Sci. Rev. 23: 851–859.

Ryan M.G. and Yoder B.J. 1997. Hydraulic limits to tree height and tree growth. Bioscience 47: 235–242.

Santrock J. and Hayes J.M. 1987. Adaption of the Unterzaucher procedure for determination of oxygen-18 in organic substances. Anal. Chem 59: 119–127.

Sauer P.E. and Sternberg L.S.L. 1994. Improved method for the determination of the oxygen isotopic composition of cellulose. Anal. Chem 66: 2409–2411.

Saurer M., Borella S., Schweingruber F. and Siegwolf R. 1997. Stable carbon isotopes in tree rings of beech: climatic versus site-related influences. Trees 11: 291–297.

Saurer M., Robertson I., Siegwolf R. and Leuenberger M. 1998b. Oxygen isotope analysis of cellulose: an interlaboratory comparison. Anal. Chem. 70: 2074–2080.

Saurer M., Schweingruber F.H., Vaganov E.A., Shiyatov S.G. and Siegwolf R. 2002. Spatial and temporal oxygen isotope trends at northern tree-line Eurasia. Geophys. Res. Lett. 29: 10–14.

Saurer M., Siegenthaler U. and Schweingruber F. 1995. The climate-carbon isotope relationship in tree rings and the significance of site conditions. Tellus 47B: 320–330.

Saurer M. and Siegenthaler U. 1989. $^{13}C/^{12}C$ isotope ratios in tree rings are sensitive to relative humidity. Dendrochronologia. 7: 9–13.

Saurer M., Siegwolf R., Borella S. and Schweingruber F. 1998a. Environmental information from stable isotopes in tree rings of *Fagus sylvatica*. In: Beniston M. and Innes J.L. (eds.), The Impacts of Climate Variability on Forests. Springer, Berlin. pp. 241–253.

Schäfer K.V.R., Oren R. and Tenhunen J.D. 2000. The effect of tree height on crown level stomatal conductance. Plant Cell Environ. 23: 365–375.

Schiegl W.E. 1974. Climatic significance of deuterium abundance in growth rings of *Picea*. Nature 251: 582–584.

Schimmellman A. 1991. Determination of the concentration and stable isotopic composition of non-exchangeable hydrogen in organic matter. Anal. Chem 63: 2456–2459.

Schimmelmann A. and DeNiro M.J. 1993. Preparation of organic and water hydrogen for stable isotope analysis – effects due to reaction vessels and zinc reagent. Anal. Chem 65: 789–792.

Schleser G.H., Frielingsdorf J. and Blair A. 1999a. Carbon isotope behaviour in wood and cellulose during artificial aging. Chem. Geol. 158: 121–130.

Schleser G.H., Helle G., Lücke A. and Vos H. 1999b. Isotope signals as climate proxies: the role of transfer functions in the study of terrestrial archives. Quaternary Sci. Rev. 18: 927–943.

Schleser G.H. and Jayasekera R. 1985. $\delta^{13}C$ variations in leaves of a forest as an indication of reassimilated CO_2 from the soil. Oecologia 65: 536–542.

Schulman E. 1958. Bristlecone pine - oldest known living thing. Nat.Geogr. 113: 355–372.

Schulze B., Wirth C., Linke P., Brand W.A., Kuhlmann I., Horna V. and Schulze E.D. 2004. Laser ablation-combustion-GC-IRMS - a new method for online analysis of intra-annual variation of delta C-13 in tree rings. Tree Physiol. 24 (11): 1193–1201.

Schweingruber F.H. 1988. Tree Rings: Basics and Applications of Dendrochronology. D. Reidel, Boston, 276pp.

Schweingruber F.H. and Briffa K.R. 1996. Tree ring density networks for climatic reconstruction. In: Jones P.D., Bradley R.S. and Jouzel J. (eds.), Climatic Variability and Forcing: Mechanisms of the last 2000 Years. Springer, Berlin, pp 43–66.

Sheu D.D., Kou P., Chiu C–H. and Chen M–J. 1996. Variability of tree ring $\delta^{13}C$ in Taiwan fir: growth effect and response to May-October temperatures. Geochim. Cosmochim. Ac. 60: 171–177.

Sonninen E. and Jungner H. 1995. Stable carbon isotopes in tree rings of a Scots pine (*Pinus sylvestris* L.) from northern Finland. Paläoklimaforschung 15: 121–128.

Sofer Z. 1980. Preparation of carbon dioxide for stable carbon isotope analysis of petroleum fractions. Anal. Chem 52: 1389–1391.

Spiker E.C. and Hatcher P.G. 1987. The effects of early diagenesis on the chemical and stable carbon isotopic composition of wood. Geochim. Cosmochim. Ac. 51: 1385–1391.

Sternberg L., De Niro M. and Savidge R. 1986. Oxygen isotope exchange between metabolites and water during biochemical reactions leading to cellulose synthesis. Plant Physiol. 82: 423–427.

Stuiver M. 1978. Atmospheric carbon dioxide and carbon reservoir changes. Science 199: 253–258.

Stuiver M. and Braziunas T.F. 1987. Tree cellulose $^{13}C/^{12}C$ isotope ratios and climate change. Nature 328: 58–60.

Stuiver M. and Reimer P.J. 1993. Extended ^{14}C data base and revised CALIB 3.0 ^{14}C age calibration program. Radiocarbon 35: 215–230.

Stump R.K. and Frazer J.W. 1973. Simultaneous determination of carbon, hydrogen and nitrogen in organic compounds. Nuclear Science Abstracts 28: 746.

Suberkropp K. and Klug M.J. 1976. Fungi and bacteria associated with leaves during processing in a woodland stream. Ecology 57: 707–719.

Suess H.E. 1970. Bristlecone-pine calibration of the radiocarbon time-scale 5,000 BC to present. In: Olsson, I.U., (ed.) Radiocarbon Variations and Absolute Chronology. John Wiley, Chichester and New York, pp. 303–311.

Switsur V.R. and Waterhouse J.S. 1998. Stable isotopes in tree ring cellulose. In: Griffiths, H., (ed.), Stable isotopes. Bios Scientific Publishing, Oxford, UK. pp. 303–321.

Switsur V.R., Waterhouse J.S., Field E.M. and Carter A.H.C. 1996. Climatic signals from stable isotopes in oak tree rings from East Anglia, Great Britain. In: Dean J.S., Meko D.,M., Swetnam T.W. (eds.), Tree Rings, Environment and Humanity. Radiocarbon, University of Arizona, Arizona, pp. 637–645.

Switsur V.R., Waterhouse J.S., Field E.M.F., Carter A.H.C., Hall M., Pollard M., Robertson I., Pilcher J.R. and Heaton T.H.E. 1994. Stable isotope studies of oak from the English Fenland and Northern Ireland. In: Funnell B.M. and Kay R.L.F. (eds.), Palaeoclimate of the last Glacial/Interglacial Cycle. Natural Environment Research Council Special Publication 94/2: 67–73.

Switsur V.R., Waterhouse J.S., Field E.M., Carter A.H.C. and Loader N.J. 1995. Stable isotope studies in tree rings from oak – techniques and some preliminary results. Paläoklimaforschung 15: 129–140.

Technical Association of the Pulp and Paper Industry (TAPPI) 1988 Test method T222 on-83. TAPPI Atlanta GA.

Tang K., Feng X. and Ettle G.J. 2000. The variations in δD of tree rings and the implications for climatic reconstruction. Geochim. Cosmochim. Ac. 64: 1663–1673.

Tang K., Feng X. and Funkhouser G. 1999. The $\delta^{13}C$ of tree rings in full-bark and strip-bark bristle cone pine trees in the White Mountains of California. Global Change Biol. 5: 33–40.

Tans P. and Mook W.G. 1980. Past atmospheric CO_2 levels and the $^{13}C/^{12}C$ ratios in tree rings. Tellus. 32: 268–283.

Tarhule A. and Leavitt S.W. 2004 Comparison of stable carbon isotope composition in the growth rings of *Isoberlinia doka*, *Daniella oliveri* and *Tamarindus indica* and west African climate. Dendrochronologia. 22: 61–70.

Taylor J.W. and Chen I. 1970. Variables in oxygen-18 analyses by mass spectrometry. Anal. Chem 42: 224.

Thompson P. and Gray J. 1977. Determination of the $^{18}O/^{16}O$ ratios in compounds containing C, H and O. Int. J. Appl. Radiat. Is. 28: 411.

Tobias H.J. and Brenna J.T. 1996. High precision D/H measurement from organic mixtures by gas chromatography continuous flow isotope ratio mass spectrometry using a palladium filter. Anal. Chem 68: 3002–3007.

Tobias H.J., Goodman K.J., Blacken C.E. and Brenna J.T. 1995. High precision D/H measurement from hydrogen gas and water by continuous flow isotope ratio mass spectrometry. Anal. Chem 67: 2486–2492.

Treydte K., Schleser G.H., Helle G., Winiger M., Frank D. and Esper J. Long term precipitation changes in western Central Asia from tree ring $\delta^{18}O$. (Pers. Comm.).

Treydte K., Schleser G.H., Schweingruber F.H. and Winiger M. 2001. The climatic significance of $\delta^{13}C$ in subalpine spruces (Lötschental, Swiss Alps). Tellus 53B: 593–611.

Verheyden A., Helle G., Schleser G.H., Dehairs F., Beeckman H. and Koedam N. 2004. Annual cyclicity in high-resolution stable carbon and oxygen isotope ratios in the wood of the mangrove tree Rhizophora mucronata Plant Cell Environ. 27: 1525–1536

von Storch H., Zorita E., Jones J., Dimitriev Y., González-Rouco F. and Tett S. 2004. Reconstructing past climate from noisy data, Science 306: 679–682.

Walcroft A.S., Silvester W.B., Whitehead D. and Kelliher F.M. 1997. Seasonal changes in stable carbon isotope ratios within annual rings of Pinus radiata reflect environmental regulation of growth processes. Aust. J. Plant Physiol. 24: 57–68.

Waterhouse J.S., Barker A.C., Carter A.H.C., Agafonov L.I. and Loader N.J. 2000. Stable carbon isotopes in Scots pine tree rings preserve a record of the flow of the river Ob. Geophys. Res. Lett. 27: 3529–3532.

Wershaw R.L., Friedman I. and Heller S.J. 1966. Hydrogen isotope fractionation in water passing through trees, In: Hobson F. and Speers M. (eds.), Advances in Organic Geochemistry. New York, Pergamon, pp. 55–67.

Werner R.A., Kornexl B.E., Roßmann A. and Schmidt H.-L. 1996. On-line determination of $\delta^{18}O$ values of organic substances. Anal. Chim. Acta 319: 159–164.

Wieser M.E. and Brand W.A. 1999. A laser extraction/combustion technique for in situ $\delta^{13}C$ analysis of organic and inorganic materials. Rapid Commun. Mass Sp. 13: 1218–1225.

Wigley T.M.L., Briffa K.R. and Jones P.D. 1984. On the average value of correlated time series, wiuth applications in dendroclimatology and hydrometeorology. J. Clim. Appl. Meteorol. 23: 201–213.

Wilson A.T. and Grinsted M.J. 1977. $^{12}C/^{13}C$ in cellulose and lignin as palaeothermometers. Nature 265: 133–135.

Wong W.W. and Klein P.D. 1986. A review of techniques for the preparation of biological samples for mass-spectrometric measurements of hydrogen-2/hydrogen-1 and oxygen-18/oxygen-16 isotope ratios. Mass Spectrom. Rev. 5: 313–342.

Wong W.W., Lee L.S. and Klein P.D. 1987. Oxygen isotope ratio measurements on carbon dioxide generated by reaction of microlitre quantities of biological fluids with guanidine hydrochloride. Anal. Chem 59: 690–693.

Yakir D. 1992. Variations in the natural abundance of oxygen-18 and deuterium in plant carbohydrates. Plant Cell Environ. 15: 1005–1020.

Yapp C.J. and Epstein S. 1982. A re-examination of cellulose carbon-bound hydrogen δD measurements and some factors affecting plant-water D/H relationships. Geochim. Cosmochim. Ac. 46: 955–965.

Zimmerman B., Schleser G.H. and Bräuning A. 1997. Preliminary results of a Tibetan stable C-isotope chronology dating from 1200 to 1994. Isot. Environ. Healt. S. 33: 157–165.

3. ISOTOPES IN BONES AND TEETH

ROBERT E.M. HEDGES (robert.hedges@rlaha.ox.ac.uk)
Research Laboratory for Archaeology and the History of Art
University of Oxford
Oxford OX1 3QJ, UK

RHIANNON E. STEVENS (rhiannon.stevens@nottingham.ac.uk)
School of Geography
University of Nottingham
Nottingham NG7 2RD, UK
and
NERC Isotope Geosciences Laboratory
British Geological Survey
Nottingham NG12 5GG, UK

PAUL L. KOCH (pkoch@es.ucsc.edu)
Department of Earth Sciences
University of California Santa Cruz
California 95064, USA

Key words: Carbon isotopes, nitrogen isotopes, oxygen isotopes, sulphur isotopes, strontium isotopes, bones, teeth, palaeodiets

Introduction

The hardness and chemical stability of bone, especially in fossil form, explains its early importance for revealing the biological history of our planet. Potentially, bone supplies a very rich array of evidence, as it is organized into a structural hierarchy, in which each level corresponds to functions that are determined by the local environment. Thus, evidence at an isotope level can be directly connected with the existence of that particular animal species at that time and place, so increasing the scope and precision of palaeoenvironmental interpretation. Most interpretations based on bone have been (palaeo)biological; that is, to do with species and their ecological preferences and behaviour, to do with demographics, and to do with health and pathological issues such as nutritional, growth and disease status. These issues can have a strong or a weak dependence on the physical environment; the relationship may be quite straightforward in some cases, or very complex in others. A very extensive literature on this has grown up over the years. In contrast, the isotope composition of bone has been studied only for the last three decades, and provides a relatively restricted range of results (at best, two or three analyses per bone). Bone is composed from what the animal eats (and drinks). While its chemical composition, which is highly complex at all chemical levels, is pretty much controlled by the homeostatic environment in which it is formed, this is not true of its isotope composition, which directly reflects what is taken up as food, subject

117

M.J. Leng (ed.), 2005. *Isotopes in Palaeoenvironmental Research*. Springer. Printed in The Netherlands.

to alterations brought about by the processes of digestion, excretion and tissue synthesis and turnover. Therefore, to a first approximation, an animal's bone isotope composition tells us (quantitatively) about the isotope composition of its diet, and to interpret the result in palaeoenvironmental terms requires insight into the isotope composition of the plants and other animals as potential food items, the availability and choice that was made from these, and fractionations associated with incorporation.

As with any archive, bone has advantages and disadvantages when compared with other archives, although the isotope information from different archives will often be complementary. Bone is ubiquitous (on land), and survives well in most, especially temperate, environments. Bone (and tooth) has a rich chemistry including several isotopes for study, in particular ^{13}C, ^{15}N, ^{2}H, ^{18}O, also ^{34}S and ^{44}Ca, and potentially for B and Mg and possibly others. Some of these isotopes occur in both organic and inorganic phases, and some of the complex organic material (mainly the protein collagen) can be subdivided into separate amino acids, adding to the possible richness of information. Bone is, at least originally, a very well defined biological entity, and the relevant isotope processes can in principle be studied today. To the extent that the environment is influenced by animals, information concerning their state and behaviour is directly relevant; on the other hand, animals respond to the environment in a variety of ways, and reconstructing palaeoenvironments from bone isotope data by itself is liable to be ambivalent. Also, animals vary individually, and the degree of isotope variability within an apparently homogeneous environment can be considerable, and may not be revealed if sampling is very limited. A major disadvantage of bones is that they rarely occur in a context enabling close association (e.g., stratigraphic) with the kind of stratified archival information so useful in palaeoenvironmental studies. Frequently their age is not known with useful precision, unless they can be separately dated by radiocarbon. Another problem with bone is the extent to which it may be altered during burial. Unfortunately, the food eaten by animals, such as plants and insects, hardly ever remains for isotope measurement, so there is a huge gap in our knowledge just where it is most needed. Finally, we are quite far from an adequate understanding of the processes responsible for isotope variation found in animal bone; so the application of bone isotopes to environmental questions is still very much a research area in which each part is learning from the other.

Isotope incorporation into bone

The mineral phase, bioapatite, of bone and teeth has the function of resisting compression and providing rigidity, and this constitutes some 75% of bone by weight, some 97% of enamel, and < 75% of tooth dentine. In enamel, bioapatite has a fairly well developed crystalline lattice, although it contains other ions such as carbonate, and can incorporate ions such as fluoride by exchange with OH, or Sr by exchange for Ca. In bone and dentine, about half the volume is taken up by the protein collagen, which is organised into structured microfibres, and which serves as a matrix for the deposition of bioapatite. The bioapatite is restricted to crystallites, intimately embedded between the collagen fibrils; on their own, the bioapatite crystallites are relatively unstable. Tooth enamel and dentine grow by accretion and preserve incremental laminae that form at a variety of timescales (daily to annual). Teeth may complete accretion well before bone growth has finished; this depends on both the species (e.g. rodent teeth grow throughout life), and on the tooth type (except for wisdom teeth (third molar, M3), the human permanent dentition is fully formed at about age 12). So a tooth can often contain a complete record of the phase of rapid growth early in an animal's life. For bone, a growth pattern of re-absorption of old bone (both collagen and bioapatite) and deposition of new bone, carried out by specialised cells, is set up, so that earlier growth is erased. At maturity a dynamic pattern of re-absorption and new deposition, within the same space is maintained (the bone is said to "turnover"). A sample of bone will therefore have a complex representation of the time when it was formed in the animal's lifetime, depending on the animal's age, type of bone, etc. This can be relevant, for example, in considering seasonality of feeding. Yet most attempts to resolve temporal variations in the isotope chemistry of mammals have exploited measurements on tooth dentine or enamel (Koch et al. (1989); Balasse et al. (2003), see figure 1).

The deposition of bioapatite involves the binding of extracellular ions of calcium, phosphate and a small amount of carbonate to a pre-existing organic matrix which is mainly collagen, although other proteins in minor amount, especially osteocalcin, play an essential part. Interest in the isotope composition of the mineral phase has concentrated on the phosphate oxygen (phosphorus has only one stable isotope unfortunately), carbonate carbon and oxygen, and strontium. In the body, phosphate-oxygen bonds are exchangeable through phosphate cycling involving phosphate-phosphate linkages (as in ADP-ATP), but this does not happen in inorganic systems. Exchange between carbonate (or bicarbonate) oxygen and water oxygen is also prevalent, while the carbonate carbon is directly related to circulating (and respired) CO_2.

As for the organic phase, collagen is assembled from circulating amino acids in specialised cells. Circulating extracellular amino acids are homeostatically controlled, and the total flux is several times that of the diet, showing that the body is highly active metabolically. Studies of protein deposits that grow without remodelling, such as hair or nails (O'Connell and Hedges 1999; Ayliffe et al. 2004; West et al. 2004) show a response to dietary change within a few days, with a new equilibrium established over several weeks to months (for humans or horses).

Relationship of bone isotope composition to an animal's diet

The pioneering work of DeNiro and Epstein (1978; 1981), based on feeding experiments of small animals, established quite early on that the isotope composition of whole bodies and most tissues directly reflected that of the diet for C and N. Since the diet can contain a mixture of components (e.g., defined as protein, carbohydrate and fat) all with possibly different isotope values, and since different tissues are known to have different isotope compositions, (e.g., the $\delta^{13}C$ value of bone carbonate is quite different from that of collagen $\delta^{13}C$), much effort has been put in to trying to define and characterise more precise relationships, and then to understand their metabolic basis. In the same way, bioapatite $\delta^{18}O$ (phosphate or carbonate) has been studied in relation to drinking water, although physiological models that can account for oxygen isotope systematics are more successful in providing an underlying theory (Kohn et al. 1996). The connection between an animal's isotope composition and the food or water it ingests is shown to be closely determined. The complexity of these connections becomes more apparent as more work is done. Furthermore, our understanding of controls on isotope patterns in modern animals has been hampered by the difficulty in establishing exactly the diets of wild animals. The size of environmental signal being considered is paramount, and it is only when small differences are being considered that the finer effects of animal physiology and nutritional status come into question.

Carbonate carbon

This represents the total dietary carbon isotope value, with an isotope offset (enrichment in the tissue) of between +9‰ and +14‰. This makes some sense in that in a steady state, the carbon isotope composition of the diet should equal that which is lost, of which most is expired CO_2, which should equilibrate with plasma and general extracellular bicarbonate. An offset of 9‰ is close to equilibrium isotope discrimination between gaseous CO_2 and calcite. The causes for values as high as 14‰ are not completely understood, though recent feeding experiments, as well as isotope mass balance models, suggest that in herbivores, where fermentation is common, loss of ^{13}C-depleted methane may contribute to ^{13}C-enrichment of the body carbonate pool (Hedges 2003; Passey et al. in press). (Reviewed by Kohn and Cerling (2002)).

Phosphate and carbonate oxygen

Phosphate oxygen is in isotope equilibrium with oxygen in body water. The most important source of oxygen to mammals is ingested water (both drinking water and water in plants), which may exhibit wide isotope variations correlated with climate. Oxygen in food is also highly variable with climate, whereas the oxygen flux into body water from the oxidation of food by atmospheric O_2 should be relatively invariant isotopically. Carbonate oxygen is equivalent to phosphate oxygen (though easier to measure, at least when enamel is available and relevant). Good empirical correlation between drinking water and bone or enamel carbonate has been demonstrated. (Reviewed by Kohn and Cerling (2002)).

Bioapatite strontium

Sr isotopes are not measurably fractionated during uptake by animals or biomineralization, so the isotope composition of bone or tooth is identical that of the source of Sr that is passed up the food web. (Reviewed by Schwarcz and Schoeninger (1991))

Collagen carbon

Typically collagen is measured as a whole, that is, all the constituent amino acids are included. In fact, the isotope values of individual amino acids vary greatly and systematically. Clearly they could supply additional information, but progress on this front will require additional development, as current techniques are not adequate as yet for routine robust measurements. Differences in amino acid content and isotope composition can account for much of the variation between different proteins. Thus collagen is generally ^{13}C-enriched (2‰ to 4‰) compared to dietary protein (Drucker and Bocherens, 2003) mainly because of its high glycine content. It is generally assumed that collagen is enriched by about 5‰ with respect to (total carbon in) diet but this is only a guide. A key issue here is whether collagen carbon comes from dietary protein only (which must be the case for indispensable amino acids) or if it contains carbon from dietary carbohydrate and lipid. This probably depends on the animal species (especially whether herbivore or omnivore), its digestive system (rumination, hind gut fermentation, etc.), and perhaps the quality of diet (e.g., protein content). Different problems require different approaches, but in the context of palaeoenvironment reconstruction, it is most useful to compare collagen differences within the same species.

Cholesterol carbon

Compound-specific methods can evaluate the very small levels of cholesterol (about 1 to 10 ppm) found in subfossil bone. Metabolically, cholesterol (which may be associated with cell membranes generally trapped in the matrix of bone) is close to the general oxidation pathway for carbon, and in practice the similarity of cholesterol to carbonate, in terms of isotope response to diet, bears this out (Howland et al. 2003; Jim et al. 2004). Cholesterol has a much faster turnover time than bone, giving it the potential for additional information (such as season of death).

Collagen nitrogen

In one sense this is simpler than collagen carbon, because almost all nitrogen comes from dietary protein. Whereas the metabolically driven isotope discrimination of carbon seems to be small compared to other discriminations, nitrogen in collagen is found to be significantly ^{15}N-enriched (about 3 to 5‰) with respect to the diet. In principle, faunal nitrogen isotope values enable the plant protein isotope values to be inferred; in practice, the magnitude and variability of isotope discrimination are not well enough documented, although work on this is proceeding. The main issues are the following. The quality of food and digestive physiology influences the balance of nitrogen excretion between urea and faeces, so that protein-poor diets tend to result in smaller enrichments (e.g., Pearson et al. (2003); Sponheimer et al. (2003a)). Water stress, while also acting on plants, seems to have an additional effect on ^{15}N-enrichment, which has been attributed to changes in urea excretion (Heaton et al. 1986; Sealy et al. 1987; Ambrose 1991; Schwarcz et al. 1999). Starvation can increase ^{15}N-enrichment. This is connected with nitrogen balance – where protein tissue is consumed it provides a source of amino acids enriched with respect to diet (Hobson et al. 1993). Where nitrogen balance is positive, as in pregnancy and lactation, a noticeable decrease in enrichment (of about 0.5‰) has been detected (Koch 1997; Fuller et al. 2004; Stevens 2004; T. O'Connell unpublished data).

Collagen hydrogen

About 75% of collagen hydrogen is bound to carbon, and therefore not directly exchangeable with body water or during burial. There are at least two steps in the Krebs cycle where H from body water bonds to a C atom. As Kreb cycle intermediates provide C-skeletons for amino acids synthesis, water H can label dispensable amino acids. The remaining (and likely more abundant) source of H in body proteins is diet (Hobson et al. 1999; Sharp et al. 2003). A broad correlation with drinking water has been shown for white-tailed deer (Cormie and Schwarcz 1994), while in regions where environmental variation in surface water isotope composition is small, such as the UK, collagen hydrogen shows a substantial trophic level enrichment recalling that of nitrogen (Birchall et al. in press). It might therefore, provide additional insight into animal diet, although this possibility has not been explored yet.

Collagen sulphur

Sulphur is present in collagen only in methionine, and so its value directly relates to sources of this essential amino acid. There is little isotope fractionation of S isotopes between diet and protein (Richards et al. 2003). Further discussion of sulphur isotope variation in foodwebs is given later on in this chapter.

Preservation of the isotope signal in bone and tooth

There is a large literature on the changes to bones and teeth during burial. These may alter the isotope composition of both the organic and the inorganic components. Collagen is broken down and leached away, and what remains can be chemically degraded state. Bone is frequently micro-excavated, probably early in its burial history, by fungi and bacteria, which locally consume all original organic material, leaving their own remnants. Also, exogenous organics, such as humic acids, can bind to and contaminate the original protein. Several protocols have been developed to optimise the extraction of the purest available collagen. Providing that more than a few percent of the original collagen remains it is generally possible to be reasonably confident that the results are valid. The main supporting evidence is the measurement of the materials' C/N ratio (a range of 2.9 and 3.6 is considered to be indicative of good collagen preservation (DeNiro 1985; Ambrose 1990), and demonstration that the amino acid composition is close to that of pristine collagen). The few percent of collagen necessary for measurement have usually been lost from bone in hot burial environments (although the conditions within caves can protect the collagen), and even in temperate regions collagen rarely survives beyond a hundred thousand years.

It is likely that the collagen matrix helps to stabilise the inorganic (bioapatite) crystalline phase. However, measurements of crystallinity, and direct electron microscopy, show that most bone reorganises its bioapatite crystallites into a lower surface area and more stable phase. This 'recrystallisation' has serious implications for the preservation of the inorganic signals – for example, it allows for carbonate exchange with ground waters. Phosphate is relatively immune to inorganic exchange of oxygen with pore fluids, but experiments have shown that when reactions are catalyzed by soil bacteria, isotope exchange of phosphate oxygen proceeds rapidly (Zazzo et al. 2004). In any case, uptake of entire phosphate or carbonate ions from pore fluid during recrystallization can reset bone isotope values. The low phosphate content in groundwater may permit isotope preservation in some settings, but this cannot be assumed for material older than several hundred thousand years. Tooth behaves more or less similarly to bone, but enamel, as already mentioned, is a much more stable and less porous form of bioapatite, and is the material of choice for phosphate and carbonate studies (as well as for trace elements, including, for example, Sr isotopes). Nevertheless, even enamel cannot be entirely relied upon (Lee Thorpe and Sponheimer 2003). Various protocols have been developed to partially dissolve powdered bone or enamel, in the hope that surfaces and newly formed crystals that have formed diagenetically can be preferentially dissolved (Koch et al. 1997). In the end any particular sample set has to be considered and tested on its own merits, and it is necessary to justify that diagenetic alteration has not corrupted the isotope ratios.

Environmental influences on isotope transport through food chains

Ultimately the ocean, atmosphere, and solid earth are the sources for O, H, C, N, S, and Sr in bones and teeth. Many subsequent transport processes can induce substantial isotope fractionation (except for Sr isotopes), although there must be branching (or reversible) points in the flow of material for this to be observable. Given the complexity of route from atmosphere to plant food, theory can guide observation, but it is mainly the results of actual field measurements that provide most of our understanding. Changes in the environment, as in climate, or in hydrology, or in soil composition, can lead to changes in the stable isotope composition of the diet of animals within the biome. While some changes can produce unmistakeable isotope signals (the shift between C3 and C4 plants being the outstanding example), most, especially within a C3 terrestrial environment such as Europe, lead to rather small effects. Faced with the fact that there is much variability within a single population, and that the faunal response incorporates the floral response at several levels, clarifying what are useful environmental signals is a challenging problem. However, it should be remembered that animals deliver a level of biological averaging over plants, and that the 3 isotope ratios from an averaged assemblage of bone provide a rich point-source of definite information even if, at present, its interpretation is tentative. The way forward in a given situation is to aim for as much corroboration as possible from other archives.

Oxygen and hydrogen

Little work has been done on bone hydrogen (already noted), so only oxygen is considered here. As mentioned above, to a good approximation bone oxygen (carbonate or phosphate) represents that of an animal's drinking water, rather than food oxygen. This cuts out much of the complicated response of oxygen in plants relative to soil water, which varies sensitively with environmental conditions (particularly humidity) and rooting depth. The variation of meteoric water $\delta^{18}O$ values with climate and geography is very well documented (e.g., Bowen and Wilkinson (2002); Longinelli and Selmo (2003); Darling and Talbot (2003); Darling et al. (this volume)) and provides good potential signals for tracking both geographical origins and climatic state. However, this may only be possible for unusually well chosen projects for several reasons. Firstly, the seasonal fluctuations in meteoritic water can be very large. Secondly, drinking water is affected by many locally specific processes, including evaporative and reservoir effects. Thirdly, when modern animals are studied in the wild (e.g., Hoppe et al. (2004)), local physiological effects contribute serious confounding factors. In any case, away from polar regions, the expected effects on meteoric water $\delta^{18}O$ values from shifts in climate will be relatively small (< 5‰), whereas the variability in bone and tooth $\delta^{18}O$ values for populations of terrestrial herbivores can be substantial (~ 2‰) (Clementz and Koch 2001). Although a few diachronic studies of bone $\delta^{18}O$ values from various sites have been made (Reinhard et al. 1996; Huertas et al. 1997), none has demonstrated a clear climatic signal. Rather, the results have been used to support a relatively general palaeoclimatic interpretation of the region. Studies of geographical movement have been more informative, although they can easily lack corroboration. Mostly these have looked at human movement (e.g., Dupras and Schwarcz (2001); Hoogewerff et al. (2001); Budd et al. (2004)) but also include animals (Bocherens et al. 2001; Balasse et al. 2002; Hoppe 2004). An area of application would be tracking migration routes, e.g. of reindeer – although there is a natural tendency for an animal to reduce the temperature extremes of its environment, and therefore to reduce the range of $\delta^{18}O$ values that it encounters. In any case, such kinds of movement are best investigated through analysis of structures of fine time resolution, such as tooth enamel (e.g., Bocherens et al. 2001). These possibilities, while already clearly demonstrated, have yet to be fully exploited.

Carbon and nitrogen

Plants are necessarily involved in the transport of C and N isotopes into bone. For carbon, the dominant isotope discrimination comes at photosynthesis. This is well explained by quantitative models (O'Leary 1988; Farquhar et al. 1989). The most critical parameters are how CO_2 is physically delivered to the site of carbon fixation, and its partial pressure within the cell. The important biochemical differences in CO_2 delivery within the plant between C3 and C4 or CAM (Crassulacean Acid Metabolism) plants are responsible for the very large difference (~ 10‰) in $\delta^{13}C$ between these different classes; and the differences in CO_2 diffusion and delivery (often involving HCO_3^-) in aquatic plants can lead to such plants having a very wide range of $\delta^{13}C$ (see Leng et al. (this volume)). Therefore species difference, in certain environments, can account for very large changes in the dietary carbon isotope values of animals. Of course, on top of this, the ecological preference of the animal must be considered. C4 plants do not form shrubs or trees, and therefore can not be 'browsed'.

Within a terrestrial C3 world there are many potential sources of subtle variation in plant $\delta^{13}C$ values offered to herbivores. First, shifts in environmental factors such as light intensity, availability of nutrients and water, temperature, salinity, and pCO_2 affect stomatal conductance of CO_2 into plants (primarily as a by-product of controlling the flux of water out of the plant) and/or the rate of enzymatic carboxylation. When stomatal conductance becomes a stronger limit on the rate of carbon fixation, plant $\delta^{13}C$ values rise; when rate of carboxylation increases as a limit on carbon fixation, plant

δ^{13}C values fall. Thus C3 plants along environmental and climatic gradients, with higher values in settings that are dry, saline, strongly illuminated, and/or nutrient deprived, and lower values where the opposite conditions apply (Heaton 1999). Second, within an ecosystem, differences in plant growth form that affect stomatal conductance (leaf shape or thickness, cuticle density) also affect δ^{13}C values. For example in boreal ecosystems, C3 plant δ^{13}C values consistently differ by more than 6‰, with highest values for conifers and lowest values for deciduous forbs (Brooks et al. 1997). Third, the local variation in the δ^{13}C value of atmospheric CO_2 under conditions of poor mixing, where there is either strong draw-down of CO_2 through very active photosynthesis, or the production of ^{12}C-enriched CO_2 through the oxidation of soil organic matter, and/or root respiration (Vogel 1978; Medina and Michin 1980; Buchmann et al. 1997). As a result of these processes and shifts in isotope discrimination at low light intensity, the δ^{13}C value of plants at ground level within a forest can be ^{13}C-depleted by 2-5‰ relative to ground plants in open environments, with values as low as −37‰ (Medina and Michin 1980). Finally, a plant is not chemically homogeneous; carbohydrates, lipids, and proteins differ in δ^{13}C value, and they will be differently distributed among the different plant tissues and maybe be selectively cropped by herbivores.

Although there is certainly the potential for recovery of environmental information from bone collagen δ^{13}C in a terrestrial C3 environment, the natural variability that registers this must first be very well documented. Measurements on collagen or apatite from contemporary populations of herbivores (red deer, black-tailed deer) show inter-individual variability from the same herd of up to 1.7‰ (standard deviation) (Clementz and Koch 2001; R. Stevens, unpublished data), whereas for herds of elephants in areas experiencing rapid flora change, within population variability was much higher (Tieszen et al. 1989; Koch et al. 1995). A few studies have demonstrated variations in faunal δ^{13}C values due to environmental parameters. Cerling et al. (2004) studied isotope variation in plants and animals in the Ituri forest (Democrat Republic of Congo) and found clear difference based on foraging position in and below the canopy. Altitudinal gradients of about 1.3‰ and 1.5‰ enrichment per 1000 metres are reported for the feather keratin of songbirds and humming birds respectively (Graves et al. 2002; Hobson et al. 2003) and correspond to gradients in plant δ^{13}C values that have been attributed to physiological adaptation of plants to changes in growing conditions, and to partial pressure of atmospheric CO_2 (Marshall and Zhang 1994; Sparks and Ehleringer 1997; Hultine and Marshall 2000; Hobson et al. 2003). Holocene mean bone δ^{13}C values were observed to be 1‰ to 2‰ more positive in warm C3-dominated regions (mainly southern European and the Middle East) compared to colder regions (mainly northern Europe) (van Klinken et al. 1994).

Potential sources of nitrogen to a plant (nitrate, ammonia, organic molecules) vary in abundance and isotope value, making it much more difficult to predict or model plant δ^{15}N values (Högberg 1997). Local differences in the soil nitrogen cycle play a fundamental role in determining the sources available to plants, and symbioses with soil microbes and fungi that are essential to plant nitrogen uptake also affect δ^{15}N values (Hobbie et al. 1999; Taylor et al. 2003). Soil nitrogen status varies on all spatial scales, and is influenced by water content and pH (which can affect the chemistry of nitrogen loss from soils), history and age of the soil, and herbivore activity. There is evidence for the broad generalisation that conditions encouraging a high N turnover rate lead to more enrichment of ^{15}N in the ecosystem (Handley et al. 1999a). This is thought to arise from the increased export of isotopically light N as volatile forms such as ammonia or nitrous oxide. Turnover is greater in hotter and drier conditions, and there is a general trend for enriched N in such ecosystems (Amundson 2003). Work in southern Africa has shown that ^{15}N-enrichment in plants correlates closely with rainfall abundance, but only for C3 plants (Swap et al. 2004). Overall, low values of $\delta^{15}N$ are often interpreted as indicating cold and/or wet conditions. Under such conditions mycorrhiza are often involved in the transport of nitrogen to plants, and these mycorrizal associations have been shown to substantially influence plant $\delta^{15}N$ (Handley et al. 1999b)

Viewed optimistically, this complexity leads to a wide range of $\delta^{15}N$ values that may be found in animal bones. Pessimistically, it may be difficult to predict or interpret faunal $\delta^{15}N$ values in terms as palaeoenvironmental indicators without substantial additional information.

Sulphur and strontium

Plants are the dominant source of sulphur and strontium to vertebrate food webs. Plants take up sulphur derived from 1) weathering of bedrock, which can be vary widely in $\delta^{34}S$ value, 2) wet atmospheric deposition (sea spray, acid rain), 3) dry atmospheric deposition (SO_2 gas), and 4) microbial processes in soils (Krouse 1989). As a consequence, the $\delta^{34}S$ value of terrestrial plants varies with location, with values ranging from -22 to $+22$‰ (Peterson and Fry 1987). In a study of grizzly bears, Felicetti et al. (2003) detected a large within-ecosystem difference in $\delta^{34}S$ value between pine-nuts and all other plant and animal foods. They offered no explanation for the strong ^{34}S-enrichment in pine nuts, but it may relate to differences in rooting depth or soil properties. In rivers and lakes, differences in the extent of anaerobic sulphate reduction (which produces sulphate extremely depleted in ^{34}S) lead to a similarly wide range of $\delta^{34}S$ values (Peterson and Fry 1987). Sulphur in marine systems is relatively uniform, with a mean value of $\sim +20$‰ (Peterson and Fry 1987).

Soil $^{87}Sr/^{86}Sr$ ratios are controlled by bedrock age and chemical composition and by atmospheric deposition of Sr as dust and precipitation. Continental rocks exhibit a large range in $^{87}Sr/^{86}Sr$ ratios that varies with rock type and age (average for rock type 0.702 to 0.716) (see review of controls on Sr isotopes in ecosystems by Capo et al. (1998)). Plants have $^{87}Sr/^{86}Sr$ ratios that match those of the soluble or available Sr in soils. The $^{87}Sr/^{86}Sr$ ratio of modern seawater (0.7092) is globally uniform because the long residence time of marine Sr, and the seawater $^{87}Sr/^{86}Sr$ ratio has fluctuated between over the last 0.7095 and 0.7070 for the last 100 million years. The $^{87}Sr/^{86}Sr$ ratios of estuarine waters are controlled by mixing. As Sr concentrations are much lower in freshwater than in seawater, the $^{87}Sr/^{86}Sr$ ratios of estuaries is quickly dominated by marine inputs (Bryant et al. 1995).

Studies of sulphur and strontium in modern animals are much more limited than studies of C, N, and O. Strontium isotope variation has largely been used to identify foraging zone and migration (e.g., Koch et al (1995); Chamberlain et al. (1997)). Sulphur isotope analysis is also used in migratory studies (e.g., Lott et al. (2003)). It is becoming an important tool in dietary studies, with great potential to detect consumption of marine foods by terrestrial animals (Knoff et al. 2001).

Application of isotope techniques to bone and teeth

Determining sheep birth seasonality by analysis of tooth enamel oxygen isotope ratios: The late Stone Age site of Kasteelberg (South Africa)

Environmental and genetic variables control the season of animal birth, but in domestic animals, humans can manipulate the timing of birth. In pastoralist subsistence economies, sustaining the availability of resources (e.g., milk and meat) throughout the year is determined by when and how many animals are born. Therefore the season of birth is crucial when attempting to understand the subsistence practices of past societies. Through intra-tooth analysis of enamel oxygen isotope ratios the distribution of birth season can be determined for herds of archaeological animals. As previously mentioned, the variation of meteoric $\delta^{18}O$ values with climate and geography is well documented. In high and middle latitudes, meteoric water $\delta^{18}O$ values are lower in the coldest months and higher in the warmest months (Gat 1980). The carbonate and phosphate oxygen in tooth enamel bioapatite precipitates in isotope equilibrium with body water, which tracks meteoric water $\delta^{18}O$ value. Thus seasonal changes in meteoric $\delta^{18}O$ are recorded within tooth enamel, and can be measured by isotope analysis of sequential enamel sub samples (e.g., Fricke and O'Neil (1996); Sharp and Cerling (1998); Kohn et al. (1998); Gadbury et al. (2000)). This relationship has been used in a study of sheep birth seasonality at the late Stone Age site of Kasteelberg (Balasse et al. 2003).

During the last 2000 years pastoralist societies have occupied the area of Kasteelberg on the Southern Cape, South Africa. Several middens have been excavated in the area, in which seal and sheep were the dominant fauna, suggesting that the site functioned as a sealing station and stock post (Klein and Cruz-Uribe 1989). Archaeological reports of the faunal evidence assumed that there was only one birthing season for the Kasteelberg sheep (Klein and Cruz-Uribe 1989; Cruz-Uribe and Schrire 1991), but reports from 18th century European travellers visiting the area suggest that European breeds had a single birthing season, whereas the indigenous breeds lambed twice a year (Mentzel 1994; Kolb 1968). Oxygen isotope analyses were performed on up to 40 sequential enamel samples (Figure 1a) from 11 prehistoric sheep teeth from Kasteelberg in order to establish whether there was one of two birthing seasons per year. Results of sequential $\delta^{18}O$ analysis on the sheep teeth show a cyclical patterns corresponding to seasonal changes in meteoric water $\delta^{18}O$ (Figure 1b). One annual cycle is observed in second molar teeth (M2) and almost two are observed in third molar teeth (M3). Two distinct cyclic patterns are seen in both M2 and M3 teeth (Figure 1b). One group of sheep have their highest (winter) $\delta^{18}O$ values in the same section of the teeth that the other group have their lowest (summer) $\delta^{18}O$ values. The significance of the bimodal isotope pattern was statistically confirmed using non-parametric (loess) regression with bootstrapped confidence intervals (Figure 1c). The most likely explanation for the bimodal distribution of $\delta^{18}O$ values is that the two groups of sheep were born at different points in the year and that their ontogenetic development started in different seasons. This interpretation presumes that the sheep analysed are contemporaneous, which is unlikely as the bone assemblages cover many centuries. However, the bimodal pattern is seen in sheep teeth from a single archaeological horizon suggesting that the pattern is not a result of analysing non-contemporaneous fauna. The dual birthing season interpretation

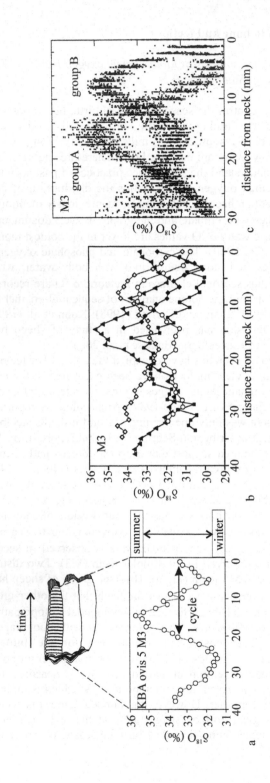

Figure 1. (a) Sequential sampling of enamel along the tooth, and intra-tooth variation in oxygen isotope ratios ($\delta^{18}O$) of enamel bioapatite. Two seasonal cycles are recorded in the third molar (M3), which corresponds to the duration of mineralization of these teeth. (b) Intra-tooth variation of oxygen isotope ratios ($\delta^{18}O$) of enamel bioapatite. The neck is defined as the point where the roots diverge. Sheep form two groups (groups A = closed symbols and B = open symbols) according to the pattern of variation of $\delta^{18}O$ values. (c) Non-parametric (loess) regression with bootstrapped confidence intervals (100 subsamples). Figure reproduced with permission from Balasse et al. (2003).

also assumes that sheep from environments with different seasonality patterns were not included in this study. Again this is unlikely, as in pastoral societies animals are often traded or exchanges over long distances. Strontium isotope signatures from some of the Kasteelberg sheep teeth suggest that a few animals previously lived in a different area (Balasse et al. 2002). However, these animals are evenly distributed between the groups with different $\delta^{18}O$ patterns, suggesting that animal movement is not the cause of the bimodal pattern.

Although the birthing periods for the two groups are offset by approximately half an annual cycle, the seasons of birth cannot be accurately established from the $\delta^{18}O$ values, as the timing of tooth development has not been fully investigated for sheep. Nevertheless, the vegetation dynamics of the area suggest that conditions in autumn and spring would be the most appropriate for birthing and nursing lambs. Two births a year can put strain on the health of the females, potentially reducing their fertility. Additionally two lambing seasons would have restricted the mobility of the human population, although would not necessarily prevent a nomadic lifestyle. The observed bimodal $\delta^{18}O$ pattern could also have been produced by three births in two years (with the herder preventing a fourth birthing season) in order to preserve the health of the sheep. Alternatively, the bimodal $\delta^{18}O$ pattern could have been produced by two groups of females giving birth at different times of the year. Regardless of which of these interpretations is correct, as a result of having two birthing season per year the period of milk availability would have been extended, suggesting sheep played a dominant role in the subsistence economy of the pastoral society at Kasteelberg.

Variation in mammal carbon and nitrogen isotope values over the last 45,000 years

Temporal variations in European faunal $\delta^{13}C$ and $\delta^{15}N$ values have only been recognised in the last ten years. G.J. Van Klinken (unpublished data) first detected variation in both carbon and nitrogen bone isotope values over the last glacial/interglacial cycle. Pleistocene faunal $\delta^{13}C$ values were noted to be slightly ^{13}C-enriched in comparison with Holocene $\delta^{13}C$ values (Figure 2) and faunal $\delta^{15}N$ values seemed relatively ^{15}N-depleted around the time of the Last Glacial Maximum (LGM). Further work in this area has confirmed and refined these trends but interpretations of the isotope values (especially carbon) have been divergent.

The drop in faunal collagen $\delta^{13}C$ values with the onset of the Holocene could relate to the expansion of forests across Northwest Europe as climate warmed, and fauna obtained more of their food from forested environments. A gradual drop in bos, red deer, and horse $\delta^{13}C$ values between 32,600 and 13,300 years BP at Paglicci Cave, southern Italy, was interpreted by Iacumin et al. (1997) to relate to the progressive development of a forest habitat. Drucker et al. (2003a) also postulated that red deer $\delta^{13}C$ values at the site of Rochedane in the Jura region of France during the Bölling / Alleröd and Younger Dryas periods were typical of an open environment, whereas the more lower red deer $\delta^{13}C$ values during the Boreal period were due to the deer consuming plants from under a forest canopy. If the observed $\delta^{13}C$ shifts were a result of consuming different types of vegetation, we would expect different isotope trends for different species depending on the ecological niche they occupy. But the change in

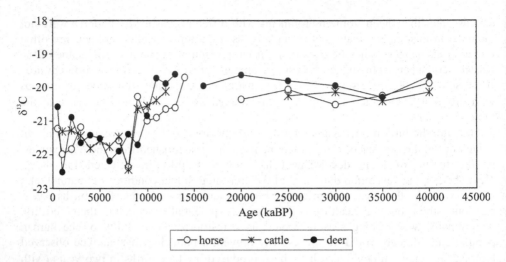

Figure 2. Carbon isotope values are plotted for equids, bovids and cervids over time (radiocarbon age). Each point represents the average of 1000 or 5000 years, with the number of samples involved varying between about 20 and 2, but usually at least 4. Samples are predominantly from the UK, but include Europe generally, north of 45° Latitude. Reproduced with permission from Hedges et al. (2004).

faunal collagen $\delta^{13}C$ values is consistent and similar among species (horse, bovids, deer), over time and between geographical regions (UK, Belgium, Germany, France, Italy) (Stevens and Hedges 2004). This suggests that a large-scale regional mechanism is the trigger for the change in faunal $\delta^{13}C$ values.

An alternative explanation for the change in faunal bone collagen $\delta^{13}C$ values observed at the Late-glacial / Holocene transition is that collectively all plant $\delta^{13}C$ values drop and the faunal $\delta^{13}C$ tracks changes in the plant $\delta^{13}C$ values. Several compilations of data from plant material (wood cellulose, bulk organic matter from lake sediment, and leaves) show similar variations in $\delta^{13}C$ over time (Krishnamurthy and Epstein 1990; Leavitt and Danzer 1991; Hammarlund 1993; van de Water et al. 1994; Beerling 1996; Prokopenko et al. 1999; Hatte et al. 2001). Firstly, changes in atmospheric CO_2 concentration have been suggested as a possible cause of variation in plant (and thus fauna) $\delta^{13}C$ values (Richards and Hedges 2003; Stevens and Hedges 2004). Analysis of CO_2 extracted from air bubbles trapped in ice cores have shown that interglacial periods had a higher CO_2 concentration of around 270 ppm in comparison to glacial periods which had a lower CO_2 concentration of around 190 ppm (Barnola et al. 1987; Jouzel et al. 1993; Smith et al. 1997). Increased atmospheric CO_2 concentration potentially could have influenced the ratio of intercellular to atmospheric CO_2 concentration (C_i/C_a), and indirectly the isotope composition of C3 plants (Arens et al. 2000). However, several studies have reported near zero-slope relationships between C_i/C_a and atmospheric CO_2 concentration (Farquhar et al. 1982; Polley et al. 1993; Ehleringer and Cerling 1995; Beerling 1996; Arens et al. 2000). Although atmospheric CO_2 concentration does not systematically influence C3 plant $\delta^{13}C$, plants may respond through physical changes their stomatal aperture and by changing the number of stomata.

Secondly, a small proportion of the change in plant $\delta^{13}C$ values could potentially be due to changes in the $\delta^{13}C$ values of atmospheric CO_2 due to input of geological carbon with a different $\delta^{13}C$ value. Geological methane sources typically have very low $\delta^{13}C$ values of around –53‰ (Quay et al. 1991; Lassey et al. 1999). At the end of the last glacial period, methane (CH_4) was released into the atmosphere from clathrates contained in sediments on continental shelves exposed due to lowered sea level and from wetlands in response to climate change (Wuebbles and Hayhoe 2002). The large-scale release of methane into the atmosphere is recorded in the ice cores as an increase in CH_4 concentration (Raynaud et al. 1988; Chappellaz et al. 1990, 1993; Blunier et al. 1993). Oxidation of CH_4 to CO_2 could potentially result in a lowering of $\delta^{13}C$ value of atmospheric CO_2. However the $\delta^{13}C$ of CO_2 from air trapped in ice cores has shown that atmospheric $\delta^{13}C$ only changed by 0.3 ± 0.2‰ over the glacial/interglacial transition (Leuenberger et al. 1992), thus it alone can not account for the observed change of around 2 to 5‰ in plant $\delta^{13}C$ values and about 2‰ in faunal $\delta^{13}C$ values. However the change in the $\delta^{13}C$ value of atmospheric CO_2 could potentially be amplified by the biosphere resulting in part or all of the observed variation in plant $\delta^{13}C$ values.

Thirdly, the change in plant $\delta^{13}C$ values over the last glacial transition may relate to availability of water. Increased relative humidity had been reported to result in a drop in plant $\delta^{13}C$ values (Ramesh et al. 1986; Stuvier and Braziunas 1987; van Klinken et al. 1994; Lipp et al. 1991; Hemming et al. 1998). Conditions in Europe became more humid from the LGM to the Holocene. This would allow C3 plants to keep their stomata open more, increasing stomatal conductance, resulting in a drop plant $\delta^{13}C$ values. The observed drop in plant and fauna $\delta^{13}C$ over the last glacial/interglacial transition most likely relates to a combination of factors including atmospheric $\delta^{13}C$, humidity, forest development, and temperature each creating minor isotope variation, and collectively (perhaps synergistically) producing the overall trend. It would, however, be very difficult to quantify the influence of these different parameters in the archaeological context from that of other environmental and climatic parameters.

Temporal and geographical variation in faunal $\delta^{15}N$ values are also observed over the last 45,000 years, but due to limited data from certain time periods and geographical regions it is currently hard to establish if variations in different regions lead or lag behind the overall trend. The overall trend over the last 45,000 years is a gradual drop in faunal $\delta^{15}N$ values, with the lowest $\delta^{15}N$ values observed at the LGM or during the late glacial period. The onset of the drop in values may be later, depending on species and geographical location (as late as 25,000 years BP), but this disparity in timing could be purely due to limited data. Major gaps in the data occur in the UK and Belgium at the LGM as fauna were not present at this time due to the harsh environmental conditions. Data are available through the LGM in Germany, and the lowest faunal $\delta^{15}N$ values occur during the Late Glacial period (Stevens 2004). Different species give conflicting trends in the South of France, with the lowest bovid and horse $\delta^{15}N$ values occurring at the LGM, and the lowest reindeer $\delta^{15}N$ values being observed during the Late Glacial period (Drucker et al. 2003b; Stevens 2004). Faunal $\delta^{15}N$ values then rapidly rise, with

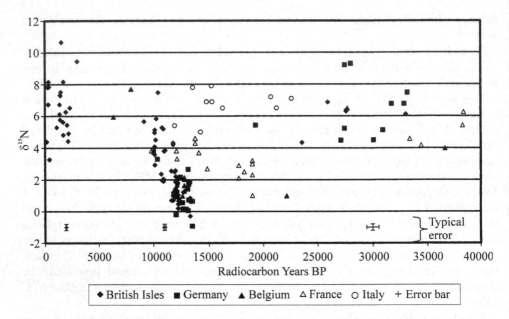

Figure 3. δ^{15}N of horse bone and tooth collagen from the British Isles, Germany, Belgium, Southern France, and Italy, over time (includes 10 horses from Drucker et al. (2000) and 11 horses from Iacumin et al. (1997)). Reproduced with permission from Stevens and Hedges (2004).

early Holocene values being similar to pre-LGM values. The duration of the rise in δ^{15}N may vary between species and geographically but without extensive radiocarbon dating of samples it is hard to establish whether or not this is the case. However, δ^{15}N values of radiocarbon dated horses from the UK rise very rapidly in under 3000 years, with values around –0.7‰ to +0.5‰ at c.13,500 years BP, rising to 4‰ to 8‰ at c. 10,000 years BP (Figure 3) (Stevens and Hedges 2004).

No single climatic or environmental parameter has been identified as the primary factor controlling faunal δ^{15}N values over the last glacial/interglacial cycle, but a number of different, inter-related factors have been identified as possible factors contributing to the drop in herbivore δ^{15}N values during cold intervals. Firstly lower herbivore δ^{15}N values could be animal based. That is, they could be physiological adaptations to the harsh environmental conditions (Richards and Hedges 2003), or they could reflect preferential consumption of plants with lower δ^{15}N values (Nadelhoffer et al. 1996). Differences in digestive physiology among species clearly lead to different plant-to-animal isotope fractionations (Sponheimer et al. 2003a), and differences in diet quality, particularly protein content, also affect this fractionation (Sponheimer et al. 2003b; Pearson et al. 2003). Still, we consider it unlikely that diet quality and/or animal physiology would change so consistently across such a broad geographic region. Secondly (and more likely), the drop in herbivore δ^{15}N values could be due to a change in plant δ^{15}N values. Plant δ^{15}N values are dependant on multiple factors including soil δ^{15}N, soil development, nutrient availability (nitrogen and phosphorus), mycorrhizal associations, nitrogen cycling (open system with preferential loss of ^{14}N versus closed systems with very limited loss of ^{14}N) (Handley et al. 1999a; 1999b; Amundson et al. 2003). Each of these factor are highly dependent on temperature and water availability, thus plant δ^{15}N values (and therefore herbivore δ^{15}N values) track climatic changes. In particular permafrost development, melting and standing water seem to play a significant role in faunal δ^{15}N values over the last glacial/interglacial cycle (Stevens and Hedges 2004). Thus the extreme climatic conditions and continuous permafrost at the LGM followed by extensive melting in northern regions could lead to dramatic depletion in faunal δ^{15}N values (Stevens 2004). In the south of France where conditions at the LGM were less harsh and permafrost was discontinuous the depletion in faunal δ^{15}N values was more limited (Figure 3)(Stevens 2004; Drucker et al. 2003b). In Italy, where the climate was much milder during the LGM and no permafrost was present, no obvious depletion in faunal δ^{15}N values occurred (Figure 3) (Iacumin et al. 1997; Stevens and Hedges 2004).

Aridity and bone collagen δ^{15}N enrichment

The quantity of rainfall can have a major influence on the δ^{15}N values of animal tissues (Heaton et al 1986). They found that modern faunal (elephants, wildebeests, giraffes and zebras) and archaeological human δ^{15}N values in South Africa and Namibia were negatively correlated with the amount of precipitation (Figure 4). For example, elephants in areas with around 800 mm rainfall per year typically had bone collagen δ^{15}N values of around +2.5‰, whereas elephants in areas with less than 200 mm of rainfall per year had values between +10‰ and +15‰. Ambrose and DeNiro (1986a; 1986b) observed a similar correlation in Kenya but the influence of rainfall was not independent from elevation in this case. Sealy et al. (1987) also reported a similar negative correlation, with herbivore collagen δ^{15}N values at Cape Point, an area with around 800 mm of precipitation per year, ranging between +3‰ and +5.5‰, whereas at Churchhaven, an area with around 250 mm of rainfall per year, herbivore collagen δ^{15}N ranging from +12‰ to +16‰ (Figure 4). Sealy et al. (1987) suggested that the correlation between amount of rainfall and faunal collagen δ^{15}N only occurs in areas with less that 400 mm rainfall per year. Cormie and Schwarcz (1996) did not observe a correlation between quantity of rainfall and white tailed deer bone collagen δ^{15}N values in North America when considering the whole population (Figure 4). However, the correlation was observed for a subset of the population that included only individuals that ate more than 10% C4 plants. Gröcke et al. (1997) also found a negative correlation between rainfall and Australian kangaroo δ^{15}N values, but did not observe a similar correlation for other marsupials (Figure 4). Ambrose (2000) observed no significant difference between bone collagen δ^{15}N of rats raised in the laboratory at 20°C and those raised at 36°C (on a controlled diet, with restricted water). The results from Gröcke et al. (1997) and Ambrose (2000) suggest that aridity may only affect the nitrogen isotope values of certain species. Gröcke et al. (1997) believe that, as the observed correlations between rainfall and herbivore bone δ^{15}N in South Africa and Australia are extremely similar, a global relationship may exist.

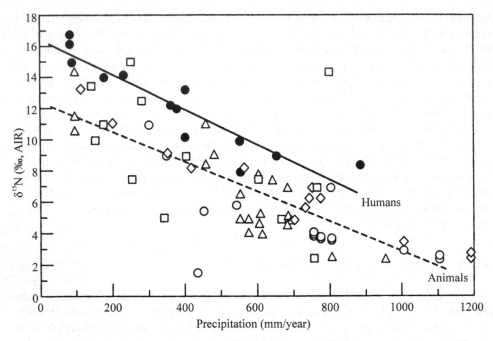

Figure 4. $\delta^{15}N$ values of human and animal bone collagen plotted against annual precipitation (mm/year). Solid symbols represent human samples, while open symbols represent animal samples. All data were extrapolated from graphs. The regression line representing human samples is a solid line ($r2=0\cdot84$; $Y=[16\cdot37\&1\cdot23]+["0\cdot01108\&0\cdot0011] x$); regression line for animals is a dashed line ($r2=0\cdot56$; $Y=[12\cdot33\&2\cdot3]+["0\cdot0094\&0\cdot00106] x$). Data for humans from Heaton et al. (1986) (closed circle) animal data from Heaton et al. (1986) (triangle), Sealy et al. (1987) (square), Cormie and Schwarcz (1996) (open circle) and Gröcke et al. (1997) (diamond). Figure reproduced with permission from Schwarcz et al. (1999).

Part of the correlation between collagen $\delta^{15}N$ values and precipitation is due to shifts in the $\delta^{15}N$ values of plants. Swap et al. (2004), describing south African plants, discuss theories offered to explain progressive ^{15}N-enrichment in plants with decrease in precipitation, which has been observed globally. Yet the correlation between collagen $\delta^{15}N$ values and amount of precipitation is too large to be fully explained by a corresponding change in the plant $\delta^{15}N$ values (Heaton et al. 1986; Sealy et al. 1987; Ambrose 1991). The mechanism that magnifies ^{15}N-enrichment in animals in more arid regions is unclear. Ambrose and DeNiro (1986a; 1986b) suggested that, as the relationship is more prominent in animals that are not obliged to drink than in obligate drinkers, higher faunal $\delta^{15}N$ values may relate to drought stress. Non-obligate drinkers reduce the loss of water from their bodies by excreting urine that is more concentrated in urea, which is ^{15}N-depleted relative to body tissues. If the fraction of body N lost as ^{15}N-depleted urea is higher in these non-obligate drinkers, body $\delta^{15}N$ values would be higher to maintain mass balance (Ambrose and DeNiro 1986a; Sealy et al. 1987; Schwarcz et al. 1999). Currently there are no studies which demonstrate that the fractional loss of body N as urea is greater in non-obligate drinkers. Sealy et al. (1987)

also suggested that reduced amount of dietary protein during period of drought could result in increased recycling of nitrogen by the animal, through conversion of urea to bacterial protein in the gut, which would then raise the trophic level as the protein is eventually consumed by the organism (Sealy et al. 1987; Schwarcz et al. 1999). This model has been criticized by Ambrose (1991) and Schwarcz et al. (1999), who argued that recycling of nitrogen should result in ^{15}N-depletion rather than enrichment as the urea consumed is ^{15}N-depleted, and that if there is not net loss of nitrogen as a result of recycling, there should be no change in overall body $\delta^{15}N$ value. Although these studies have attempted to establish the mechanism by which the relationship between $\delta^{15}N$ values and aridity occurs, a full understanding has not yet been achieved.

Isotope evidence for diet, migration, and herd structure in North American proboscideans

Proboscideans (mastodons, mammoths, gomphotheres) are among the most common fossils in Pleistocene deposits in North America, South America, Eurasia, and Africa. Today, elephants can be keystone species, maintaining and structuring the diversity of plants and smaller animals through their roles as agents of floral disturbance, their propensity to dig wells during droughts, and other landscape transforming activities (Owen-Smith 1988). In addition, while late Pleistocene kill and butcher sites containing extinct large mammals are not common in North or South America, those that do exist frequently have proboscideans (Grayson and Meltzer 2002). As a consequence, proboscideans are central players in the debate over the cause of late Pleistocene extinctions in the New World (Fisher 1996). If the palaeoecology of proboscideans can be constrained through isotope, faunal, growth increment, or other types of analysis, the plausibility of different scenarios for extinction and the role of proboscideans as agents of ecosystem change can be assessed (Fisher 1987; Zimov et al. 1995).

Carbon isotope analysis of sympatric and allopatric populations of mastodons and mammoths has shown that they had different habitat and dietary preferences. Mastodons were browsers, consuming exclusively C3 vegetation. Where mastodons co-occur with deer and tapir, mastodons consistently have higher $\delta^{13}C$ values, suggesting that they browsed in more open habitats, whereas deer and tapir were foraging in ^{13}C-depleted, deep-forest habitats (Koch et al. (1998), Table 1). Mammoths, in contrast, were grazers, though when they co-occurred with bison (committed hyper-grazers), they frequently had lower $\delta^{13}C$ values, suggesting a small component of C3 browse in mammoth diets (Connin et al. (1998); Koch et al. (1998); (2004), Table 1).

Some models for late Pleistocene extinction suggest that preferred habitat patches became increasingly more isolated, or that prolonged drought forced animals to congregate in remaining well-watered regions (Owen-Smith 1988; Haynes 1991). In addition, proboscideans may have altered their migratory patterns or ranging behavior to avoid human predators. By analogy with modern mammals, we might predict that grassland adapted mammoths might have had greater home ranges or might have undertaken greater migratory loops than forest-dwelling mastodons. To understand the ranging behavior of mammoths and mastodons, Hoppe et al. (1999) studied the Sr isotope chemistry of animals in Florida. The Plio-Pleistocene marine carbonate bedrock

Table 1. Carbon isotope values for tooth enamel from Late Pleistocene. For each region, we report the mean ± one standard deviation, with the number of specimens indicated below in parentheses. The scale at the bottom can be used to convert $\delta^{13}C$ values into %C4 in the diet, assuming end-member $\delta^{13}C$ values for C3 and C4 plants in the late Pleistocene a diet-to-enamel fractionation of 14‰. Data are from Koch et al. (1998; 2004), Connin et al. (1998) and Koch (unpublished data).

Species	Texas	Florida	Missouri	Southwest
Mastodon	−10.5±0.6	−11.0±0.9	−11.4±0.6	
(*Mammut*)	(25)	(41)	(37)	
Mammoth	−2.4±1.4	−1.6±1.8	−1.6±0.4	−4.2±3.4
(*Mammuthus*)	(64)	(8)	(6)	(18)
Bison	−1.0±1.5	−0.7±2.4	−0.7±1.9	−0.6±2.5
(*Bison*)	(22)	(8)	(2)	(14)
Tapir	−11.2±0.5	−13.6±1.4	−12.8±0.8	
(*Tapirus*)	(7)	(15)	(2)	
Deer	−12.2±1.2	−13.6±1.4	−14.4±0.3	−10.8
(*Odocoileus*)	(17)	(15)	(2)	(1)

$\delta^{13}C$	<−12	−12	−10	−8	−6	−4	−2	0	2
%C4	Canopy	2	15	29	42	55	69	82	95

of the southern part of the state and the sedimentologically more diverse Cretaceous-Miocene marine bedrock in the northern and central part of the state, have a relatively uniform and low $^{87}Sr/^{86}Sr$ ratio. Hoppe et al. (1999) characterized Florida $^{87}Sr/^{86}Sr$ ratios by measuring the isotope composition of modern plants (Figure 5). To the north, Palaeozoic sediments and igneous and metamorphic rocks in the Appalachian mountains have more variable and higher $^{87}Sr/^{86}Sr$. They characterized environmental isotope compositions in this region by measuring modern rodents. At sites in southern Florida, mammoths, mastodons, and co-occurring deer and tapir all had invariant $^{87}Sr/^{86}Sr$ ratios identical to Plio-Pleistocene marine bedrock. At sites in central Florida, mammoths had values circumscribed by Florida plants, whereas many mastodons had higher values. At the Page-Ladson site on the Florida panhandle, mammoths, deer and tapir all had low values consistent with residence in Florida. All the mastodons had higher values than Florida plants, and tooth enamel microsamples from one individual indicated that it repeatedly migrated between areas with high and low $^{87}Sr/^{86}Sr$ ratios. Thus by examining sites arrayed along a strong environmental gradient in $^{87}Sr/^{86}Sr$ ratios, they demonstrated that forest-dwelling mastodons undertook migrations of at least 120 to 300 km, but probably not more than 700 km. The uniform ratios observed in mammoths suggest that these animals did not regularly range more than a few

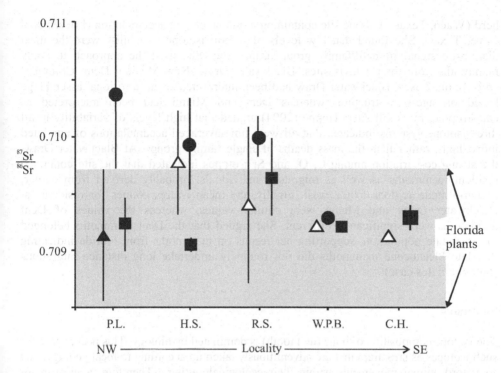

Figure 5. Average Sr isotope ratios of bulk samples of mastodons (circles), mammoths (squares), and deer and /or tapirs (triangles). Lines represent calculated 1 standard deviation from average values for each population of more than three individuals, or range of samples values when only two individuals were measured. P.L is Page-Ladson-Aucilla River, H.S. is Hornsby springs, R.S. us rock springs, W.P.B is West Palm Beach, and C.H. is Cutler Hammock. Reproduced with permission from Hoppe et al. (1999).

hundred kilometers; they did not undertake transcontinental migration, as suggested by some authors (Churcher 1980).

Some fossil localities containing the remains of multiple mammoths have been interpreted as representing the mass death of an entire herd, or family group (Haynes 1991). Assuming that mammoths had a matriarchal social structure similar to modern African and Asian elephants, the age-sex structure of these associations has been interpreted as representing a catastrophic death assemblage. In the southwestern U.S., a number of these sites are associated with human artifacts, raising the possibility that Paleoindians were capable of harvesting entire family groups (Saunders 1992). Hunting elephants is a dangerous activity, and confronting an entire family group would require a remarkable level of skill and cooperative behavior. The interpretation of these sites remains controversial, however, because it is difficult to demonstrate that all the dead individuals died simultaneously. Hoppe (2004) argued if Pleistocene mammoths traveled together in small family groups, then mammoths from sites that represent family groups should have lower variability in carbon, oxygen, and strontium isotope values than mammoths from sites containing unrelated individuals. She tested this idea by comparing the isotope variability among mammoths from one mass death of a single herd (Waco, Texas) and one site containing a time-averaged accumulation (Friesenhahn Cave, Texas). She found that low levels of carbon isotope variability were the most diagnostic signal of herd/family group association. She used the approach to study mammoths from three Clovis sites: Blackwater Draw, New Mexico; Dent, Colorado; and Miami, Texas. Blackwater Draw had been interpreted as an attritional assemblage based on age-sex structure, whereas Dent and Miami had been interpreted as catastrophic, mass-kill sites. Hoppe (2004) argued that high levels of variability in all three isotope systems indicated that all were time-averaged accumulations of unrelated individuals, rather than the mass deaths of single family groups. At Blackwater Draw, the strong co-variation among C, O, and Sr isotopes indicated that the site contained resident mammoths as well as migratory individuals, probably derived from cooler, higher-altitude regions to the west. Finally, the mean isotope values for residents at Blackwater Draw and Miami were similar values, whereas the values of Dent mammoths were significantly different. She argued that the Dent mammoths belonged to a separate population, supporting her results on mammoths from Florida indicating that late Pleistocene mammoths did not routinely undertake long distance migrations (600 km in this case).

Summary

The isotope composition of bone (and tooth) is dominated by biology. This is even true for such isotopes as strontium that are adventitiously taken up according to local geology, and so record animal movements within their ecological setting. Therefore, a measure of understanding of the main biological features in the subject of study (such as food web structure) is needed before environmental inferences can be successful. Such inferences may be coarse-grained, or rather subtle and highly resolved in time. The need to take account of biology makes the study of bone isotopes somewhat specialised. A further difference is the sporadic and somewhat acontextual deposition of bone, rarely enabling its data to be directly integrated with complementary data from the same site. Integration must be obtained by working on a larger, more general scale. On the other hand, as the chapter shows, as archives, bone and teeth are rich in the different kinds of isotope signal they can incorporate. Furthermore, some of these signals may be very sensitive – the non-linear change in C3/C4 plant abundance (and therefore with $\delta^{13}C$ in the bone collagen of herbivorous grazers) with local climate is a very clear case. The chapter has given the background to understanding how isotope changes in bone are caused, and shown how these changes may be understood, though not always, in terms of environmental effects. We hope it also shows that there is a wealth of data "out there" still to be systematically explored, in order to provide more penetrating insight into the interaction between environment, vertebrate biology, and the isotope chemistry of an archive.

References

Ambrose S.H. 2000. Controlled diet and climate experiments on nitrogen isotope ratios of rats. In: Ambrose S.H. and Katzenberg M.A. (eds), Biogeochemical approaches to paleodietary analysis. Advances in Archaeological and Museum Science. Kluwer Academic / Plenum, New York. pp 243–259.

Ambrose S.H. 1990. Preparation and characterization of bone and tooth collagen for isotopic analysis. J. Archaeol. Sci. 17: 431–451.

Ambrose S.H. 1991. Effects of diet, climate and physiology on nitrogen isotope abundances in terrestrial foodwebs. J. Archaeol. Sci. 18: 293–317.

Ambrose S.H. and DeNiro M.J. 1986a. The isotopic ecology of East African mammals. Oecologia 69: 395–406.

Ambrose S.H. and DeNiro M.J. 1986b. Reconstruction of African human diet using bone collagen carbon and nitrogen isotope ratios. Nature 319: 321–324.

Amundson R., Austin A.T., Schuur E.A.G., Yoo K., Matzek V., Kendall C., Uebersax, A., Brenner D. and Baisden W.T. 2003. Global patterns of the isotopic composition of soil and plant nitrogen. Global Biogeochem. Cy. 17: 1031.

Arens N.C., Jahren A.H. and Amundson R. 2000. Can C_3 plants faithfully record the carbon isotopic composition of atmospheric carbon dioxide? Paleobiology 26: 137–164.

Ayliffe L.K., Cerling T.E., Robinson T., West A.G., Sponheimer M., Passey B.H., Hammer J., Roeder B., Dearing M.D. and Ehleringer J.R. 2004. Turnover of carbon isotopes in tail hair and breath CO_2 of horses fed an isotopically varied diet. Oecologia 139: 11–22.

Balasse M., Smith A.B., Ambrose S.H. and Leigh S.R. 2003. Determining sheep birth seasonality by analysis of tooth enamel oxygen isotope ratios: The late stone age site of Kasteelberg (South Africa). J. Archaeol. Sci. 30: 205–215.

Balasse M., Ambrose S. H., Smith A. B. and Price T. D. 2002. The seasonal mobility model for prehistoric herders in the south-western Cape of South Africa assessed by isotopic analysis of sheep tooth enamel. J. Archaeol. Sci. 29: 917–932.

Barnola J.M., Raynaud D., Kortkcvich Y.S. and Lorius C. 1987. Vostok ice core provides 160,000-year record of atmospheric CO_2. Nature 329: 408–414.

Beerling D.J. 1996. C-13 discrimination by fossil leaves during the late-glacial climate oscillation 12-10 ka BP: Measurements and physiological controls. Oecologia 108: 29–37.

Birchall J., O'Connell T.C., Heaton T.H.E. and Hedges R.E.M. In press. Hydrogen isotope ratios in animal body protein reflect trophic level. J. Anim. Ecol.

Blunier T., Chappellaz J.A., Schwander J. Barnola J.M., Desperts T., Stauffer B., and Raynaud D. 1993. Atmospheric methane record from a Greenland ice core over the last 1000 years. Geophys. Res. Lett. 20: 2219–2222.

Bocherens H., Mashkour M., Billiou D., Pelle E. and Mariotti A. 2001 A new approach for studying prehistoric herd management in arid areas: intra-tooth isotopic analyses of archaeological caprine from Iran. Cr. Acad. Sci. IIA. 332: 67–74.

Bowen G.J. and Wilkinson B. 2002. Spatial distribution of delta O-18 in meteoric precipitation. Geology 30: 315–318.

Brooks J.R., Flanagan L.B., Buchmann N. and Ehleringer J.R. 1997. Carbon isotope composition of boreal plants: Functional grouping of life forms. Oecologia 110: 301–311.

Bryant J.D., Jones D.S. and Mueller P.A. 1995. Influence of fresh water flux on $^{87}Sr/^{86}Sr$ chronostratigraphy in marginal marine environments and dating of vertebrate and invertebrate faunas. J. Paleontol. 69: 1–6.

Buchmann N., Guehl J.M., Barigah T.S. and Ehleringer J.R. 1997. Interseasonal comparison of CO_2 concentrations, isotopic composition, and carbon dynamics in an Amazonian rainforest (French Guiana). Oecologia 110: 120–13.

Budd P., Millard A., Chenery C., Lucy S. and Roberts C. 2004. Investigating population movement by stable isotope analysis: a report from Britain. Antiquity 78:127–141.

Capo R.C., Stewart B.W. and Chadwick O.A. 1998. Strontium isotopes as tracers of ecosystem processes: Theory and methods. Geoderma 82: 197–225.

Cerling T.E., Hart J.A. and Hart T.B. 2004. Stable isotope ecology of the Ituri Forest. Oecologia 138: 5–12.

Chamberlain C.P., Blum J.D., Holmes R.T., Feng X.H., Sherry T.W. and Graves G.R. 1997. The use of isotope tracers for identifying populations of migratory birds. Oecologia 109: 132–141.

Chappellaz J.A., Blunier T., Raynaud D., Barnola J.M., Schwander J. and Stauffer B. 1993. Synchronous changes in atmospheric CH_4 and Greenland climate between 40 and 8 kyr B.P. Nature 366: 443–445.

Chappellaz J.A. 1990. Etude du methane atmospherique au cours du dernier cycle climatique a partir d l'analyse de l'air piege dans la glace antarctique, PhD thesis, University of Grenoble.

Churcher C.S. 1980. Did North American mammoths migrate? Canadian Journal of Anthropology 1: 103–105.

Clementz M.T. and Koch P.L. 2001. Differentiating aquatic mammal habitat and foraging ecology with stable isotopes in tooth enamel. Oecologia 129: 461–472.

Connin S.L., Betancourt J. and Quade J. 1998. Late Pleistocene C4 plant dominance and summer rainfall in the southwestern United States from isotopic study of herbivore teeth. Quaternary Res. 50: 179–193.

Cormie A.B. and Schwarcz H.P. 1994. Stable isotopes of nitrogen and carbon of North American white-tailed deer and implications for paleodietary and other food web studies. Palaeogeogr. Palaeocl. 107: 227–241.

Cruz-Uribe K. and Schrire C. 1991. Analysis of faunal remains from Oudepost 1, an early outpost of the Dutch East India Company, Cape Province. S. Afr. Archaeol. Bull. 46, 92–106.

Darling W.G., Bath A.H., Gibson J.J. and Rozanski K. This volume. Isotopes in Water. In: M.J.Leng (ed.). Isotopes in Palaeoenvironmental Research. Springer, Dordrecht, The Netherlands.

Darling W.G. and Talbot J.C. 2003. The O & H stable isotopic composition of fresh waters in the British Isles. Hydrol. Earth Syst. Sc. 7: 163–181.

DeNiro M.J. 1985. Postmortem preservation and alteration of in vivo bone collagen isotope ratios in relation to palaeodietary reconstruction. Nature 317: 806–809.

DeNiro M.J. and Epstein S. 1978. Influence of diet on distribution of carbon isotopes in animals. Geochim. Cosmochim. Ac. 42: 495–506.

DeNiro M.J. and Epstein S. 1981. Influence of diet on distribution of nitrogen isotopes in animals. Geochim. Cosmochim. Ac. 45: 341–351.

Drucker D., Bocherens H., Bridault A. and Billiou D. 2003a. Carbon and nitrogen isotopic composition of red deer (*Cervus elaphus*) collagen as a tool for tracking palaeoenvironmental change during the Late-Glacial and Early Holocene in the northern Jura (France). Palaeogeogr. Palaeocl. 195: 375–388.

Drucker D.G., Bocherens H. and Billiou D. 2003b. Evidence for shifting environmental conditions in Southwestern France from 33 000 to 15 000 years ago derived from carbon-13 and nitrogen-15 natural abundances in collagen of large herbivores. Earth Planet. Sc. Lett. 216: 163–173.

Bocherens H. and Drucker D. 2003. Trophic level isotopic enrichment of carbon and nitrogen in bone collagen: Case studies from recent and ancient terrestrial ecosystems. Int J Osteoarchaeol 13, 46–53.

Dupras T.L. and Schwarcz H.P. 2001. Strangers in a strange land: Stable isotope evidence for human migration in the Dakhleh Oasis, Egypt. J. Archaeol. Sci. 28: 1199–1208.

Ehleringer J.R. and Cerling T.E. 1995. Atmospheric CO_2 and the ratio of intercellular to ambient CO_2 concentrations in plants. Tree Physiol. 15: 105–111.

Farquhar G.D., Ehleringer J.R. and Hubick K.T. 1989. Carbon isotope discrimination and photosynthesis. Annu. Rev. Plant Phy. 40: 503–537.

Farquhar G.D., O'Leary M.H. and Berry J.A. 1982. On the relationship between carbon isotope discrimination and the intercellular carbon dioxide concentration in leaves. Aust. J. Plant Physiol. 9:121–137.

Felicetti L.A., Schwartz C.C., Rye R.O., Haroldson M.A., Gunther K.A., Phillips D.L. and Robbins C.T. 2003. Use of sulfur and nitrogen stable isotopes to determine the importance of whitebark pine nuts to Yellowstone grizzly bears. Can. J. Zool. 81: 763–770.

Fisher D.C. 1987. Mastodont procurement by Paleoindians of the Great Lakes Region: hunting or scavenging? In: Nitecki M.H. and Nitecki D.V. (eds), The Evolution Of Juman Hunting. Plenum Publishing Corporation, Chicago, pp. 309–421.

Fisher D.C. 1996. Extinction of proboscideans in North America. In: Shoshani J. and Tassy P. (eds), The Proboscidea: Evolution and palaeoecology of elephants and their Relatives. Oxford University Press, New York, pp. 141–160.

Fricke H.C. and O'Neil J.R. 1996. Inter- and intra-tooth variation in the oxygen isotope composition of mammalian tooth enamel phosphate: Implications for palaeoclimatological and palaeobiological research. Palaeogeogr. Palaeocl. 126: 91–99.

Fuller B.T., Fuller J.L., Sage N.E., Harris D.A., O'Connell T.C. and Hedges R.E.M. 2004. Nitrogen balance and delta N-15: why you're not what you eat during pregnancy. Rapid Commun. Mass Sp. 18: 2889–2896.

Gadbury C., Todd L., Jahren A.H. and Amundson R. 2000. Spatial and temporal variations in the isotopic composition of bison tooth enamel from the Early Holocene Hudson-Meng Bone Bed, Nebraska. Palaeogeogr. Palaeocl 157: 79–93.

Gat J.R. 1980. The isotopes of hydrogen and oxygen in precipitation. In: Fritz P. and Fontes J.C (eds), Handbook of environmental isotope geochemistry volume 1: The terrestrial environment. Elsevier, Amsterdam, pp 21–42.

Graves G.R., Romanek C.S. and Navarro A.R. 2002. Stable isotope signature of philopatry and dispersal in a migratory songbird. P. Natl. Acad. Sci. USA 99: 8096–8100.

Grayson D.K. and Metlzer D.J. 2002. Clovis hunting and large mammal extinction: A critical review of the evidence. J. World Prehist. 16: 313–359.

Gröcke D.R., Bocherens H. and Mariotti A. 1997. Annual rainfall and nitrogen-isotope correlation in macropod collagen: application as a palaeoprecipitation indicator. Earth Planet.Sc. Lett. 153: 279–285.

Hammarlund D. 1993. A distinct $\delta^{13}C$ decline in organic lake sediments at the Pleistocene-Holocene transition in Southern Sweden. Boreas 22: 236–243.

Handley L.L., Austin A.T., Robinson D., Scrimgeour C.M., Raven J.A., Heaton T.H.E., Schmidt S. and Stewart G.R. 1999a The N-15 natural abundance (delta N-15) of ecosystem samples reflects measures of water availability. Aust. J. Plant Physiol. 26: 185–199.

Handley L.L., Azcon R., Lozano, J.M.R. and Scrimgeour C.M. 1999b. Plant delta N-15 associated with arbuscular mycorrhization, drought and nitrogen deficiency. Rapid Commun. Mass Sp. 13: 1320–1324.

Hatte C., Antoine P., Fontugne M., Lang A., Rousseau D.D. and Zoller L. 2001. delta C-13 of loess organic matter as a potential proxy for paleoprecipitation. Quaternary Res. 55: 33–38.

Haynes G. 1991. Mammoths, mastodonts, and elephants: Biology, behaviour, and the fossil record. Cambridge University Press, Cambridge, 413 pp.

Heaton T.H.E., Vogel J.C., von la Chevallerie G. and Collett G. 1986. Climatic influence on the isotopic composition of bone collagen. Nature 322: 822–823.

Heaton T.H.E. 1999. Spatial, species, and temporal variations in the $^{13}C/^{12}C$ ratios of C$_3$ plants: Implications for palaeodiet studies. J. Archaeol. Sci. 26: 637–649.

Hedges R.E.M. 2003. On bone collagen - apatite-carbonate isotopic relationships. Int. J. Osteoarchaeol. 13: 66–79.

Hemming D.L., Switsur V.R., Waterhouse J.S., Heaton T.H.E. and Carter A.H.C. 1998. Climate variation and the stable carbon isotope composition of tree ring cellulose: an intercomparison of Quercus robur, Fagus silvatica and Pinus silvestris. Tellus B. 50: 25–33.

Hobbie E.A., Macko S.A. and Shugart, H.H. 1999. Insights into nitrogen and carbon dynamics of ectomycorrhizal and saprotrophic fungi from isotopic evidence. Oecologia 118: 353–360.

Hobson K.A., Atwell L. and Wassenaar L.I. 1999. Influence of drinking water and diet on the stable hydrogen isotope ratios of animal tissues. P. Natl. Acad. Sci. USA 96: 8003–8006.

Hobson K.A., Alisauskas R.T. and Clark R.G. 1993. Stable-nitrogen isotope enrichment in avian issues due to fasting and nutritional stress: implications for isotopic analyses of diet. Condor 95: 388–394.

Hobson K.A., Wassenaar L.I., Mila B., Lovette I., Dingle C. and Smith, T.B. 2003. Stable isotopes as indicators of altitudinal distributions and movements in an Ecuadorean hummingbird community. Oecologia 136: 302–308.

Högberg P. 1997. ^{15}N natural abundance in soil-plant systems. New Phytol. 137: 179–203.

Hoogewerff J., Papesch W., Kralik M., Berner M., Vroon P., Miesbauer H., Gaber O., Kunzel K.H. and Kleinjans J. 2001. The last domicile of the Iceman from Hauslabjoch: A geochemical approach using Sr, C and O isotopes and trace element signatures. J. Archaeol. Sci. 28: 983–989.

Hoppe K.A. 2004. Late Pleistocene mammoth herd structure, migration patterns, and Clovis hunting strategies inferred from isotopic analyses of multiple death assemblages. Paleobiology 30: 129–145.

Hoppe K.A., Stover S.M., Pascoe J.R. and Amundson R. 2004. Tooth enamel biomineralization in extant horses: implications for isotopic microsampling. Palaeogeogr. Palaeocl. 206: 355–365.

Hoppe K.A., Koch P.L., Carlson R.W. and Webb S.D. 1999. Tracking mammoths and mastodons: reconstruction of migratory behaviour using strontium isotope ratios. Geology 27: 439–442.

Howland M.R., Corr L.T., Young S.M.M., Jones V., Jim S., van der Merwe N.J., Mitchell A.D. and Evershed R.P. 2003. Expression of the dietary isotope signal in the compound-specific δ^{13}C values of pig bone lipids and amino acids. Int. J. Osteoarchaeol. 13: 54–65.

Huertas A.D., Iacumin P. and Longinelli A. 1997. A stable isotope study of fossil mammal remains from the Paglicci cave, southern Italy, 13 to 33 ka BP: Palaeoclimatological considerations. Chem. Geol. 141: 211–223.

Hultine K.R. and Marshall J.D. 2000. Altitude trends in conifer leaf morphology and stable carbon isotope composition. Oecologia 123: 32–40.

Iacumin P., Bocherens H. and Huertas A.D. 1997. A stable isotope study of fossil mammal remains from the Paglicci cave, Southern Italy. N and C as palaeoenvironmental indicators. Earth Planet. Sc. Lett. 148: 349–357.

Jim S., Ambrose S.H. and Evershed R.P. 2004. Stable carbon isotopic evidence for differences in the dietary origin of bone cholesterol, collagen and apatite: Implications for their use in palaeodietary reconstruction. Geochim. Cosmochim. Ac. 68: 61–72.

Jouzel J., Barkov N.I., Barnola J.M., Bender M., Chappellaz J., Genthon C., Kotlyakov V.M., Lipenkov V., Lorius C., Petit J.R., Raynaud D., Raisbeck G., Ritz C., Sowers T., Stievenard M., Yiou F. and Yiou P. 1993. Extending the Vostok ice-core record of paleoclimate to the penultimate glacial period. Nature 364: 407–412.

Klein R.G. and Cruz-Uribe K. 1989. Faunal evidence for pre-historic herder-forager activities at Kasteelberg, western Cape Province, South Africa. S. Afr. Archaeol. Bull. 44: 82–97.

Knoff A.J., Macko S.A. and Erwin R.M. 2001. Diets of nesting Laughing Gulls (*Larus atricilla*) at the Virginia Coast Reserve: Observations from stable isotope analysis. Isot. Environ. Healt. S. 37: 67–88.

Koch P.L., Fisher D.C. and Dettman D. 1989. Oxygen isotope variation in the tusks of extinct proboscideans - a measure of season of death and seasonality. Geology 17: 515–519.

Koch P.L., Heisinger J., Moss C., Carlson R.W., Fogel M.L. and Behrensmeyer A.K. 1995. Isotopic tracking of change in diet and habitat use in African elephants. Science 267: 1340–1343.

Koch P.L., Hoppe K.A. and Webb S.D. 1998. The isotopic ecology of late Pleistocene mammals in North America - Part 1. Florida. Chem. Geol. 152: 119–138.

Koch P. 1997. Reproductive status, nitrogen balance, and the isotopic ecology of horses. 5th Advanced Seminar on Palaeodiet, France.

Koch P.L., Tuross N. and Fogel M.L. 1997. The effects of sample treatment and diagenesis on the isotopic integrity of carbonate in biogenic hydroxylapatite. J. Archaeol. Sci. 24: 417–429

Kohn M.J., Schoeninger M.J. and Valley J.W. 1996. Herbivore tooth oxygen isotope compositions: Effects of diet and physiology. Geochim. Cosmochim. Ac. 60: 3889–3896

Kohn M.J., Schoeninger M.J. and Valley J.W. 1998. Variability in oxygen isotope compositions of herbivore teeth: reflections of seasonality or developmental physiology? Chem. Geol.152: 97–112.

Kohn M.J. and Cerling T.E. 2002. Stable isotope compositions of biological apatite, phosphates. In: Kohn M.L., Rakovan J. and Hughes J.M. (eds), Geochemical, geobiological, and materials importance. Reviews in Mineralogy and Geochemistry, 48: 455–488.

Kolb P. 1968. The present state of Cape of Good-Hope. Weston La Barre, New York.

Krishnamurthy R.V. and Epstein S. 1990. Glacial-Interglacial excursion in the concentration of atmospheric CO_2 - Effect in the C-13/C-12 ratio in wood cellulose. Tellus B 42: 423–434.

Krouse H.R. 1989. Sulfur isotope studies of the pedosphere and biosphere. In: Nagy K.A. (ed.), Stable isotopes in ecological research. Springer-Verlag, New York, pp. 424–444.

Lassey K.R, Lowe D.C. and Manning M.R. 1999. The trend in atmospheric methane $\delta^{13}C$ and implications for isotopic restraints on the global methane budget. Global Biogeochem. Cy. 14: 41–49.

Leavitt S.W. and Danzer S.R. 1991. Chronology from plant matter. Nature 352: 671.

Lee-Thorp J. and Sponheimer M. 2003. Three case studies used to reassess the reliability of fossil bone and enamel isotope signals for paleodietary studies. J. Anthropol. Archaeol. 22: 208–216.

Leng M.J., Lamb A.L., Heaton T.H.E., Marshall J.D., Wolfe B.B., Jones M.D, Holmes J.A. and Arrowsmith C. This volume. Isotopes in lake sediments. In: M.J.Leng (ed.). Isotopes in Palaeoenvironmental Research. Springer, Dordrecht, The Netherlands.

Leuenberger M., Siegenthaler U. and Langway C.C. 1992. Carbon isotope composition of atmospheric CO_2 from an Antarctic ice core. Nature 357: 488–490.

Lipp J., Trimborn P., Fritz P., Moser H., Becker B. and Frenzel B. 1991. Stable isotopes in tree ring cellulose and climate change. Tellus B. 43: 322–330.

Longinelli A. and Selmo E. 2003. Isotopic composition of precipitation in Italy: a first overall map J. Hydrol. 270: 75–88.

Lott C.A., Meehan T.D. and Heath J.A. 2003. Estimating the latitudinal origins of migratory birds using hydrogen and sulfur stable isotopes in feathers: influence of marine prey base. Oecologia 134: 505–510.

Marshall J.D. and Zhang J.W. 1994. Carbon isotope discrimination and water-use efficiency in native plants of the North-Central Rockies. Ecology 75: 1887–1895.

Medina E. and Michin P. 1980. Stratification of $\delta^{13}C$ in Amazonian rainforests. Oecologia 45 337–378.

Mentzel O.F. 1944. A geographical and topographical description of the Cape of Good Hope. Part Three. Cape Town: The Van Riebeeck Society no. 25.

Nadelhoffer K., Shaver G., Fry B., Giblin A., Johnson L. and Mackane R. 1996. ^{15}N natural abundances and N use by tundra plants. Oecologia 107: 386–394.

O'Connell T.C. and Hedges R.E.M. 1999. Investigations into the effect of diet on modern human hair isotopic values. Am. J. Phys. Anthropol. 108: 409–425.

O'Leary M. H. 1988. Carbon isotopes in photosynthesis. Bioscience 38: 328–336. Owen-Smith R.N. 1988. Megaherbivores. Cambridge University Press, Cambridge, 369 pp.

Passey B.H., Robinson T.F., Ayliffe L.K., Cerling T.E., Sponheimer M., Dearing M.D., Roeder B.L. and Ehleringer J.R. In press. Carbon isotopic fractionation between diet, breath, and bioapatite in different mammals. J. Archaeol. Sci.

Pearson S.F., Levey D.J., Greenberg C.H. and del Rio C.M. 2003. Effects of elemental composition on the incorporation of dietary nitrogen and carbon isotopic signatures in an omnivorous songbird. Oecologia 135: 516–523.

Peterson B.J. and Fry B. 1987. Stable isotopes in ecosystem studies. Annu. Rev. Ecol. Syst. 18: 293–320.

Polley H.W., Johnson H.B., Marino B.D. and Mayeux H.S. 1993. Increase in C_3 plant water-use efficiency and biomass over glacial to present CO_2 concentrations. Nature 361: 61–64.

Prokopenko A.A., Williams D.F., Karabanov E.B. and Khursevich G.K. 1999. Response of Lake Baikal ecosystem to climate forcing and $pCO_{(2)}$ change over the last glacial/interglacial transition. Earth Planet. Sci. Lett. 172: 239–253.

Quay P.D., King S.L., Stutsman J., Wilbur D.O., Steele L.P., Fung I., Gammon R.H., Brown T.A., Farwell G.W., Grootes P.M. and Schmidt F.H. 1991. Carbon isotopic composition of atmospheric CH_4: Fossil and biomass burning source and strengths. Global Biogeochem. Cy. 5, 25–47.

Ramesh R., Bhattacharya S.K. and Gopalan K. 1986. Stable isotope systematics in tree cellulose as palaeoenvironmental indicators - a review. J. Geol. Soc. India 27: 154–167.

Reinhard E., de Torres T. and O'Neil J. 1996. O-18/O-16 ratios of cave bear tooth enamel: A record of climate variability during the Pleistocene. Palaeogeogr. Palaeocl. 126: 45–59.

Raynaud D., Chappellaz J.A., Barnola J.M., Korotkevich Y.S., and Lorius C. 1988. Climatic and CH_4 cycle implications of glacial-interglacial CH_4 change in the Vostok ice core. Nature 333: 655–657.

Richards M.P., Fuller B.T., Sponheimer M., Robinson T. and Ayliffe L. 2003. Sulphur isotopes in palaeodietary studies: A review and results from a controlled feeding experiment. Int. J. Osteoarchaeol. 13: 37–45.

Richards M.P. and Hedges R.E.M. 2003. Variations in bone collagen $\delta^{13}C$ and $\delta^{15}N$ values of fauna from Northwest Europe over the last 40,000 years. Palaeogeogr. Palaeocl. 193: 261–267.

Saunders J. J. 1992. Blackwater Draw: mammoths and mammoth hunters in the terminal Pleistocene. In: Fox J.W., Smith C.B. and Wilkins K.T. (eds), Proboscidean and paleoindian Interactions. Baylor University Press, Waco, Texas, pp. 123–147.

Schwarcz H.P. and Schoeninger M.J. 1991. Stable isotope analyses in human nutritional ecology. Yearb. Phys. Anthropol. 34: 283–321.

Schwarcz H.P., Dupras T.L. and Fairgrieve S.I. 1999. ^{15}N enrichment in the Sahara: in search of a global relationship. J. Archaeol. Sci. 26: 629–636.

Sealy J.C., van der Merwe N.J., Lee-Thorp J.A. and Lanham J.L. 1987. Nitrogen isotopic ecology in Southern Africa: implications for environmental and dietary tracing. Geochim. Cosmochim. Ac. 51: 2707–2717.

Sharp Z.D., Atudorei V., Panarello H.O., Fernandez J. and Douthitt, C. 2003. Hydrogen isotope systematics of hair: archeological and forensic applications. J. Archaeol. Sci. 30: 1709–1716.

Sharp Z.D. and Cerling T.E. 1998. Fossil isotope records of seasonal climate and ecology: straight from the horse's mouth. Geology 26: 219–222.

Smith H.J., Wahlen M., Mastroianni D. and Taylor K.C. 1997. The CO_2 concentration of air trapped in GISP2 ice from the last glacial maximum-Holocene transition. Geophys. Res. Lett. 24: 1–4.

Sparks J.P. and Ehleringer J.R. 1997. Leaf carbon isotope discrimination and nitrogen content for riparian trees along elevational transects. Oecologia 109: 362–367.

Sponheimer M., Robinson T., Ayliffe L., Roeder B., Hammer J., Passey B., West A., Cerling T., Dearing D. and Ehleringer J. 2003a. Nitrogen isotopes in mammalian herbivores: Hair $\delta^{15}N$ values from a controlled feeding study. Int. J. Osteoarchaeol. 13: 80–87.

Sponheimer M., Robinson T.F., Roeder B.L., Passey B.H., Ayliffe L.K., Cerling T.E., Dearing M.D. and Ehleringer J.R. 2003b. An experimental study of nitrogen flux in llamas: is ^{14}N preferentially excreted? J. Archaeol. Sci. 30: 1649–1655.

Stevens R.E. 2004. Establishing links between climate/environment and both modern and archaeological hair and bone isotope values: Determining the potential of archaeological bone collagen $\delta^{13}C$ and $\delta^{15}N$ as palaeoclimatic and palaeoenvironmental proxies. D.Phil thesis, University of Oxford.

Stevens R.E. and Hedges R.E.M. 2004. Carbon and nitrogen stable isotope analysis of Northwest European horse bone and tooth collagen, 40,000 BP - present: Palaeoclimatic interpretations. Quaternary Sci. Rev. 23: 977–991.

Stuiver M. and Braziunas T.F. 1987. Tree cellulose $^{13}C/^{12}C$ isotope ratios and climatic change. Nature 328: 58–60.

Swap R.J., Aranibar J.N., Dowty P.R., Gilhooly W.P. and Macko S.A. 2004. Natural abundance of ^{13}C and ^{15}N in C_3 and C_4 vegetation of southern Africa: patterns and implications. Glob. Change Biol. 10: 350–358.

Taylor A.F.S., Fransson P.M., Högberg P., Högberg M.N. and Plamboeck A.H. 2003. Species level patterns in ^{13}C and ^{15}N abundance of ectomycorrhizal and saprotrophic fungal sporocarps. New Phytol. 159: 757–774.

Tieszen L.L, Boutton T.W, Ottichilo W.K., Nelson D.E. and Brandt D.H. 1989. An assessment of long-term food habits of Tsavo elephants based on stable carbon and nitrogen isotope ratios of bone collagen. Afr. J. Ecol. 27: 219–226.

van de Water P.K., Leavitt S.W. and Betancourt J.L. 1994. Trends in stomatal density and $^{13}C/^{12}C$ ratios of Pinus flexilis needles during last glacial-interglacial cycle. Science 264: 239–243.

van Klinken G.J., van der Plicht J. and Hedges R.E.M. 1994. Bone $^{13}C/^{12}C$ ratios reflect (palaeo) climatic variations. Geophys. Res. Lett. 21: 445–448.

Wuebbles D.J. and Hayhoe K. 2002. Atmospheric methane and global change. Earth Sci. Rev. 57: 177–210.

Vogel J.C. 1978. Isotopic assessment of the dietary habits of ungulates. S. Afr. J. Sci. 74: 298–301.

West A.G., Ayliffe L.K., Cerling T.E., Robinson T.F., Karren B., Dearing M.D. and Ehleringer J.R. 2004. Short-term diet changes revealed using stable carbon isotopes in horse tail-hair. Func. Ecol. 18: 616–624.

Zazzo A., Lecuyer C. and Mariotti A. 2004. Experimentally-controlled carbon and oxygen isotope exchange between bioapatites and water under inorganic and microbially-mediated conditions. Geochim. Cosmochim. Ac. 68: 1–12.

Zimov S.A., Chuprynin V.I., Oreshko A.P., Chapin F.S. III., Reynolds J. F. and Chapin M.C. 1995. Steppe tundra transition: a herbivore-driven biome shift at the end of the Pleistocene. Am. Nat. 146: 765–794.

4. ISOTOPES IN LAKE SEDIMENTS

MELANIE J. LENG (mjl@bgs.ac.uk)
NERC Isotope Geosciences Laboratory
British Geological Survey
Nottingham NG12 5GG, UK
and
School of Geography
University of Nottingham
Nottingham NG7 2RD, UK

ANGELA L. LAMB (alla@bgs.ac.uk)
TIMOTHY H.E. HEATON (theh@bgs.ac.uk)
NERC Isotope Geosciences Laboratory
British Geological Survey
Nottingham NG12 5GG, UK

JAMES D. MARSHALL (isotopes@liverpool.ac.uk)
Department of Earth Sciences
University of Liverpool
Liverpool L69 3GP, UK

BRENT B. WOLFE (bwolfe@wlu.ca)
Department of Geography and Environmental Studies
Wilfrid Laurier University
Waterloo Ontario N2L 3C5, Canada

MATTHEW D. JONES (matthew.jones@nottingham.ac.uk)
School of Geography
University of Nottingham
Nottingham NG7 2RD, UK

JONATHAN A. HOLMES (j.holmes@ucl.ac.uk)
Department of Geography
University College London
London WC1H OAP, UK

CAROL ARROWSMITH (caar@bgs.ac.uk)
NERC Isotope Geosciences Laboratory
British Geological Survey
Nottingham NG12 5GG, UK

Key words: Oxygen isotopes, carbon isotopes, hydrogen isotopes, nitrogen isotopes, lakes, lake-sediments, palaeolimnology.

147

M.J. Leng (ed.), 2005. *Isotopes in Palaeoenvironmental Research.* Springer. Printed in The Netherlands.

Introduction

Palaeolimnology, the interpretation of past environment/climate conditions from sediments accumulated in lake basins, is a rapidly expanding area of research (Last and Smol 2001a; 2001b; Smol et al. 2001a; 2001b). This is largely due to the explosion of research into climate change over recent decades, and also because lakes provide us with widespread, continuous records of terrestrial environmental change. Stable isotopes have become an essential part of palaeolimnology, since the work of McCrea (1950) and Urey et al. (1951) highlighted the potential for oxygen isotope compositions to be used for palaeotemperature reconstruction. The technique has been applied to both lacustrine sediments and fossils, and stratigraphic changes in $\delta^{18}O$ are commonly attributed to changes in lake water isotope composition which can be dependent on temperature, air mass source area and/or precipitation/evaporation ratio. Pioneering works include: Stuiver (1970), Fritz and Poplawski (1974), Fritz et al. (1975), Eicher and Siegenthaler (1976), and reviews have been provided by Buchardt and Fritz (1980), Siegenthaler and Eicher (1986) and more recently by Ito (2001), Schwalb (2003) and Leng and Marshall (2004).

Typically we can measure several stable isotope ratios ($^{2}H/^{1}H$, $^{13}C/^{12}C$, $^{15}N/^{14}N$, $^{18}O/^{16}O$) from a lake sediment, depending on what material becomes incorporated into the lake deposits (Figure 1). Detrital grains incorporated into the lake from the catchment are referred to as allochthonous, while minerals precipitated within the lakes are referred to as authigenic (and include biogenic minerals). Many lakes produce primary mineral precipitates, which are either authigenic or biogenic. In general authigenic refers to calcite (marl) precipitated in response to algal and macrophyte photosynthesis, while biogenic includes organic matter but generally refers to skeletal structures such as ostracod and mollusc shells as well as diatom frustules. Bulk carbonates may comprise several different fractions (e.g., marl, molluscs, ostracods, detrital carbonates) but also different carbonate minerals (calcite, aragonite and dolomite are the most common) - these different fractions need to be separated. The carbonate mineralogy must be identified by methods including petrography, SEM or XRD; since carbonate minerals have different mineral-water isotope fractionation effects (e.g., Friedman and O'Neil (1977); Rosenbaum and Sheppard (1986); also see Leng and Marshall (2004)). It is relatively easy to separate marl from biogenic carbonates, but difficult to separate different carbonate minerals and authigenic from detrital carbonates. In general carbonates have isotope values ($\delta^{18}O$ and $\delta^{13}C$) that are related to the temperature and isotope composition of the lake-waters from which they precipitated. Snail and ostracod shells may have microhabitat or inter-species fractionation differences (e.g., Abell and Williams (1989)) so single species are preferred. Diatom frustules are normally too small to separate individual species so bulk samples are used. Exceptions to this are where there is a particularly large size distribution of frustules (Shemesh et al. 1995).

Most, if not all, lakes preserve organic matter which originates from a mixture of lipids, carbohydrates, proteins and other organic matter produced mostly by plants which grow in and around the lake. Bulk organic matter is most frequently analysed for carbon and nitrogen isotope ratios. Where specific organic fractions can be separated by

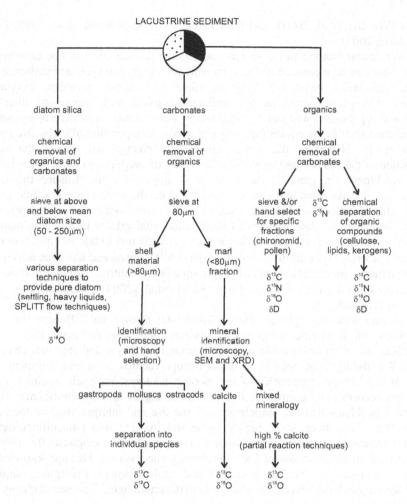

Figure 1. Flow diagram showing the most common components within a lake sediment and treatments to isolate them prior to isotope analysis.

physical (e.g., pollen, chironomid head capsules) or chemical techniques (cellulose, specific organic compounds such as lipid biomarkers (e.g., Sauer et al. (2001); Huang et al. (2002); (2004)) these may be preferable for carbon and nitrogen but can also be analysed for oxygen and hydrogen isotopes. The bulk $\delta^{13}C$ and $\delta^{15}N$ analysis of organic matter can tell us about the source of organic matter, past productivity and changes in nutrient supply (Meyers and Teranes 2001). The dynamics of nitrogen biochemical cycling are more complicated than those of carbon, thereby making interpretation of sedimentary $\delta^{15}N$ more difficult (Talbot 2001). More recently, the oxygen and hydrogen isotope composition of organic compounds have been used to deduce changes in lake water isotope composition, in particular aquatic cellulose (Edwards and McAndrews 1989; Edwards et al. 1996; 2004; Wolfe et al. 2001a), chitin from chironomid head capsules (Wooller et al. 2004), and bulk kerogen (Krishnamurthy et al. 1995; Hassan and Spalding 2001).

However, using isotope ratios in environmental reconstruction is not easy because variables that can be measured in the sediments (commonly oxygen and carbon isotope ratios in carbonates, oxygen in biogenic silica, and carbon, nitrogen, oxygen and hydrogen in organic materials) are influenced by a wide range of interlinked environmental processes and not a single factor. For example, a change in temperature will produce a shift in the equilibrium oxygen isotope composition of carbonate forming in lake water. However, the same temperature change will affect the isotope composition of the rainfall and may also affect rates of evaporation, both in the lake and in the catchment. In general, therefore, it is impossible to measure the isotope composition of carbonates and silicates and translate the values into absolute or even relative climate variation without making some very significant assumptions. In addition there may be disequilibrium (often called vital effects in biogenic materials) and microenvironmental effects, so that not all minerals precipitate in equilibrium with their environments. Disequilibrium mostly affects $\delta^{18}O$ values and has been attributed to rates of mineral precipitation, pH through speciation control, metabolic fluids, and microenvironmental differences (see Leng and Marshall (2004)). Values of $\delta^{13}C$ can be affected by respired CO_2.

This chapter seeks to highlight the environmental factors that influence the isotope composition of lacustrine carbonates, silicates and organic materials, and the assumptions that need to be made in using isotope variation to indicate past changes in climate. We highlight the need to calibrate isotope records from lake sediments using studies of the isotope systematics of the modern lake system. Such studies can often ensure that records can be interpreted with a much higher level of confidence. Despite the various problems that are associated with the use and interpretation of the isotope records, there have been some highly successful insights into palaeolimnology and palaeoenvironmental change using isotope techniques, and we emphasis this through a series of case study examples of Late Quaternary successions. Through knowledge of isotope systematics, calibration exercises and undertaking a multiproxy approach, isotope techniques have become invaluable to understanding environmental change.

Oxygen isotope systematics in inorganic materials

Palaeoclimate studies commonly use stratigraphic changes in the oxygen isotope compositions of lacustrine carbonates or silicates. The composition of either biogenic, or authigenic mineral precipitates, can be used to infer changes in either temperature or the oxygen isotopic composition of lake water. These may be an important proxy for climate change where the isotopes reflect changes in the source of water to the lake or climate changes such as temperature or the precipitation/evaporation ratio.

Figure 2 indicates the factors that can influence the oxygen isotope composition of a lacustrine carbonate precipitate ($\delta^{18}O_{carb}$). An almost identical diagram could be drawn for the oxygen isotope composition of silicates. For equilibrium mineral precipitation, the isotope composition is predictable by thermodynamics. In this case the oxygen isotope composition of the mineral is controlled only by the temperature and by the isotope composition of the lake water from which the mineral precipitated. In principle,

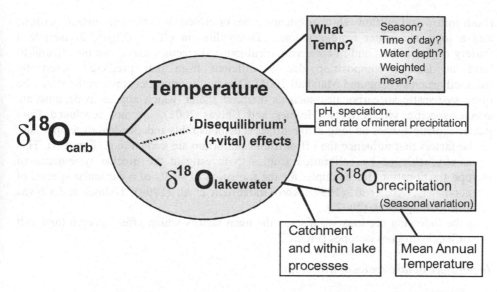

Figure 2. Controls on the oxygen isotope composition of lacustrine carbonates ($\delta^{18}O_{carb}$). If carbonate is precipitated in isotopic equilibrium, $\delta^{18}O_{carb}$ depends entirely on temperature and the isotopic composition of the lake water ($\delta^{18}O_{lakewater}$). Disequilibrium effects, commonly known as 'vital effects' in biogenic precipitates, caused by local changes in microenvironment or rate of precipitation, can induce systematic or non-systematic offsets in the $\delta^{18}O_{carb}$ signal. Interpretation of $\delta^{18}O_{carb}$ signal in terms of temperature depends on a knowledge of the environment and time of year in which a particular type of authigenic or biogenic carbonate forms. For example authigenic carbonates typically form in response to summer photosynthesis in surface waters whereas some benthic ostracods are known to calcify in cold deep waters. Quantitative interpretation similarly demands an understanding of the local controls on the isotopic composition of lake water ($\delta^{18}O_{lakewater}$). In non-evaporating lakes there is a simple relationship to $\delta^{18}O_{precipitation}$ but in others the water composition is strongly influenced by evaporation within the catchment and within the lake itself. $\delta^{18}O_{precipitation}$ is increasingly being shown to be an important indicator of climate change: it typically changes with mean annual temperature but non-linear responses to climate change are increasingly being recognised. See text for more discussion of this figure. Reproduced with permission from Leng and Marshall (2004).

if it can be shown that a particular type of carbonate or silicate forms in isotope equilibrium, fractionation equations (often known as palaeotemperature equations) can be used to estimate past temperatures and their changes if there is no concurrent change in the oxygen isotope composition of the lake water. The interpretation of oxygen isotope compositions in practice, however, is complicated because both temperatures and water compositions can be affected by changes in climate.

Knowledge of the factors that may have influenced the isotope composition of the lake water ($\delta^{18}O_{lakewater}$ on Figure 2) is obviously vital to the interpretation of the $\delta^{18}O_{carb}$ signal. Some of the processes that influence lake water composition are discussed in the following sections. It should be emphasised that there is a need to understand the basic hydrology (assuming this is essentially unchanging) of any lake that is to be used for palaeoclimate reconstruction. Perhaps the most critical is knowing if the lake is open (both inflow and outflow, short residence time) or closed (no effective surface outflow, and a long lake water residence time). Disequilibrium effects (Figure 2) include a variety of rate effects and microenvironment-induced changes that cause the mineral to have an isotope composition that is different from that predicted purely by thermodynamics (Leng and Marshall 2004). Disequilibrium processes can, however, be quite systematic so carbonates such as ostracod shells, which appear to demonstrate species-specific vital effects (Holmes and Chivas 2002), are not precluded from palaeoclimatic studies as long as the offset is demonstrably independent of temperature or the factors that influence the offset from equilibrium are known (e.g., Figure 3). This is usually achieved by calibration studies to investigate the precise systematics of isotope fractionation, for example, for the formation of shells of a particular species of ostracod (e.g., Xia et al. (1997); von Grafenstein et al. (1999); Holmes and Chivas (2002); Keatings et al. (2002)).

In the following sections we discuss the main factors which affect oxygen (and will also affect hydrogen) isotopes (Figure 2).

Importance of lake (palaeo) hydrology

The oxygen isotope composition of water ($\delta^{18}O_{lakewater}$ on Figure 2) in hydrologically open lakes, subject to minimal evaporation, usually reflects the isotope composition of precipitation (both rain and snowfall) received by the lake ($\delta^{18}O_{precipitation}$ and usually referred to as δp). An extended account of the controls on the isotope composition of precipitation can be found elsewhere (see Clark and Fritz (1997); Darling et al. (this volume)) but many studies have shown that the oxygen isotope composition of mean annual precipitation varies globally and the covariation in $\delta^{18}O_{precipitation}$ and δD ($\delta^2 H$) defines a global meteoric water line (Craig (1961) and Figure 4a). Outside the tropics, where 'amount' effects (relating to intense precipitation events) are common, δp varies systematically with mean annual temperature (Clark and Fritz 1997 and the IAEA-WMO GNIP database) and will broadly correlate with the latitude and altitude of a site (Bowen and Wilkinson 2002, and refs therein). The global relationship between changes in δp with temperature ($d\delta p/dT \sim 0.2$ to $0.7\permil/^{\circ}C$) is referred to as the 'Dansgaard relationship' after the seminal early compilation of meteoric water values (Dansgaard 1964).

However, it is not safe to assume that either the modern, or past, oxygen isotope composition of lake water ($\delta^{18}O_{lakewater}$) reflects that of mean annual precipitation. (For example see Figure 5 where $\delta^{18}O_{lakewater}$ is related to the amount of precipitation). The residence time of water in the lake and modification of water composition by catchment and lake processes are particularly important to consider, as evaporation will affect the water composition. The size of a lake in comparison to its catchment is important because the isotope composition of rain and snowfall are very variable on short timescales: a lake therefore needs to be big enough and well enough mixed for its isotope composition to 'average out' the short term variation and reflect mean annual precipitation. The greatest degree of variation in the isotope composition of rainfall occurs on the time scale of hours to days (Darling et al. this volume). Within individual storm events, $\delta^{18}O$ can vary by many ‰ but this is unlikely to have any significance for the sediment record - except perhaps for the isotope composition of some species of

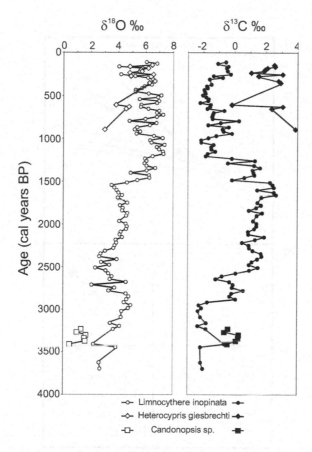

Figure 3. Late Holocene oxygen and carbon isotope variations from Kajemarum Oasis (pit K1), a presently dry lake in the West African Sahel, based on multiple-shell samples of monospecific ostracods. Values of $\delta^{18}O$ for *Limnocythere inopinata* suggest an abrupt shift from higher effective precipitation prior to about 1500 years BP to a drier and more variable hydrological regime. Making assumptions about isotope fractionation and temperatures, lake water $\delta^{18}O$ values are estimated to be around +9‰ for recent ostracods - a value consistent with other lakes in the area - whereas values between 3400 and 2700 BP would have been about +5‰. Independent palaeosalinity values (discussed in Street-Perrott et al. (2000)) suggest that variations in $\delta^{18}O_{precipitation}$ values may explain some of this variability. *Candonopsis* have lower $\delta^{18}O$ and higher $\delta^{13}C$ than *Limnocythere*. Moreover, because the vital offset in the Candoninae is larger (up to +2‰) than in *Limnocythere*, the water in which the *Candonopsis* specimens calcified must have had a lower $\delta^{18}O$ than the values for *Limnocythere* imply. This difference in $\delta^{18}O$ and $\delta^{13}C$ values for the two taxa could reflect the fact that the *Candonopsis* specimens were living during intervals of especially high effective precipitation, a suggestion consistent with the fact that *Candonopsis* tends to prefer fresher water than *Limnocythere inopinata*. Differences between *Limnocythere inopinata* and *Heterocypris giesbrechtii* (for which the vital offset is currently unknown) are more complex: at times the two species show parallel $\delta^{18}O$ records whereas elsewhere they differ. The three ostracod species also show differences in their $\delta^{13}C$ values, probably related to microhabitat effects. (Data from Holmes, Street-Perrott unpublished).

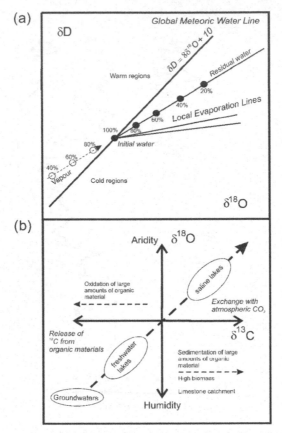

Figure 4. (a) Major controls on the $\delta^{18}O$ v δD of precipitation and lake waters. The Global Meteoric Water Line (GMWL) is an average of many local or regional meteoric water lines whose slopes and intercepts may differ from the GMWL due to varying climatic and geographic parameters. In general however, warmer, tropical rains have higher δ values, while colder, polar precipitation has lower δ values Lake waters plotting on or close to the GMWL are isotopically the same as precipitation, whereas lake waters that plot off the GMWL on a local evaporation line (LEL) have undergone kinetic fractionation. Molecular diffusion from the water to the vapour is a fractionating process due to the fact that the diffusivity of $^1H_2^{16}O$ in air is greater than $^2H^1H^{16}O$ or $^1H_2^{18}O$. With evaporation the isotopic composition of the residual water in the lake and the resulting water vapour become progressively more enriched, in both cases the kinetic fractionation of ^{18}O exceed that of ^2H. In general, groundwater-fed open lake waters should have a $\delta^{18}O$ and δD composition similar to mean weighted values for precipitation, and fall on a MWL. Evaporating lakes will have $\delta^{18}O$ and δD values which lie on a LEL with a slope determined by local climate (see Clarke and Fritz (1997)). (b) $\delta^{13}C$ v $\delta^{18}O$ of lakewaters. Hydrologically closed lakes often show $\delta^{13}C_{TDIC}$ v $\delta^{18}O$ covariance, the high values reflect different degrees of equilibration of the TDIC with atmospheric CO_2 and preferential evaporative loss of the ^{16}O respectively. Groundwaters and river waters have $\delta^{13}C_{TDIC}$ values that are typically low, values are generally between –10 to –15 ‰ from plant respiration and production of CO_2 in the catchment soils. Reproduced with permission from Leng and Marshall (2004).

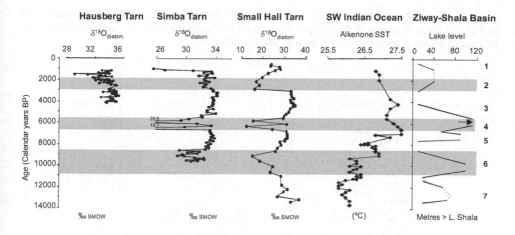

Figure 5. Biogenic silica is especially useful in acidic lakes with no authigenic or biogenic carbonates. In three high-altitude, freshwater lakes on Mount Kenya the oxygen isotope composition of biogenic silica is useful because the lakewater oxygen isotope composition records lake moisture balance. The diatom oxygen isotope data from Simba Tarn, Small Hall Tarn, and Hausberg Tarn (Mount Kenya) are compared to lake levels in the Ziway-Shala Basin, Ethiopia, derived from ^{14}C dated shorelines. Abrupt shifts of up to 18‰ in diatom silica are thought to represent lake moisture balance. Episodes of heavy convective precipitation dated ~11.1-8.6, 6.7-5.6, 2.9-1.9, and <1.3 thousand cal. BP, were linked to enhanced soil erosion, neoglacial ice advances, and forest expansion on Mt. Kenya. Numbers 1 to 7 down right hand margin refer to wet-dry phases. Zones 2, 4 and 6 refer to high lake levels. Reproduced with permission from Barker et al. (2001).

organisms such as gastropods and ostracods which can inhabit the most ephemeral of pools.

Seasonal variation in precipitation is likely to be much more significant in small lakes with short residence times, as these tend to have $\delta^{18}O_{lakewater}$ values which are regularly displaced by later precipitation. At temperate latitudes monthly mean rainfall $\delta^{18}O$ values typically have an overall range of between 2-8‰, the range increasing with continentality of the site. At such sites, winter rainfall has significantly more negative $\delta^{18}O$ than its summer equivalent. If a lake is very small in relation to its catchment (with a residence time of < 1 year), winter rainfall will be physically displaced by summer rainfall and thus $\delta^{18}O_{lakewater}$ will be influenced by seasonal variation. The precise isotope composition of lake water at any time will depend on the amount of rain in the different seasons and the degree of mixing of winter and summer rainfall. Even in lakes with relatively long overall residence times, if the water becomes stratified (as often happens in summer) surface waters may have isotope compositions that reflect summer rainfall rather than mean annual precipitation.

Evaporation has a major influence on the isotope composition of any standing body of water. Closed (terminal) lakes, particularly those in arid regions where water loss is mainly through evaporation, have waters with variable oxygen isotope composition and more positive values than regional rainfall. Measured $\delta^{18}O$ (and δD) values are always

Table 1. Generalised features in lakes that are likely to produce seasonality, temperature, δp and/or P/E reconstructions from isotopic compositions of authigenic components within the lake. Reproduced with permission from Leng and Marshall (2004).

Lake-water volume	Very small	Small-medium open lakes	Small-medium closed lakes	Large
Residence time	< 1 year ('open' lake)	≥ 1 year	10's years	100's years ('closed' lake)
Predominant forcing	S, T, δP	T, δP	P/E	P/E
δ18O ranges through the Holocene	Often –ve values, small range of 1-2‰, possibly large range in ‰ for materials precipitated in different seasons e.g., [1]Lake Chuma (Kola Penninsula), [2]Lake Abisko (Sweden)	Often –ve values, small range of 1-2‰ e.g., [3]Hawes Water (UK), [4]Lake Ammersee (Germany)	-ve to +ve values, large swings (5 to >10‰) e.g., [5]Greenland lakes, [6]Lake Tilo (Ethiopia).	+ve values, subdued signal homogenised by buffering of large lake volume e.g., [8]Lake Malawi, [9]Lake Turkana (Kenya)

S = seasonality, T = temperature, δP = isotope composition of precipitation, P/E = amount of precipitation relative to evaporation [1]V.Jones et al. (2004), [2]Shemesh et al. (2001), [3]Marshall et al. (2002), [4]von Grafenstein et al. (1999), [5]Anderson and Leng (2004), [6]Lamb et al. 2000), Ricketts and Johnson (1996), [8]Ricketts and Anderson (1998).

higher than those of ambient precipitation as the lighter isotope ^{16}O (and ^{1}H) is preferentially lost to evaporation. Isotope records from the sediments in such lakes show large swings in composition as the ratio of the amount of precipitation to evaporation (P/E) changes with climate. In extreme circumstances evaporation in soil zones prior to water entering the lake and from the surface of a lake can lead to significantly elevated $\delta^{18}O_{lakewater}$ values. Any interpretation of the isotope records from a lake must take into account the hydrology of the lake and likely changes in hydrology that may have occurred in the past. A summary of lake types and the type of information isotopes may provide is given in Table 1.

A more quantitative approach can be taken in modelling the isotope hydrological balance in lakes. Although this mass balance modelling approach has been used relatively often in the studies of contemporary lake systems (e.g., Dinçer (1968); Gibson et al. (1996); (2002)) it has only been used sporadically for the interpretation of palaeoisotope records (Ricketts and Johnson 1996; Lemke and Sturm 1997; Benson and Paillet 2002). The following sections describe in more detail the controls on $\delta^{18}O_{lakewater}$ and the isotope hydrology balance model that can lead to a more quantitative interpretation of these records.

Open lakes

Palaeotemperature equations
In many open lake systems $\delta^{18}O_{lakewater}$ changes are related to temperature or air mass change. Where there are unlikely to be changes in the source of precipitation, then palaeotemperature equations can be used. These equations are derived from equilibrium isotope fractionation, which is related directly to the thermodynamics of the mineral

precipitation reaction. The increase in vibrational, and other, energy associated with increased temperature leads to a decrease in isotope fractionation between the water and the mineral that precipitates from it. For equilibrium carbonates, for example, the mineral isotope composition decreases by about 0.24‰ for each 1°C increase in temperature (Craig 1965), whereas for diatom silica empirical data suggest decreases of between 0.2 to 0.5‰/°C (Brandriss et al. 1998; Shemesh et al. 2001).

A number of studies have attempted to determine the empirical relationship between temperature, the oxygen isotope composition of different carbonate minerals and the composition of the water from which they formed. Many so-called palaeotemperature equations utilise the empirical relationship of Epstein et al. (1953) which was turned into a palaeotemperature equation by Craig (1965). More recent attempts to determine fractionation behaviour for inorganic and biogenic calcites (Grossman and Ku 1986; Kim and O'Neil 1997) involving careful laboratory experiments, have yielded more precise equilibrium relationships which have gradients that are very similar to the original Craig data. There are a number of palaeotemperature equations for the equilibrium precipitation of carbonates from solution. Many workers use a version of the 'Craig' equation, although Kim and O'Neil (1997) propose an equation based on new experimental data, which was re-expressed in a more convenient form by Leng and Marshall (2004):

$$T°C = 13.8 - 4.58(\delta c - \delta w) + 0.08(\delta c - \delta w)^2 \qquad (1)$$

In this equation δc is $\delta^{18}O_{carb}$ of the carbonate compared to the VPDB international standard and δw is $\delta^{18}O_{water}$ of the water compared to the VSMOW international standard. The new equation of Kim and O'Neil gives 'palaeotemperatures' that are lower than those calculated using the original Craig equation or its derivatives.

Most palaeotemperature equations are based on calcite, although there are many types of carbonate minerals that can be precipitated from a lake water and this is largely dependent on lake water chemistry. Aragonite and magnesian calcites precipitated at equilibrium generally have $\delta^{18}O$ values that are higher than those for low-magnesium calcites. Aragonite $\delta^{18}O$ values are typically around +0.6‰ higher than equivalent calcite (Grossman 1982; Abell and Williams 1989), and magnesium calcites have $\delta^{18}O$ values that are elevated by 0.06‰/mol% $MgCO_3$ (Tarutani et al. 1969). Dolomite has $\delta^{18}O$ values 3‰ higher than equivalent calcite (Land 1980). For both aragonite and high-magnesium calcite the offset from the calcite value appears to be independent of temperature (cf., Figure 1 in Kim and O'Neil (1997)).

The temperature dependence of oxygen isotope fractionation between diatom silica and water is still controversial. The fractionation has been estimated previously from analyses of diatoms from marine sediments, coupled with estimates of the temperatures and isotope compositions of coexisting waters during silica formation (Labeyrie 1974; Juillet-Leclerc and Labeyrie 1987; Matheney and Knauth 1989). Realistic fractionations vary somewhat because data from calibration studies are limited, and based on bulk samples (Labeyrie and Juillet 1982; Wang and Yeh 1985; Juillet-Leclerc and Labeyrie 1987; Shemesh et al. 1995). Published estimates of the average temperature dependence for typical ocean temperatures range from −0.2 to −0.5 ‰/°C (Juillet-Leclerc and

Labeyrie 1987; Shemesh et al. 1992) and these estimates are often used in lake-based studies. This was partially addressed by analysis of diatoms cultured in the laboratory (Brandriss et al. 1998), which showed a diatom-temperature coefficient of –0.2‰/°C.

In lake studies it is often assumed, though not proven, that carbonates and silicates form in isotope equilibrium and hence that the palaeotemperature equations can be used with confidence. However, many studies lack the detailed investigation of contemporaneous materials to prove this. Ideally any palaeoclimate study of sediments or fossils from a lacustrine sequence should include a study of the contemporary lake water and the modern mineral precipitates. The presence of a systematic relationship between temperature, water isotope composition and mineral composition for the modern lake enables quantitative interpretation of the sediment record. It is particularly useful to know when and where the carbonate/silicate precipitates in the modern environment. This may tell us whether the sediment record reflects seasonal or year round conditions (e.g., Leng et al. (2001)).

Assuming that equilibrium precipitation has occurred, the interpretation of oxygen isotope data in terms of palaeotemperatures requires an understanding of two temperature effects that have opposing effects on the composition of a carbonate or silicate precipitate. At intermediate and high latitudes the oxygen isotope composition of mean annual precipitation correlates directly with change in temperature with a gradient of approximately +0.6‰/°C (Dansgaard 1964). Opposing this effect from being transferred directly into the sediment record, the equilibrium isotope fractionation between carbonate and water has a gradient of around –0.24‰/°C. For many lake records the carbonate response to temperature will be dominated by the change in the isotope composition of precipitation and effectively 'damped' by the opposing effect of mineral-water fractionation. In this case the measured carbonate values will covary with temperature - with an increase of about 0.36‰/°C. Eicher and Siegenthaler (1976) were amongst the first to use this relationship in the interpretation of a lacustrine record and the approach has been followed by many others. This is reasonable for the palaeoclimatic interpretation of many lakes but it implicitly assumes that dδp/dT always changes according to the Dansgaard relationship. Recent work has shown that this is not necessarily the case and where there are independent estimates of temperature change, $\delta^{18}O$ can be used to assess δp (see below).

Palaeo airmass (δp)

As outlined above, the oxygen isotope composition of modern global precipitation (δp) varies with latitude and altitude. In detail δp values are a function of the isotope composition of the moisture source (generally the ocean), air-mass trajectory and the condensation temperature of the rain or snow. The basic physics of the processes controlling δp are relatively well understood and it has been incorporated as an output variable in several numerical climate models (e.g., Hoffmann et al. (2000)). During episodes of climate change, δp at a particular site will commonly change, and most significantly changes may be independent of changes in temperature.

If the δp signal from a lake-sediment record can be resolved from the local hydrological and temperature effects that affect the isotope composition of lake water and thus the composition of the mineral precipitates, it can provide an independent monitor of changes in meteorology during episodes of climate change and, independently, be used to test the performance of climate models. However, recent studies have shown that past δp at a site is not always related to local temperature change. For example, distant changes in the temperature or isotope composition of the seawater undergoing evaporation or changes in the long-distance trajectory of air-masses may both affect δp in the absence of temperature change at the lake site. Significant short-term deviation from the Dansgaard relationship in some sediment records have highlighted important shifts in the Holocene climate system (Edwards et al. (1996); Teranes and McKenzie (2001); Hammarlund et al. (2002), Figure 6). Isotope studies of lake sediments that attempt to interpret oxygen isotope records solely in terms of temperature change at the site potentially ignore the possible effects of changes in δp that are related to more distant climatic processes.

The oxygen isotope composition of sedimentary (authigenic and biogenic) lacustrine carbonates or silicates has the potential to be used as a proxy for δp but extreme care must be taken before this interpretation is made. Specifically, it is necessary to know the relationship between $\delta^{18}O_{lakewater}$ and δp. In lakes where evaporation effects are significant, or where the residence times are too short for the lake water to be representative of mean precipitation, then $\delta^{18}O_{lakewater} \neq \delta p$.

In deep lakes where the isotope composition of the lake water can be demonstrated to reflect mean annual precipitation, it has been demonstrated that a δp signal can be determined by the analysis of biogenic calcite from the shells of benthic ostracods which live below the thermocline and are thus not affected by seasonal shifts in temperature or evaporation. A high-resolution δp record from Ammersee, southern Germany (von Grafenstein et al. 1996), parallels the $\delta^{18}O$ record from the Greenland ice-cores but also identifies significant quantitative differences in the records that are attributed to both changes in the source of the water in the North Atlantic, and changes related to changes in storm tracks across north western Europe.

To gain a fuller understanding of past climate, and test climate models, it will become increasingly necessary to derive global and regional maps of δp. Suitable deep lakes are rare so techniques are required to derive reliable δp records from other archives. Sediments and fossils from small lakes or those formed in the surface waters of large lakes, may be useful in this respect but quantitative records of changes in lake-water composition, and by inference δp, may only be possible where certain conditions apply. For instance, calibration studies of carbonate precipitation are essential to determine the relationships between contemporary $\delta^{18}O_{carb}$, δp, $\delta^{18}O_{lakewater}$, and temperature. Where carbonates can be shown to precipitate in equilibrium, or with a known vital offset, the sediment isotope record can be derived in terms of δp using independent biological or elemental palaeotemperature proxies. Mg/Ca ratios may provide an indicator of temperature change but major uncertainties in partitioning, particularly in very low Mg lake waters, may lead to considerable uncertainty in temperature determination (Figure 2 in Holmes and Chivas (2002)). Numerical transfer functions based on modern populations (beetles, chironomids or pollen), and extended into the past using the nearest living relative approach, have a potential precision of 1°C thus enabling calculations of past water composition to within 0.3‰. However, care must be taken as each proxy and temperature reconstruction has its own problems. For example, when reconstructing temperature from pollen-based records, vegetation recovery after a major

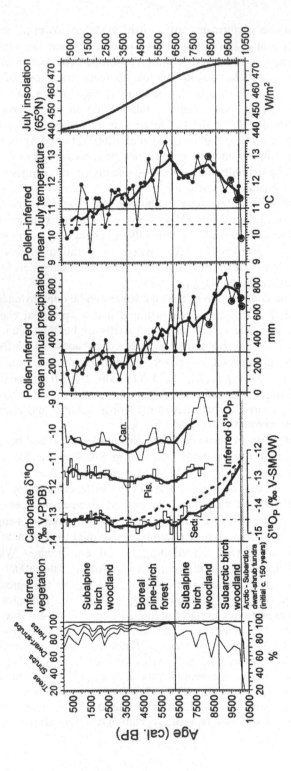

Figure 6. Carbonate-based isotope records from Lake Tibetanus, a small open, groundwater-fed lake in Northern Sweden, show that there are differences in the normal δ¹⁸O-temperature relationship which are attributed to changes in oceanography at the source of the precipitation. The transition from relatively moist, maritime conditions in the early Holocene to a much drier climate after 6500 cal. BP is also reflected by major changes in forest coverage and type as recorded in the pollen and macrofossil data. At the time of maximum influence of westerly air-mass circulation c. 9500 cal. BP, brought about by high summer insolation and enhanced meridional pressure gradients, δ¹⁸Op (= δ¹⁸Oprecip) at Lake Tibetanus was about 2‰ higher than would be predicted by the modern isotope-temperature relationship. The occurrence of long-term changes in δ¹⁸Op-temperature relations, which are more sensitive measures of palaeoclimate than either δ¹⁸Op or temperature alone, is potentially an important consideration when interpreting δ¹⁸O records. Sed = fine grained sedimentary carbonate (mainly Chara), Pis = *Pisidium* sp. Mollusc aragonite, Can = adult *Candona candida* ostracod calcite. Reproduced with permission from Hammarlund et al. (2002).

climate shift often lags behind δp changes and the amount of lag can be hundreds to thousands of years. Faunal reconstructions, in general, often face environmental variable problems. For example, in some circumstances the distribution and abundance of chironomid assemblages may be influenced by pH, total phosphorus, dissolved oxygen and lake depth, more strongly than by temperature, and this may compromise the reliability of chironomid-inferred temperature reconstruction (Brooks 2003).

Closed lakes

In closed lakes, $\delta^{18}O_{lakewater}$ depends on the balance between the isotope composition of inputs (including the source and amount of precipitation, surface runoff and groundwater inflow) and outputs (evaporation and groundwater loss). Unless there is significant groundwater seepage, most closed lakes will lose water primarily through evaporation, the rate of which is controlled by wind speed, temperature and humidity (Hostetler and Benson 1994). The phase change of evaporation results in the lighter isotope of oxygen, ^{16}O, being preferentially evaporated from water bodies leaving water that is relatively enriched in the heavier isotope, ^{18}O (see Darling et al. (this volume)). Any effects of varying precipitation source or temperature on $\delta^{18}O_{lakewater}$ are often small in comparison to evaporative concentration, and measured $\delta^{18}O_{lakewater}$ (and $\delta D_{lakewater}$) values become elevated above those of ambient precipitation (Gat 1980; Gasse and Fontes 1992). In extreme circumstances, evaporation in soil zones and from the surface of a lake can lead to significantly elevated $\delta^{18}O_{lakewater}$ values. $\delta^{18}O_{lakewater}$ and $\delta D_{lakewater}$ values from closed lakes in arid and semi-arid regions plot to the right of the Global Meteoric Water Line, often on slopes of 4.5 to 5.5, indicative of highly evaporated systems (Craig et al. 1963) (Figure 4).

Large swings in $\delta^{18}O$ (which can be >10‰ e.g., Lamb et al. (2000)) can often be directly attributed to precipitation/evaporation balances when the other variables affecting $\delta^{18}O_{lakewater}$ have been considered. Palaeoclimate records inferred from $\delta^{18}O$ have been produced from many closed lakes in arid and semi-arid regions, where evaporation dominates over other fractionation controls. Closed lakes on the African continent have recorded significant changes in $\delta^{18}O_{lakewater}$ in response to widely fluctuating lake-levels since the last glacial maximum (see Gasse (2000) for review). For example, $\delta^{18}O_{carb}$ measured in cores from Lake Turkana (NW Kenya) show that the level of the lake fell from ca. 4,000 BP, interrupted by brief humid periods at 1800-2000 BP and towards the end of the 19th century (Johnson et al. 1991). Similar climatic fluctuations have been recorded across East Africa, West Africa, Central Asia and Central America (Street-Perrott et al. 1989; Gasse and van Campo 1994).

Lake $\delta^{18}O$ variations have also been successfully linked to precipitation/evaporation balances of closed lakes in the Polar regions (Noon et al. 2003; Anderson and Leng 2004), the Mediterranean (Frogley et al. 1999; Leng et al. 1999), North America (Fritz et al. 1975; Li et al. 1997), South America (Abbott et al. 2003) and Australia (Chivas et al. 1993). For example, variations in $\delta^{18}O_{carb}$ from Lake Qinghai, the largest inland lake in China are shown to respond to changes in the strength of the Asian monsoon system (Lister et al. (1991); Henderson et al. (2003), Figure 7). Instrumental records of lake-level change at Lake Mono, California enabled Li et al. (1997) to compare high-

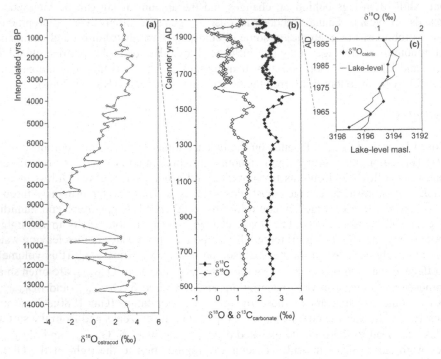

Figure 7. Lake Qinghai is a large closed basin-lake which undergoes seasonal evaporation, leaving its waters with high $\delta^{18}O$. Variations in $\delta^{18}O_{ostracod}$ over the last 14,500 years (a, redrawn from Lister et al. (1991)) record bottom-water $\delta^{18}O$ and are thus less affected by temperature variations, recording variations in P/E. The broad trends in the data correspond to long-term changes in monsoon strength since the last glacial, whereas the short-term trends record unusually wet or dry periods. High values before ~12500 BP reflect an arid, monsoon-weak climate; after which monsoon circulation begins to strengthen. River inflow increases after 10,800 BP at the onset of the Holocene, which shows several periods of rising lake-levels. A progressive increase in $\delta^{18}O$ occurs during the climatic optimum of 6800-3000 BP, as the lake evaporates under steady-state conditions. This results in relatively low short-term fluctuations in $\delta^{18}O$, characteristic of large lakes experiencing stable environmental conditions (Lister et al. 1991). $\delta^{18}O$ and $\delta^{13}C$ values from authigenic calcite from a core during the last 1500 years (b, Henderson (2004)) are relatively stable. Between 1500 AD and 1920 AD there is increased isotopic variability attributed to changes in effective precipitation. Three prominent negative excursions in $\delta^{18}O$ indicate increasing effective moisture and a freshening of the lake causing higher lake-levels (concurrent with the northern hemisphere Little Ice Age). This has been attributed to significant decreases in evaporation as a result of reduced air temperature and prolonged ice cover. From 1920 AD a large positive shift is seen indicating increased evaporation in the catchment, reducing lake-levels (coincident with a well-documented drought in NW China). The similarity in the $\delta^{18}O$ and $\delta^{13}C$ records is typical of closed lakes. By comparing $\delta^{18}O_{calcite}$ with lake-level observations over the last 35 years at Qinghai, Henderson et al. (2003) were able to compare the sensitivity of $\delta^{18}O$ further. The $\delta^{18}O_{calcite}$ curve shows falling values from 1960-1995, in line with falling lake-levels, indicating that the $\delta^{18}O$ record reflects reduced effective moisture.

resolution $\delta^{18}O_{calcite}$-inferred lake-level changes with measured changes since 1912. On annual to decadal timescales $\delta^{18}O_{calcite}$ recorded changes in precipitation amount, reflected in lake-levels, that matched the instrumental record. Where instrumental records do not exist, $\delta^{18}O$ variations are often compared to other palaeoenvironmental proxies (such as diatoms and pollen) and such multiproxy approaches are now routine in stable isotope studies (e.g., Roberts et al. (2001)).

The degree to which evaporation will increase $\delta^{18}O_{lakewater}$ depends on the residence time of the lake (lake volume/throughput rate). Closed lakes will have increased residence times relative to open systems, lengthening their exposure to evaporation (Hostetler and Benson 1994). Changes to a lake's residence time, caused by changes in basin hydrology and varying groundwater fluxes, will also influence the degree of enrichment, as will changes in the nature of catchment vegetation and soils. Groundwater fluxes, as opposed to riverine fluxes, are relatively difficult to estimate in lake budgets, especially if they vary through time. Deposition of sediments within a lake can reduce groundwater losses through time. Lakes that receive a significant part of their input from groundwater sources are also influenced by the residence time of the groundwater in the aquifer. If rainwater recharge is local, and the aquifer responds quickly to changes in climate, the lake may still be sensitive to climate change (e.g., Gasse and Fontes (1989)). However, water may be stored in groundwater aquifers for many years, thus acting as a buffer to climatic change. Street (1980) proposed that such lakes are unlikely to register climatic fluctuation with amplitudes of less than thousands of years. Rice Lake in the Northern Great Plains (USA) is a good example as it receives variable amounts of groundwater, recharged by isotopically light snowmelt, affecting the lake's response to climate (Yu et al. 2002). Additionally, groundwater may transcend drainage divides, disturbing closed-basin hydrology (see Ito (2001)). Further complications can arise in highly evaporated systems, whereby varied carbonate mineralogies often occur with increasing salinity (e.g., Lake Bosumtwi, Talbot and Kelts (1986)).

An understanding of the modern hydrology of a lake is key for interpreting records of $\delta^{18}O$ change but any interpretation must also take into account past changes in hydrology that may have occurred. The use of $\delta^{18}O_{carb}$ and $\delta^{13}C_{carb}$ (also applicable to the modern system via $\delta^{18}O_{lakewater}$ and $\delta^{13}C_{TDIC}$) is useful for identifying periods when a lake basin is or was hydrologically closed (Figure 4b). Details of controls on $\delta^{13}C_{carb}$ are described in McKenzie (1985), Talbot (1990), Li and Ku (1997), and Leng and Marshall (2004). Closed-basin lakes commonly show co-variation between $\delta^{18}O_{carb}$ and $\delta^{13}C_{carb}$, characteristically where $r \geq 0.7$ (e.g., Johnson et al. (1991)). Talbot (1990) suggested that this was probably due to hydrological effects associated with dilution and concentration balances. When there is a large freshwater input into a lake, $\delta^{18}O_{carb}$ and $\delta^{13}C_{carb}$ values both become lower. Likewise, when a lake becomes closed, and lake volume begins to fall, $\delta^{18}O_{carb}$ and $\delta^{13}C_{carb}$ are likely to increase via evaporation and vapour exchange. Li and Ku (1997) suggest that the relationship is inevitably more complex than this. For example, the $\delta^{13}C$ composition of total dissolved inorganic carbon (TDIC) is known to be affected by high carbonate alkalinity and CO_2 concentrations. Li and Ku (1997) argue that if freshwater inflow were to enter an alkaline lake, the existing levels of alkalinity and/or total CO_2 in the lake would dampen the $\delta^{13}C$ dilution effect. Likewise, if $\delta^{18}O_{carb}$ and $\delta^{13}C_{carb}$ approach isotopic steady state and their values remain constant, co-variation will be weak. It has also been recognised that $\delta^{18}O_{carb}$ and $\delta^{13}C_{carb}$ co-variation is only true for lakes that have attained hydrological closure over extended time frames (Drummond et al. 1995; Li and Ku 1997). For further discussion of $\delta^{18}O$ and evaporation in closed lakes see Craig and Gordon (1965), Gonfiantini (1986) and Gat (1995).

Modelling can be used to understand the sensitivity of a lake system, find the most significant control on lake $\delta^{18}O$ records and give some quantitative idea about changes in climate state over longer timescales (also see Darling et al (this volume)). However, high-resolution quantification of individual climatic parameters can be difficult due to the large number of variables controlling the lake isotope system. The following section describes the basis of isotope hydrological modelling in lake basins.

Lake type and hydrological modelling

We can estimate the type of information that we may get from a particular lake from its location, size, whether it has an inflow and an outflow etc. Given that there is an infinite continuum between all lake types and that no two lakes are the same - even if located in the same area - it is difficult to broadly categorise lake types (see Table 1). However, Table 2 attempts to demonstrate the type of information we might get from a range of lake types from large to small and open to closed hydrologies. In a general sense we can consider what type of information a lake sediment might record.

To fully understand factors controlling $\delta^{18}O$ in a particular system we need to turn to modelling both the contemporary and palaeolake hydrology. The hydrological budget of a lake can provide information on whether a lake is open or closed and whether this has changed through time.

The hydrological budget of a lake can be explained by the sum of the inflows and outflows from the system (e.g., Ricketts and Johnson (1996); Gibson et al. (1999)), for example:

$$dV = P + S_i + G_i - E - S_o - G_o \tag{2}$$

where d is the change in V, lake volume; P, precipitation on the lake surface; S_i, surface inflow from rivers and/or overland runoff; G_i, groundwater inflow; E, evaporation from the lake surface; S_o, surface outflow; G_o, groundwater outflow. (V, P, S_i, G_i, E, S_o and G_o are measured in the same units over a given time). Some of these values can be measured directly e.g., P, S_i, others have to be calculated (e.g., E) or estimated (e.g., G_o).

As well as the water balance equation above (Equation 2) the stable isotope values of the lake hydrological system must also balance such that:

$$dV.\delta_l = P\delta_P + S_i\delta_{Si} + G_i\delta_{Gi} - E\delta_E - S_o\delta_{So} - G_o\delta_{Go} \tag{3}$$

where the values δ_l (= $\delta^{18}O_{lakewater}$ described elsewhere), δ_p, δ_{Si}, δ_{Gi}, δ_E, δ_{So}, δ_{Go} are the isotope values (either $\delta^{18}O$ or δD) of the lake waters, lake surface precipitation, surface

Table 2. Controls on parameters controlling δ_l. See chapter text and Darling et al. (this volume) for more detailed discussion and typical values for some of these parameters.

Parameter	Controls
V	P, S_i, G_i, E, S_o, G_o
P	P, A_l
S_i	P, A_c, k, E_c
G_i	A_l, P, E_c
E	T_a, T_l, u, RH, I
S_o	V
G_o	V
δ_P	T_a, source area, P etc.
δ_{Si}	δ_P, E_c
δ_{Gi}	δ_P, E_c, t
δ_E	δ_l, δ_A, T_l, T_a, RH, u
δ_A	δ_P

P = precipitation; A_l = lake area; A_c = catchment area; k = runoff coefficient; T_a = air temperature; T_l = lake surface temperature; u = wind speed; RH = relative humidity; I = insolation; E_c = catchment evaporation. All other symbols are defined in the text. δ_P is dependent on δ_A, not vice versa. However δ_A is rarely measured and so values of δ_A are usually estimated from δ_P ($\delta_A = \delta_P - \epsilon^*$) (Gibson et al. 1999); see Darling et al. (this volume) for discussion of ϵ^*).

inflow, groundwater inflow, lake surface evaporation, surface outflow and groundwater outflow respectively over a given time (Gibson et al. 1999; Benson and Paillet 2002).

δ_l, δ_{Si}, δ_{So} are easy to measure directly from water samples collected from the lake and catchment. δ_P can be measured if rainwaters are sampled near the lake site; alternatively precipitation isotope values can be found in the GNIP database. Samples for measurement of δ_{Gi} and δ_{Go} are difficult to collect. δ_{Gi} can be estimated by measuring groundwater in or near the catchment that may come from wells or springs, δ_{Go} is lake water and therefore is usually taken to equal δ_l.

δ_E is very difficult to measure directly and is usually calculated based on the Craig-Gordon model of evaporation (Craig and Gordon 1965). Most authors (e.g., Gonfiantini (1986); Gibson et al. (1999)) use the equation as discussed by Darling et al. (this volume; Equation 9) although an alternative equation for calculating δ_E was derived from observations at Pyramid Lake, Nevada by Benson and White (1994) and has been used for modelling palaeolake variability (Ricketts and Johnson 1996; Benson and Paillet 2002).

Many of the hydrological components in Equations 2 and 3 are dependent on climatic parameters and all these factors therefore control recorded values of δ_l ($\delta^{18}O_{lakewater}$) and values of $\delta^{18}O$ recorded in sedimentary archives. Table 2 outlines the controlling factors behind all the parameters in Equations 2 and 3. By using these values and equations to model contemporary lake systems the dominant control on changes in $\delta^{18}O$ can be established and used to interpret records of past change.

The mass balance modelling approach has been used only sporadically for the interpretation of palaeoisotope records (e.g., Ricketts and Johnson (1996); Lemke and Sturm (1997); Benson and Paillet (2002)). For example, detailed isotopic modelling of the $\delta^{18}O$ record from Lake Turkana demonstrated that some of the $\delta^{18}O$ variability was due to the change of inputs and outputs and the isotope composition of inflowing water, as well as lake-level shifts (Ricketts and Johnson 1996). When the inputs and outputs to a closed lake are equal, δ_l will increase by evaporation until a steady state is reached ($\delta^{18}O_{evaporation} = \delta^{18}O_{input}$; e.g., Lister et al. (1991)). At steady state, δ_l will remain stable regardless of the volume (depth) and only changes in the rate of inputs and outputs will alter this. Thus fluctuations in δ_l (or $\delta^{18}O_{lakewater}$) in closed lakes are not necessarily linked to lake-level fluctuations.

Carbon, isotopes in lacustrine organic matter

Carbon isotopes in lacustrine organic matter

Organic matter within lake sediments can provide important information on the palaeoenvironment of a lake and its catchment. Lacustrine organic matter comes predominantly from plant materials and to a lesser extent organisms that live in and around the lake (Figure 8). The concentration of total organic carbon (%TOC) represents the amount of organic matter preserved after sedimentation, and so depends on both on the initial production and the degree of degradation (Meyers and Teranes 2001). The amount of sedimentary organic matter that originates from aquatic and terrestrial plants can be sometimes distinguished by C/N ratios (% organic C versus % total N) and to a lesser extent by $\delta^{13}C$. Organic nitrogen occurs preferentially in proteins and nucleic acids which are relatively abundant in lower plants such as aquatic phytoplankton (Talbot and Johannessen 1992). Organic material derived solely from lacustrine phytoplankton therefore has a characteristically low C/N ratio, typically <10. The C/N ratios of bulk surface sediments from several lakes, including Lake Michigan and Lake Ontario in North America as well as Lake Baikal in Siberia, have low C/N (8-11) which has been attributed to organic material predominantly from mixed phytoplankton (Table II in Meyers and Teranes (2001)). In contrast, organic material derived from terrestrial plants usually has higher C/N ratios for example, 17-163 in Meyers and Teranes (2001). Where there is a mixed contribution, it is difficult to distinguish inputs and interpret $\delta^{13}C$.

The carbon isotope signal ($\delta^{13}C$) from lacustrine organic matter acts as a tracer for past changes in the terrestrial and aquatic carbon cycles. If the amount of organic material from the terrestrial environment can be assessed from other proxies (TOC, C/N, magnetic susceptibility) and if it is limited, then aquatic $\delta^{13}C$ can provide information on within-lake processes provided there has been limited post depositional changes. Lacustrine algae and submergent macrophytes utilise dissolved CO_2 and HCO_3^-. Variations in the isotope composition of the dissolved CO_2 and HCO_3^- are due to many factors but of particular note are changes in the $\delta^{13}C$ of the C supplying the organic matter, and changes in $\delta^{13}C$ related to productivity and nutrient supply (these are described in more detail below). On a global scale, it is also worth noting that $\delta^{13}C$ in organic materials will be affected by the changing amount and composition of atmospheric CO_2 and climate (Meyers and Teranes 2001). In European lacustrine sediments, for example, a number of studies have attributed increased $\delta^{13}C$ values in

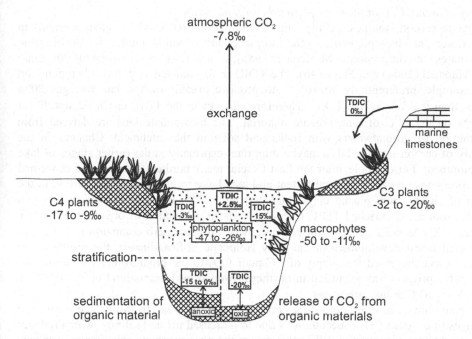

Figure 8. Carbon isotope values for the major sources of carbon into lakes and examples of the range of resulting $\delta^{13}C_{TDIC}$. The $\delta^{13}C_{TDIC}$ is higher (around −3‰ as opposed to −15‰) when groundwater flows in an area dominated by C4 plants. Where lake water stratification occurs the TDIC pool may have relatively high $\delta^{13}C_{TDIC}$ as the onset of permanent anoxic bottom waters slows organic oxidation and allows the preservation of greater amounts of organic material, effectively removing ^{12}C from the system. In highly organic lakes (especially if a portion of the organic material is derived from the terrestrial environment) oxidation of organic matter leads to lower $\delta^{13}C_{TDIC}$ as ^{12}C is liberated. Values of $\delta^{13}C_{TDIC}$ can be significantly higher than +2.5‰ shown here in evaporating brines that generate methane. The range of $\delta^{13}C_{TDIC}$ values are from Rau (1978), Coleman and Fry (1991), Keeley and Sandquist (1992), Meyers and Teranes (2001). Modified from Leng and Marshall (2004).

organic sediments to the transition from the Late Glacial Interstadial to the Younger Dryas due to lower atmospheric CO_2 concentrations and reduced annual air temperatures during the Late Glacial (Hammarlund 1993; Turney 1999; Nuñez et al. 2002).

If the amount of organic material can be shown to be predominantly from catchment vegetation then $\delta^{13}C$ can provide information on changes in vegetation. For example, a small crater lake in the Ethiopian Rift Valley has high C/N throughout the Holocene suggesting that organic input to the lake was at least in part from terrestrial plants and that the corresponding $\delta^{13}C$ values provide a proxy for changes in catchment vegetation that are supported by pollen data (Lamb et al. 2004).

Changes in the $\delta^{13}C$ of the C supplying the organic matter

Inorganic carbon isotopes, mainly from TDIC (Total Dissolved Inorganic Carbon) in lake waters, are incorporated into the inorganic carbon within authigenic and biogenic carbonates (see discussion in McKenzie (1985); Li and Ku (1997); Talbot (1990); Leng and Marshall (2004) and Figure 4b). The TDIC is also utilised by plants. Phytoplankton for example, preferentially utilise ^{12}C to produce organic matter that averages 20‰ lower than the $\delta^{13}C$ of the TDIC. Therefore changes in the TDIC can have significant effects on the $\delta^{13}C$ of bulk organic material. The bicarbonate ions are derived from interaction of groundwaters with rocks and soils in the catchment. Changes in the supply of carbon into the lakes maybe important especially in the earlier stages of lake development. For example, after the Last Glacial many northern European lakes formed in depressions created by glacial retreat and may have received inorganic carbon in the form of HCO_3^- from glacial clay which has high $\delta^{13}C$. Incorporation of carbon into plants from a ^{13}C-enriched TDIC reservoir leads to production of organic matter with higher $\delta^{13}C$. As soils develop, for example, when pioneer herb communities gave way to boreal forest development in the early Holocene in Scandinavia, the resultant soil development increased the supply of ^{13}C-poor CO_2 to the groundwater recharging the lake. This process was identified in northern Sweden by Hammarlund et al. (1997) as decreasing $\delta^{13}C$ in the carbonates.

During climate amelioration, drier conditions can also lead to lower lake levels and changes in aqueous CO_2 concentrations due to increased pH and salinity, which reduces dissolved CO_2 concentration. When this occurs the photosynthetic aquatic plants change from CO_2^- to HCO_3^- metabolism (Smith and Walker 1980; Lucas 1983), and plants become relatively enriched in ^{13}C (Smith and Walker 1980; Beardall et al. 1982), producing higher $\delta^{13}C$ values. Plant $\delta^{13}C$ values up to –9‰ have been recorded during these types of conditions (Meyers and Teranes 2001). These values may be increased further by exchange of the TDIC with isotopically heavier atmospheric carbon, which is a common feature in long residence time, closed lakes (Li and Ku (1997), Figure 4b).

During wetter periods the lake carbon pool will have been refreshed from a greater influx of soil-derived CO_2, which would have much lower $\delta^{13}C$. In Western Europe, for example, $\delta^{13}C_{DIC}$ values in ground- and river waters entering lakes are generally between –10 to –15‰ and –6 to –12‰ respectively (Andrews et al. 1993; 1997).

Changes in $\delta^{13}C$ related to productivity and nutrient supply

In lakes the dissolved carbon pool is often changed by biological productivity, mainly by preferential uptake of ^{12}C by aquatic plants during photosynthesis. During periods of enhanced productivity, or in lakes with a large biomass, the carbon pool in the water becomes enriched in ^{13}C and consequently has a higher $\delta^{13}C_{TDIC}$. Changes in plant photosynthesis and respiration can be seasonal: for example at Malham Tarn (NW England, UK) it has been proposed that photosynthesis by the calcareous algae Chara during the summer, causes the remaining TDIC to be enriched in ^{13}C (Coletta et al. 2001). Summer stratification in the water column in response to photosynthesis and organic production in the surface waters occurs in many lakes and leads to significant differences in the carbon isotope composition of the total dissolved organic carbon, which will be reflected in the carbon isotope composition of carbonates precipitated at different depths in the water column. Another example is Lake Greifen, NE Switzerland (McKenzie 1985; McKenzie and Hollander 1993) where mixing of the water column and very low photosynthesis in winter and early spring causes $\delta^{13}C_{TDIC}$ in the water column to remain relatively homogeneous. In the late spring and early summer, temperature increase and lake water overturn brings a supply of nutrients to the surface and causes phytoplankton blooms. These phytoplankton preferentially utilise the ^{12}C and so there is a significant increase in the $\delta^{13}C$ value of the TDIC in the surface waters, peaking around $-7.5‰$, while the bottom water TDIC drops to around $-13‰$ due to oxidation of the sinking organic matter. In highly organic shallow lakes, preferential oxidation of ^{12}C in organic matter can lead to the formation of very low $\delta^{13}C_{TDIC}$.

Compound specific organic carbon

The introduction of gas chromatography-isotope ratio mass spectrometry (GC-IRMS), which permits carbon isotope measurements to be made on individual compounds of known origin (biomarkers), has enabled the terrestrial and aquatic signals to be disentangled with greater confidence. Initially, compound-specific isotope analysis was carried out on biomarkers extracted from marine cores (Jasper and Hayes 1990; Freeman et al. 1990). Huang et al. (1995) were amongst the first to apply this technique to lacustrine sediments of Late Quaternary age, as part of a multidisciplinary study of past changes in climate and carbon cycling on Mt. Kenya, East Africa, and other studies are following (Street-Perrott et al. (2004), Figure 9). A full account of compound specific organic carbon is beyond the scope of this chapter but more details can be found in Huang et al. (1999) and references therein.

Nitrogen isotopes in lacustrine organic matter

In addition to carbon isotopes, nitrogen isotopes can also provide information on palaeoenvironment but are less commonly used in lacustrine isotope studies. In the past this was partly due to the analytical difficulties associated with the need for much larger sample sizes: $^{15}N/^{14}N$ analysis typically requires at least 20 times more sediment than $^{13}C/^{12}C$ analysis. Additionally, there is the greater complexity (relative to carbon) of the nitrogen inputs to lakes, and N isotope fractionations within lakes. Both of these lead to a greater difficulty in unambiguously interpreting $^{15}N/^{14}N$ data. Inputs to lakes include those from the atmosphere (deposition of N compounds, and fixation of atmospheric N_2) and from surface and subsurface waters (rivers, groundwater).

Atmospheric N deposition largely comprises dissolved inorganic nitrogen (DIN) in the form of nitrate and ammonium, and dissolved organic N (DON), with $\delta^{15}N$ values showing significant diurnal and seasonal variation, but usually averaging in the range -6 to $+6‰$ (Heaton 1987; Cornell et al. 1995; Kendall 1998; Russell et al. 1998; Heaton et al. 2004). Atmospheric N_2 (which has $\delta^{15}N = 0‰$ by definition) is fixed by cyanobacteria, with relatively little isotope fractionation, to produce organic matter with $\delta^{15}N$ typically -3 to $+1‰$ (Fogel and Cifuentes 1993). Water-borne inputs include particulates and DON derived from plants and soils, and DIN chiefly as nitrate or occasionally as ammonium.

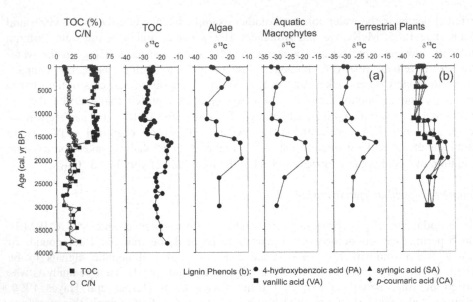

Figure 9. Sacred Lake is one of several well-studied, high altitude freshwater lakes located on the northeast flank of Mount Kenya. A 40,000 year-record has been analysed for total organic carbon (TOC), C/N ratios, $\delta^{13}C_{TOC}$ and weighted-average $\delta^{13}C$ values of algal, aquatic-macrophyte and terrestrial-plant biomarkers of which (a) are long-chain *n*-alkanes and (b) lignin phenols. The ability to analyse individual biomarkers allows a more detailed understanding of the changes to emerge than the bulk organic $\delta^{13}C$ curve alone. The core TOC has C/N ratios of between 10 and 27 which normally indicate a mixture of algal and terrestrial plant input into a lake, however, high levels of the algae *Botryococcus braunii* within the sediment means the C/N curve is difficult to interpret as this algae has unusually high C/N ratios. There are swings in all the $\delta^{13}C$ values at the last glacial maximum (LGM) (23,000 – 14,300 cal. BP), the late-glacial interstadial (14,300 – 12,300 cal. BP), the Younger Dryas (~11,800 cal. BP), and the mid-to-late Holocene. The similarities between the terrestrial and aquatic biomarkers, suggests the changes are due to pCO_2. Lower pCO_2 during the LGM combined with an arid climate resulted in much lower available carbon levels in the lake and catchment ecosystems. Superimposed on this were changes in the relative importance of the major carbon sources; high $\delta^{13}C_{TOC}$ and $\delta^{13}C$ of the biomarkers indicating high levels of C_4 graminoids, green algae and certain diatoms. Reproduced with permission from Street-Perrott et al. (2004).

Over most of the globe, plant $\delta^{15}N$ values typically fall in the range –5 to +8‰, and soil N about 0 to +8‰, with the values at a particular location being partly determined by climate (Heaton 1987; Handley et al. 1999). Surface and groundwaters in non-polluted areas may contribute DIN and DON with $\delta^{15}N$ between 0 to +10‰ (Kendall 1998); with the major forms of pollutant N contributing nitrate and ammonium with $\delta^{15}N$ between –4 to +4‰ for fertilizers, and as high as +8 to +18‰ for human and animal sewage waste (Heaton (1986), unpublished data). Terrestrial plant- and soil-derived particulate organic matter typically has higher C/N ratios and lower $\delta^{13}C$ values than lacustrine-derived organic matter. It is therefore common to plot sediment $\delta^{15}N$ values against C/N and $\delta^{13}C$. Whilst correlations are often observed, they may go 'either way' - i.e., higher C/N and $\delta^{13}C$ may correlate with lower $\delta^{15}N$ in some sequences and with higher $\delta^{15}N$ in others (Talbot 2001; Watanabe et al. 2004). This simply reflects the wide global range of $\delta^{15}N$ values for terrestrial plants and soils (Handley et al. 1999).

Within the lake, the more resistant types of plant- and soil-derived particulate matter may settle out to form allochthonous organic matter in the sediment. This material, commonly rich in cellulose and lignin with high C/N ratios, will tend to have less impact on the $\delta^{15}N$ value of the sediment than on the $\delta^{13}C$ value. The·remaining inputs will contribute to the lake's internal N-cycle: the assimilation (uptake) of DIN (and some DON) by phytoplankton and macrophytes to form new organic matter, and the decay (mineralisation and oxidation) of organic matter to release DIN and DON.

Within-lake processes
As ^{14}N is preferentially incorporated into phytoplankton, phytoplankton N can have $\delta^{15}N$ values several ‰ lower than that of the DIN, so the $\delta^{15}N$ value of the remaining DIN will increase. Two other processes are also associated with significant fractionation effects: denitrification (reduction of nitrate to gaseous N_2 under anoxic conditions) and ammonia volatilization (loss of ammonium as gaseous NH_3 under alkaline conditions). These again favour reaction of ^{14}N molecules, and can lead to substantial increases in the $\delta^{15}N$ value of the remaining nitrate and ammonium (Heaton 1986). Thus the DIN inputs to a lake may be substantially modified by subsequent processes within the lake, and the organic matter formed by assimilation of this DIN may have a $\delta^{15}N$ different from (usually lower than) that of the DIN. In the case of assimilation, however, if almost all of the DIN is consumed the final $\delta^{15}N$ value of the phytoplankton will be very similar to that of the initial DIN (a simple requirement of isotope mass balance). As a general rule, N-rich systems, in which DIN is in excess, or more complex lakes where anoxia or high pH may develop, may provide an environment where substantial changes in $\delta^{15}N$ of DIN and organic matter are possible. In N-limited systems, containing only low DIN concentrations, and where anoxia and high alkalinity are absent, there is a greater likelihood that the $\delta^{15}N$ values of autochthonous organic matter reflect that of the inputs. In a survey of N-limited upland lakes in Britain, R. Jones et al. (2004) found that there was a positive correlation between the $\delta^{15}N$ value of sediment organic N and the $\delta^{15}N$ value of the total dissolved N (DIN + DON).

When organic matter settles to form sediments it may undergo bacterial and diagenetic alteration which can potentially alter its $\delta^{15}N$ (and $\delta^{13}C$) value (Lehmann et al. 2002). Most studies of these effects have been on oceanic sediments, and have yielded conflicting results, but several studies suggest that lacustrine sediments do faithfully record the $\delta^{15}N$ value of deposited organic matter (Meyers and Ishiwatari (1993); Teranes and Bernasconi (2000); Ogawa et al. (2001), Figure 10).

Meromixis (stratification)
Lake stratification can result in low oxygen concentrations or reduced nutrient supply, and have important consequences for $^{15}N/^{14}N$ ratios. Anoxia can lead to inhibition of nitrification, and hence build-up of ammonium, or denitrification. These are conditions which can result in an increase in the $\delta^{15}N$ value of the remaining DIN (see above), and have been cited as reasons for high $\delta^{15}N$ (> +10‰) for sediment organic matter in a

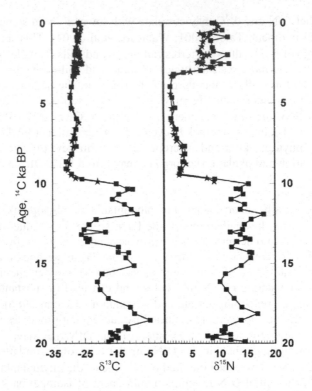

Figure 10. Lake Bosumtwi is a meteorite crater lake in Ghana, West Africa. The sediment record is a high-resolution archive of the last one million years of environmental change in this region (Talbot et al. 1984; Talbot and Johannessen 1992). Within the organic matter in the core there is some terrestrial material but this is low compared to the phytoplankton, so the δ^{15}N record is thought to largely reflect changes in Lake Bosumtwi's N cycle. Between 25000-10000 BP the sediment is typical of a shallow and alkaline lake with abundant organic matter. This type of lake would have favoured the production of abundant ammonia and its subsequent loss through volatilisation involving a strong fractionation and would lead to the development of lake waters with a very high δ^{15}N$_{DIN}$. Assimilation of this strongly enriched DIN pool is a likely explanation for the corresponding high δ^{15}N of the organic matter. An alternative explanation is that the high δ^{15}N in the organic matter reflects biological drawdown of a limited DIN pool, a consequence of reduced flushing of soil N during periods of aridity (Meyers and Lallier-Verges 1999). From 10,000 to 3,300 BP, δ^{15}N are low, and the organic matter in the core contains abundant filaments of the cyanobacteria *Anabaena*, an N-fixing phytoplankton. After 3300 BP the lake is thought to have deepened, and seasonal mixing which would have allowed the N-fixing phytoplankton to be replaced by forms, adapted to assimilating dissolved nitrate and ammonium. Reproduced with permission from Talbot (2001).

number of lakes (Collister and Hayes 1991; Talbot and Johannessen 1992; Ogawa et al. 2001). Alternatively, build-up of high ammonium concentrations due to meromixis might result in a larger isotope fractionation associated with ammonium assimilation, causing a decrease in the δ^{15}N values of organic matter (Teranes and Bernasconi 2000).

Lack of mixing may also reduce the supply of DIN, formed by mineralisation of organic matter near the sediment boundary, to surface waters. The depletion in DIN in the surface waters could then favour N_2-fixing algae, again causing a decrease in organic $\delta^{15}N$.

Oxygen and hydrogen isotopes in lacustrine organic matter

As well as carbon and nitrogen isotopes, oxygen and hydrogen can be measured in lacustrine organic matter (Edwards and McAndrews 1989; Edwards et al. 1996; 2004; Wolfe et al. 2001a). It is assumed that the oxygen or hydrogen isotope composition of autochthonous organic matter (and this may need to be investigated) deposited in sediments preserves the oxygen and hydrogen isotope composition of the lake water from which it formed (Figure 11). Equation (4) can be used to quantitatively relate the oxygen (or hydrogen) isotope composition of the lake sediment organic compound to the lake water source (also see Edwards and McAndrews (1989); Wolfe et al. (2001b)):

$$\delta^{18}O_{org} = \delta^{18}O_{precip} + \varepsilon^{18}_{hydro} + \varepsilon^{18}_{org\text{-}lakewater} \qquad (4)$$

where $\delta^{18}O_{org}$ is the oxygen isotope composition of the organic compound, $\delta^{18}O_{precip}$ is the oxygen isotope composition of local meteoric water, ε^{18}_{hydro} is the oxygen isotope enrichment between lake water and meteoric water due to hydrologic factors and which is commonly dominated by evaporation (= $\varepsilon^{18}_{lake\text{-}precip}$), and $\varepsilon^{18}_{org\text{-}lakewater}$ is the oxygen isotope enrichment (or depletion as is generally the case for $\varepsilon^{2}_{org\text{-}lakewater}$) that occurs during synthesis of the organic compound. Thus, decoding palaeoenvironmental information from stratigraphic records of $\delta^{18}O_{org}$ (or $\delta^{2}H_{org}$) depends on the ability to characterise or constrain the relative importance of $\delta^{18}O_{precip}$ ($\delta^{2}H_{precip}$), ε^{18}_{hydro} (ε^{2}_{hydro}), and $\varepsilon^{18}_{org\text{-}lakewater}$ ($\varepsilon^{2}_{org\text{-}lakewater}$).

With the exception of aquatic cellulose, in which numerous field and laboratory studies have shown that $\varepsilon^{18}_{cellulose\text{-}lakewater}$ averages between +27 and +28‰, and is independent of water temperature and plant species (see Sternberg (1989); Yakir (1992); Wolfe et al. (2001b)), $\varepsilon_{org\text{-}lakewater}$ values for other organic compounds are generally less well known. Recent field studies, however, have produced promising results. For example, Wooller et al. (2004) report that $\delta^{18}_{chironomid\text{-}lakewater}$ is similar to that of cellulose (28‰) based on relationships between chironomid $\delta^{18}O$ measurements in surface sediments and estimated $\delta^{18}O_{precip}$ from four lakes across a broad climatic gradient assumed to have minimal ε^{18}_{hydro}. The apparent similarity between $\varepsilon^{18}_{cellulose\text{-}lakewater}$ and $\varepsilon^{18}_{chironomid\text{-}lakewater}$ suggests common biochemical reactions govern cellulose and chitin oxygen isotope fractionation. A similar 'surface sediment-lake water' approach has been used to estimate $\varepsilon^{2}_{palmitic\ acid\text{-}lakewater}$ from a number of lakes in eastern North America with results ranging from –213 to –145‰ (mean = –171‰ ±15‰; Huang et al. (2002)). Recently reported follow-up studies from the same field sites explored the usefulness of other lipid biomarkers and confirmed the strong palaeoenvironmental potential of ester-bound palmitic acid (Huang et al. 2004). As recognized by Wooller et al. (2004), controlled laboratory experiments would likely increase the sensitivity of these emerging organic oxygen and hydrogen isotope archives. Scatter in apparent $\varepsilon_{org\text{-}lakewater}$ values estimated from field studies can largely

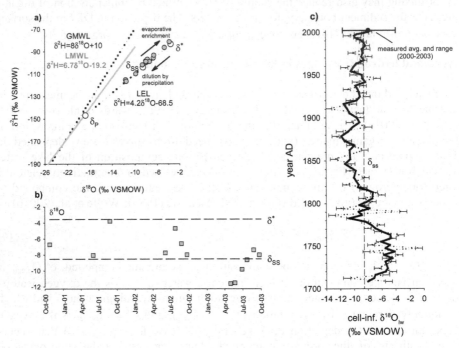

Figure 11. Hydrological research in the Peace-Athabasca Delta, Canada, a northern boreal ecosystem, is being undertaken by the use of isotope tracers to assess present and past water balance in wetlands and lakes (Wolfe et al. 2002). Results from stable isotope analyses of lake water collected over three years (October 2000 to September 2003) from Spruce Island Lake, a shallow (1 m) terminal wetland, are tightly constrained along the Local Evaporation Line (LEL) reflecting strong sensitivity to the opposing effects of evaporative enrichment and dilution by precipitation (a,b). Clustering of samples (shaded circles and squares) collected between the months of June and October beyond the predicted terminal basin steady-state value (δ_{SS}) indicates pronounced non-steady-state evaporation (i.e., net evaporative loss) during the open-water season. Extreme enrichment approaching the limiting value ($\overset{*}{\delta}$) occurred in summer 2001, reflecting particularly dry conditions, whereas late summer rain events curtailed the effects of evaporative isotopic enrichment during 2002. Further lowering of lake water isotopic composition due to input of isotopically-depleted snowmelt is indicated by samples obtained in early spring 2003. A 300-yr lake water $\delta^{18}O$ history for Spruce Island Lake is inferred from oxygen isotope analysis of aquatic plant cellulose using a $\varepsilon^{18}_{cellulose-lakewater}$ value of 28‰ (c). Both raw (error bars represent analytical uncertaintly of +/- 0.51‰) and three-point running mean profiles are shown. Also shown is the average and range of lake water $\delta^{18}O$ values from a) and, for reference, the contemporary terminal basin steady-state value (δ_{SS}). High cellulose-inferred $\delta^{18}O_{lw}$ ($=\delta^{18}O_{lakewater}$) values (cell-inf. $\delta^{18}O_{lw}$ on c)) from the early 1700s to ~1770, which exceed contemporary δ_{SS}, are indicative of very dry conditions. A decline in cellulose-inferred $\delta^{18}O_{lw}$ values after ~1820 to the mid- to late 1800s is interpreted to reflect wetter conditions. This is followed by a subsequent trend towards the present to values that are close to δ_{SS}. Overall, the cellulose-inferred $\delta^{18}O_{lw}$ record clearly demonstrates that both wetter and drier conditions have occurred at Spruce Island Lake over the past 300 years when compared to recent decades (Wolfe et al. unpublished).

be due to insufficient characterisation of seasonal and annual variability in lake water oxygen and hydrogen isotope composition (cf., Wolfe and Edwards (1997); Wolfe et al. (2001a)).

In cases where $\varepsilon_{org-lakewater}$ can be defined or constrained, stratigraphic interpretation of organic-inferred $\delta^{18}O$ (or δ^2H) lake water profiles generally involves separating primary isotope effects caused by shifts in the isotope composition of source water (i.e., $\delta^{18}O_{precip}$, δ^2H_{precip}) from those associated with secondary hydrological processes such as evaporative enrichment (i.e., ε^{18}_{hydro}, ε^2_{hydro}). Various approaches have been utilized to separate lake water $\delta^{18}O$ (and δ^2H) histories into $\delta^{18}O_{precip}$ and ε^{18}_{hydro} (ε^2_{hydro}) components, such as incorporation of independently derived records of palaeoprecipitation isotope composition history (e.g., Wolfe et al. (2000); (2001a)) and comparison of lake water isotope records from several lakes in the same region with varying hydrological sensitivities to evaporation (e.g., Edwards et al. (1996)). Information from other isotope records (i.e., $\delta^{13}C$, $\delta^{15}N$) or other proxies (diatoms, pollen, mineralogy etc.), as well as assessment of modern isotope hydrology of the study lake can also be used to constrain interpretations (e.g., Wolfe et al. (1996); Huang et al. (2002), Figure 11). In this context, coupling of oxygen- and hydrogen-isotope archives has tremendous potential for direct quantitative deconvolution of lake water $\delta^{18}O$ and δ^2H histories.

Summary

This chapter has attempted to demonstrate that the oxygen, carbon, nitrogen and hydrogen isotope composition of lacustrine sedimentary materials can yield a wide range of useful palaeoclimate information. However, with all these isotope systems a full interpretation of the data from the various components within a lake sediment requires a detailed knowledge of the processes that control and modify the signal. This must be largely determined for an individual lake system, although typical responses are assumed.

Oxygen isotopes are the main isotopes used in palaeoenvironmental studies and can be obtained from a large range of materials. However, even within the most established materials (i.e., carbonates) the interpretation is not easy. For example, a change in temperature will produce a shift in the equilibrium oxygen isotope composition of carbonate forming in lake water. However, the same temperature change will affect the isotope composition of the rainfall and may also affect rates of evaporation, both in the lake and in the catchment. One way of resolving this might be the analysis of the oxygen isotope composition of lacustrine organic materials, for example aquatic cellulose, which is thought to have $\delta^{18}O$ values that are independent of water temperature and plant species. However, there are different biochemical fractionations within plants, and separating specific plant compounds is not always easy.

Other palaeoenvironmental information can be achieved from organic materials in lakes. With bulk organic matter the interpretation is critically dependent on whether the majority of the material originates from plants growing within or around the lake. Bulk organic matter is most frequently analysed for carbon and nitrogen isotope ratios and the data can tell us about the source of organic matter, past productivity and changes in nutrient supply. The dynamics of nitrogen biochemical cycling are more complicated than those of carbon, thereby making interpretation of sedimentary $\delta^{15}N$ more difficult. A better approach is to analyse specific organic fractions (e.g., pollen, chironomid head capsules, cellulose, specific organic compounds), although separating them is not always easy.

Whatever is analysed from a lake sediment it is important, if possible, to carry out a calibration exercise investigating the basic systematics of isotope variation in the modern lake environment to establish the relationship between the measured signal and the isotope composition of the host waters. A robust calibration is seldom easy as the materials may not occur in the contemporary lake, the lake may be in an isolated geographical region or have dried out, making a rigorous contemporary study impossible. Where such a calibration is not possible assumptions have to be made, but these should be based on evidence from a multiproxy approach using isotope signals from different materials as well as other palaeolimnological techniques.

References

Abbott M.B., Wolfe B.B., Wolfe A.P., Seltzer G.O., Aravena R., Mark B.G., Polissar P.J., Rodbell D.T., Rowe H.D. and Vuille M. 2003. Holocene paleohydrology and glacial history of the central Andes using multiproxy lake sediment studies. Palaeogeogr. Palaeocl. 194: 123–138.

Abell P.I. and Williams M.A.J. 1989. Oxygen and carbon isotope ratios in gastropod shells as indicators of paleoenvironments in the Afar region of Ethiopia. Palaeogeogr. Palaeocl. 74: 265–278.

Anderson N.J. and Leng M.J. 2004. Increased aridity during the early Holocene in West Greenland inferred from stable isotopes in laminated-lake sediments. Quaternary Sci. Rev. 23: 841–849.

Andrews J.E., Riding R. and Dennis P.F. 1993. Stable isotope compositions of recent freshwater cyanobacterial carbonates from the British Isles: local and regional environmental controls. Sedimentology 40: 303–314.

Andrews J.E., Riding R. and Dennis P.F. 1997. The stable isotope record of environmental and climatic signals in modern terrestrial microbial carbonates from Europe. Palaeogeogr. Palaeocl. 129: 171–189.

Barker P.A., Street-Perrott F.A., Leng M.J., Greenwood P.B., Swain D.L., Perrott R.A., Telford J. and Ficken K.J. 2001. A 14 ka oxygen isotope record from diatom silica in two alpine tarns on Mt Kenya. Science 292: 2307–2310.

Beardall J.H., Griffiths J.A. and Raven J.A. 1982. Carbon isotope discrimination and the CO_2 accumulation mechanism in *Chlorella emersoni*. J. Exp. Bot. 33: 729–737.

Benson L. and Paillet F. 2002. HIBAL: a hydrologic-isotopic-balance model for application to paleolake systems. Quaternary Sci. Rev. 21: 1521–1539.

Benson L.V. and White J.W.C. 1994. Stable isotopes of oxygen and hydrogen in the Truckee River-Pyramid Lake surface-water system. 3. Source of water vapour overlying Pyramid Lake. Limnol. Oceanogr. 39: 1954–1958.

Bowen G.J. and Wilkinson B. 2002. Spatial distribution of $\delta^{18}O$ in meteoric precipitation. Geology 30: 315–318.

Brandriss M.E., O'Neil J.R., Edlund M.B. and Stoermer E.F. 1998. Oxygen isotope fractionation between diatomaceous silica and water. Geochim. Cosmochim. Ac. 62: 1119–1125.

Brooks S.J. 2003. Chironomid analysis to interpret and quantify Holocene climate change. In: Mackay A.W., Battarbee R.W., Birks H.J.B. and Oldfield F. (eds), Global Change in the Holocene. Arnold, London, pp. 328–341.

Buchardt B. and Fritz P. 1980. Environmental isotopes as environmental and climatological indicators. In: Fritz P and Fontes J.Ch. (eds), Handbook of Environmental Isotope Geochemistry. Elsevier, Amsterdam, pp. 473–504.

Chivas A.R., De Dekker P., Cali J.A., Chapman A., Kiss E. and Shelly J.M.G. 1993. Coupled stable isotope and trace element measurements of lacustrine carbonates as palaeoclimatic indicators. In: Swart P.K., Lohmann K.C., McKenzie J.A. and Savin S. (eds), Climate Change in Continental Isotopic Records. Geophysical Monograph 78, pp. 113–122.

Clark I. and Fritz I. 1997. Environmental Isotopes in Hydrogeology. Lewis Publishers, Boca Raton, New York, 328 pp.

Coleman D.C. and Fry B. 1991. Carbon Isotope Techniques. Academic Press, London, 274 pp.

Colleta P., Pentecost A. and Spiro B. 2001. Stable isotopes in charophyte incrustations: relationships with climate and water chemistry. Palaeogeogr. Palaeocl. 173: 9–19.

Collister J.W. and Hayes J.M. 1991. A preliminary study of the carbon and nitrogen isotope biogeochemistry of lacustrine sedimentary rocks from the Green River Mountain Formation, Wyoming, Utah and Colorado. U.S.G.S. Bulletin, 1973-A-G: C1–C12.

Cornell S., Rendell A. and Jickells T. 1995. Atmospheric inputs of dissolved organic nitrogen to the oceans. Nature 376: 243–246.

Craig H. 1961. Isotopic variations in meteoric waters. Science 133: 1702–1703.

Craig H. 1965. The measurement of oxygen isotope palaeotemperatures. In: Tongiorgi E. (ed.), Stable Isotopes in Oceanographic Studies and Palaeotemperatures. Pisa, Consiglio Nazionale delle Ricerche Laboratorio di Geologia Nucleare, pp. 161–182.

Craig H. and Gordon L.I. 1965. Deuterium and oxygen-18 variation in the ocean and marine atmosphere. In: Tongiorgi E. (ed.), Stable Isotopes in Oceanographic Studies and Palaeotemperatures. Pisa, Consiglio Nazionale delle Ricerche Laboratorio di Geologia Nucleare, pp. 277–373.

Craig H., Gordon L. and Horibe Y. 1963. Isotopic exchange effects in the evaporation of water. 1. low-temperature experimental results. J. Geophys. Res. 68: 5079–5087.

Dansgaard W. 1964. Stable isotopes in precipitation. Tellus 16: 436–468.

Darling W.G., Bath A.H., Gibson J.J. and Rozanski K. This volume. Isotopes in Water. In: Leng M.J. (ed.), Isotopes in Palaeoenvironmental Research. Springer, Dordrecht, The Netherlands.

Dinçer T. 1968. The use of Oxygen 18 and Deuterium concentrations in the water balance of lakes. Water Resour. Res. 4: 1289–1306.

Drummond C.N., Patterson W.P. and Walker J.C.G. 1995. Climatic forcing of carbon-oxygen isotopic covariance in temperate-region marl lakes. Geology 23: 1031–1034.

Edwards T.W.D. and McAndrews J.H. 1989. Paleohydrology of a Canadian Shield lake inferred from [18]O in sediment cellulose. Can. J. Earth Sci. 26: 1850–1859.

Edwards T.W.D., Wolfe B.B., Gibson J.J. and Hammarlund D. 2004. Use of water isotope tracers in high-latitude hydrology and paleohydrology. In: Pienitz R., Douglas M.S.V. and Smol J.P. (eds), Long-Term Environmental Change in Arctic and Antarctic Lakes. Springer, Dordrecht, The Netherlands.

Edwards T.W.D., Wolfe B.B. and MacDonald G.M. 1996. Influence of changing atmospheric circulation on precipitation δ^{18}O-temperature relations in Canada during the Holocene. Quaternary Res. 46: 211–218.

Eicher U. and Siegenthaler U. 1976. Palynological and oxygen isotope investigations on Late-Glacial sediment cores from Swiss lakes. Boreas 5: 109–117.

Epstein S., Buchsbaum R., Lowenstam H.A. and Urey H.C. 1953. Revised carbonate water isotopic temperature scale. Geol. Soc. Am. Bull. 64: 1315–1326.

Fogel M.L. and Cifuentes L.A. 1993. Isotope fractionation during primary production. In: Engel, M.H. and Macko S.A. (eds), Organic Geochemistry. Plenum Press, New York, pp. 73–98.

Freeman K.H., Hayes J.M., Trendel J.M. and Albrecht P. 1990. Evidence from carbon-isotope measurements for diverse origin of sedimentary hydrocarbons. Nature 343: 254–256.

LENG ET AL

Friedman I. and O'Neil J.R. 1977. Compilation of stable isotope fractionation factors of geochemical interest. In: Fleischer M. (ed.), Data of Geochemistry. Sixth edition. Geological Survey professional paper 440-KK, US government printing office, Washington, pp. 1–40.

Fritz P. and Poplawski S. 1974. ^{18}O and ^{13}C in the shells of freshwater molluscs and their environments. Earth Planet. Sc. Lett. 24: 91–98.

Fritz P., Anderson T.W. and Lewis C.F.M. 1975. Late-Quaternary climatic trends and history of Lake Erie from stable isotope studies. Science 190: 267–269.

Frogley M.R., Tzedakis P.C. and Heaton T.H.E. 1999. Climate variability in Northwest Greece during the last Interglacial. Science 285: 1886–1889.

Gasse F. 2000. Hydrological changes in the African tropics since the Last Glacial Maximum. Quaternary Sci. Rev. 19: 189–211.

Gasse F. and Fontes J-C. 1989. Palaeoenvironments and palaeohydrology of a tropical closed lake (Lake Asal, Djibouti) since 10,000 yr B.P. Palaeogeogr. Palaeocl. 69: 67–102.

Gasse F. and Fontes J-C. 1992. Climatic changes in northwest Africa during the last deglaciation (16 - 7 ka B.P.) In: Bard E. and Broeker W. (eds), The last deglaciation: Absolute and radiocarbon chronologies. NATO ASI series 12, Kluwer Academic Publishers, Dordrecht, The Netherlands, pp. 295–325.

Gasse F. and van Campo E. 1994. Abrupt post-glacial climate events in West Asia and North Africa monsoon domains. Earth Planet. Sc. Lett. 126: 435–456.

Gat J.R. 1980. Isotope hydrology of very saline lakes. In: Nissenbaum A. (ed.), Hypersaline Brines and Evaporitic Environments. Elsevier, Amsterdam, pp. 1–8.

Gat J.R. 1995. Stable isotopes of fresh and saline lakes. In: Lerman A., Imboden D.M. and Gat J.R. (eds), Physics and Chemistry of Lakes. Springer, Berlin, pp. 139–165.

Gibson J.J., Edwards T.W.D. and Prowse T.D. 1996. Development and validation of an isotopic method for estimating lake evaporation. Hydrol. Process. 10: 1369–1382.

Gibson J.J., Edwards T.W.D. and Prowse T.D. 1999. Pan-derived isotopic composition of atmospheric water vapour and its variability in northern Canada. J. Hydrol. 217: 55–74.

Gibson J.J., Prepas E.E. and McEachern P. 2002. Quantitative comparison of lake throughflow, residency, and catchment runoff using stable isotopes: modelling and results from a regional survey of Boreal lakes. J. Hydrol. 262: 128–144.

Gonfiantini R. 1986. Environmental isotopes in lake studies. In: Fritz P. and Fontes J-C. (eds), Handbook of Environmental Isotope Geochemistry. Elsevier, Amsterdam, pp. 113–168.

Grossman E. 1982. Stable isotope analysis in live benthic foraminifera from the Southern California Borderland. PhD thesis, University of Southern California, Los Angeles.

Grossman E.L. and Ku T.L. 1986. Oxygen and carbon isotope fractionation in biogenic aragonite – temperature effects. Chem. Geol. 59: 59–74.

Hammarlund D. 1993. A distinct delta-C decline in organic lake sediments at the Pleistocene-Holocene transition in Southern Sweden. Boreas 22: 236–243.

Hammarlund D., Aravena R., Barnekow L., Buchardt B. and Possnert G. 1997. Multi-component carbon isotope evidence of early Holocene environmental change and carbon-flow pathways from a hard-water lake in northern Sweden. J. Paleolimnol. 18: 219–233.

Hammarlund D., Barnekow L., Birks H.J.B., Buchardt B. and Edwards T.W.D. 2002. Holocene changes in atmospheric circulation recorded in the oxygen-isotope stratigraphy of lacustrine carbonates from northern Sweden. Holocene 12: 339–351.

Handley L.L., Austin A.T., Robinson D., Scrimgeour C.M., Raven J.A., Heaton T.H.E., Schmidt S. and Stewart G.R. 1999. The ^{15}N natural abundance ($\delta^{15}N$) of ecosystem samples reflects measures of water availability. Aust. J. Plant Physiol. 26: 185–199.

Hassan K.M. and Spalding R.F. 2001. Hydrogen isotope values in lacustrine kerogen. Chem. Geol. 175: 713–721.

Heaton T.H.E. 1986. Isotopic studies of nitrogen pollution in the hydrosphere and atmosphere: a review. Chem. Geol. 59: 87–102.

Heaton T.H.E. 1987. The $^{15}N/^{14}N$ ratios of plants in South Africa and Namibia: relationship to climate and coastal/saline environments. Oecologia 74: 236–246.

Heaton T.H.E., Wynn P. and Tye A. 2004. Low $^{15}N/^{14}N$ ratios for nitrate in snow in the High Arctic (79°N). Atmos. Environ. 38: 5611–5621.

Henderson A.C.G. 2004. Late Holocene environmental change on the NE Tibetan Plateau: a palaeolimnological study of Lake Qinghai and Lake Gahai, based on stable isotopes. PhD thesis, University of London, London.

Henderson A.C.G., Holmes J.A., Zhang J.W., Leng M.J. and Carvalho L.R. 2003. A carbon- and oxygen-isotope record of recent environmental change from Qinghai Lake, NE Tibetan Plateau. Chinese Sci. Bull. 48:1463–1468.

Hoffmann G., Jouzel J. and Masson V. 2000. Stable water isotopes in atmospheric general circulation models. Hydrol. Process. 14: 1385–1406.

Holmes J.A. and Chivas A.R. 2002. Ostracod shell chemistry – overview. In: Holmes J.A. and Chivas A.R. (eds), The Ostracoda: Applications in Quaternary Research. Geophysical Monograph 131, American Geophysical Union, Washington D.C. pp. 185–204.

Hostetler S.W. and Benson L.V. 1994. Stable isotopes of oxygen and hydrogen in the Truckee River-Pyramid Lake surface-water system. 2. A predictive model of $\delta^{18}O$ and δ^2H in Pyramid Lake. Limnol. Oceanogr. 39: 356–364.

Huang Y., Freeman K.H., Eglinton T.I. and Street-Perrott F.A. 1999. $\delta^{13}C$ analyses of individual lignon phenols in Quaternary lake sediments: A novel proxy for deciphering past terrestrial vegetation changes. Geology 27: 471–474.

Huang Y., Shuman B., Wang Y. and Webb III T. 2002. Hydrogen isotope ratios of palmitic acid in lacustrine sediments record late Quaternary climate variations. Geology 30: 1103–1106.

Huang Y., Shuman B., Wang Y. and Webb III T. 2004. Hydrogen isotope ratios of individual lipids in lake sediments as novel tracers of climatic and environmental change: a surface sediment test. J. Paleolimnol. 31: 363–375.

Huang Y., Street-Perrott F.A., Perrott R.A. and Eglinton G. 1995. Molecular and carbon isotopic stratigraphy of a glacial/interglacial sediment sequence from a tropical freshwater lake: Sacred Lake, Mt. Kenya. In: Grimalt J.O. and Dorronsoro C. (eds), Organic Geochemistry: Developments and Applications to Energy, Climate, Environment and Human History. Pergamon, Oxford, pp. 826–829.

IAEA/WMO 2001. Global Network of Isotopes in Precipitation. The GNIP Database. Accessible at: http://isohis.iaea.org

Ito E. 2001. Application of stable isotope techniques to inorganic and biogenic carbonates. In: Last W. M. and Smol J.P. (eds), Tracking Environmental Change Using Lake Sediments. Volume 2: Physical and Geochemical Techniques. Kluwer Academic Publishers, Dordrecht, The Netherlands, pp. 351–371.

Jasper J.P. and Hayes J.M. 1990. A carbon isotope record of CO_2 levels during the late Quaternary. Nature 347: 462–464.

Johnson T.C., Halfman J.D. and Showers W.J. 1991. Paleoclimate of the past 4000 years at Lake Turkana, Kenya, based on the isotopic composition of authigenic calcite. Palaeogeogr. Palaeocl. 85: 189–198.

Jones R.I., King L., Dent M.M., Maberly S.C. and Gibson C.E. 2004. Nitrogen stable isotope ratios in surface sediments, epilithon and macrophytes from upland lakes with different nutrient status. Freshwater Biol. 49: 382–391.

Jones V.J., Leng M.J., Solovieva N., Sloane H.J. and Tarasov P. 2004. Holocene climate on the Kola Peninsula; evidence from the oxygen isotope record of diatom silica. Quaternary Sci. Rev. 23: 833–839.

Juillet-Leclerc A. and Labeyrie L. 1987. Temperature dependence of the oxygen isotopic fractionation between diatom silica and water. Earth Planet. Sc. Lett. 84: 69–74.

Keatings K.W., Heaton T.H.E. and Holmes J.A. 2002. Carbon and oxygen isotope fractionation in non-marine ostracods: results from a 'natural culture' environment. Geochim. Cosmochim. Ac. 66: 1701–1711.

Keeley J.E. and Sandquist D.R. 1992. Carbon – freshwater plants. Plant Cell Environ. 15: 1021–1035.

Kendall C. 1998. Tracing nitrogen sources and cycling in catchments. In: Kendall C. and McDonnell J.J. (eds), Isotope Tracers in Catchment Hydrology. Elsevier, Amsterdam, pp. 519–576.

Kim S.T. and O'Neil J.R. 1997. Equilibrium and nonequilibrium oxygen isotope effects in synthetic carbonates. Geochim. Cosmochim. Ac. 61: 3461–3475.

Krishnamurthy R.V., Syrup K.A., Baskaran M. and Long A. 1995. Late glacial climate record of midwestern United States from the hydrogen isotope ratio of lake organic matter. Nature 269: 1565–1567.

Labeyrie L.D. 1974. New approach to surface seawater palaeotemperatures using $^{18}O/^{16}O$ ratios in silica of diatom frustules. Nature 248: 40–42.

Labeyrie L.D. and Juillet A. 1982. Oxygen isotopic exchangeability of diatom valve silica; interpretation and consequences for paleoclimatic studies. Geochim. Cosmochim. Ac. 46: 967–975.

Lamb A.L., Leng M.J., Lamb H.F., Telford R.J. and Mohammed M.U. 2004. Holocene climate and vegetation change in the Main Ethiopian Rift Valley, inferred from the composition (C/N and $\delta^{13}C$) of lacustrine organic matter. Quaternary Sci. Rev. 23: 881–891.

Lamb A.L., Leng M.J., Lamb H.F. and Umer M. 2000. A 9,000-year oxygen and carbon isotope record of hydrological change in a small Ethiopian crater lake. Holocene 10: 167–177.

Land L.S. 1980. The isotopic and trace element geochemistry of dolomite: the state of the art. In: Zenger D.H., Dunham J.B. and Ethington R.L. (eds), Concepts and Models of Dolomitisation. Society of Economic Paleontologists and Mineralogists Special Publication No. 28, pp. 87–110.

Last W.M. and Smol J.P. (eds). 2001a. Tracking Environmental Change Using Lake Sediments. Volume 1: Basin Analysis, Coring, and Chronological Techniques. Kluwer Academic Publishers, Dordrecht, The Netherlands, 548 pp.

Last W.M. and Smol J.P. (eds). 2001b. Tracking Environmental Change Using Lake Sediments. Volume 2: Physical and Geochemical Methods. Kluwer Academic Publishers, Dordrecht, The Netherlands, 504 pp.

Lehmann M.F., Bernasconi S.M., Barbieri A. and McKenzie J.A. 2002. Preservation of organic matter and alteration of its carbon and nitrogen isotope composition during simulated and in situ early sedimentary diagenesis. Geochim. Cosmochim. Ac. 66: 3573–3584.

Lemke G. and Sturm M. 1997. $\delta^{18}O$ and trace element measurements as proxy for the reconstruction of climate changes at Lake Van (Turkey): Preliminary results. In: Dalfes N.D. (ed.), Third Millennium BC Climate Change and Old World Collapse, NATO ASI Series. 149, 653–678.

Leng M.J., Barker P., Greenwood P., Roberts N. and Reed J.M. 2001. Oxygen isotope analysis of diatom silica and authigenic calcite from Lake Pinarbasi, Turkey. J. Paleolimnol. 25: 343–349.

Leng M.J., Lamb A.L., Lamb H.F. and Telford R.J. 1999. Palaeoclimatic implications of isotopic data from modern and early Holocene shells of the freshwater snail Melanoides tuberculata, from lakes in the Ethiopian Rift Valley. J. Paleolimnol. 21: 97–106.

Leng M.J. and Marshall J.D. 2004. Palaeoclimate interpretation of stable isotope data from lake sediment archives. Quaternary Sci. Rev. 23: 811–831.

Li H.-C. and Ku T.-L. 1997. $\delta^{13}C-\delta^{18}O$ covariance as a paleohydrological indicator for closed-basin lakes. Palaeogeogr. Palaeocl. 133: 69–80.

Li H.-C., Ku T.-L., Stott L.D. and Anderson R.F. 1997. Stable isotope studies on Mono Lake (California). 1. $\delta^{18}O$ in lake sediments as proxy for climatic change during the last 150 years. Limnol. Oceanogr. 42: 230–238.

Lister G.S., Kelts K., Zao C.K., Yu J.K. and Niessen K. 1991. Lake Qinghai, China: closed basin lake levels and the oxygen isotope record for ostracoda since the Late Pleistocene. Palaeogeogr. Palaeocl. 84: 141–162.

Lucas W.J. 1983. Photosynthetic assimilation of exogenous HCO_3^- by aquatic plants. Ann. Rev. Plant Physio. 34: 71–104.

McCrea J.M. 1950. On the isotopic chemistry of carbonates and palaeo-temperature scale. J. Chem. Phys. 18: 849–857.

McKenzie J.A. 1985. Carbon isotopes and productivity in the lacustrine and marine environment. In: W. Stumm (ed.), Chemical Processes in Lakes. Wiley, New York, pp. 99–118.

McKenzie J.A. and Hollander D.J. 1993. Oxygen-isotope record in recent carbonate sediments from Lake Greifen, Switzerland (1750-1986): application of continental isotopic indicator for evaluation of changes in climate and atmospheric circulation patterns. In Swart P.K., Lohmann J., McKenzie J. and Savin S. (eds), Climate Change in Continental Isotopic Records. Geophysical Monograph 78, pp. 101–111.

Matheney R.K. and Knauth L.P. 1989. Oxygen-isotope fractionation between marine biogenic silica and seawater. Geochim. Cosmochim. Ac. 53: 3207–3214.

Marshall J.D., Jones R.T., Crowley S.F., Oldfield F., Nash S. and Bedford A. 2002. A high resolution late glacial isotopic record from Hawes Water Northwest England. Climatic oscillations: calibration and comparison of palaeotemperature proxies. Palaeogeogr. Palaeocl. 185: 25–40.

Meyers P.A. and Ishiwatari R. 1993. Lacustrine organic geochemistry – an overview of indicators of organic matter sources and diagenesis in lake sediments. Org. Geochem. 20: 867–900.

Meyers P.A. and Lallier-Verges E. 1999. Lacustrine sedimentary organic matter records of Late Quaternary paleoclimates. J. Paleolimnol. 21: 345–372.

Meyers P.A. and Teranes J.L. 2001. Sediment organic matter. In: Last W. M. and Smol J.P. (eds), Tracking Environmental Change Using Lake Sediments. Volume 2: Physical and Geochemical Techniques. Kluwer Academic Publishers, Dordrecht, The Netherlands, pp. 239-269.

Noon P.E., Leng M.J. and Jones V.J. 2003. Oxygen isotope ($\delta^{18}O$) evidence of hydrological change through the mid- to late- Holocene (c. 6000 ^{14}C BP to present day) at Signey Island, maritime Antarctic. Holocene 13: 153-160.

Nuñez R., Spiro B., Pentecost A., Kim A. and Coletta P. 2002. Organic-geochemical and stable isotope indicators of environmental change in a marl lake, Malham Tarn, North Yorkshire, U.K. J. Paleolimnol. 28: 403–417.

Ogawa N.O., Koitabashi T., Oda H., Nakamura T., Ohkouchi N. and Wada E. 2001. Fluctuations of nitrogen isotope ratio of gobiid fish (Isaza) specimens and sediments in Lake Biwa, Japan, during the 20th century. Limnol. Oceanogr. 46: 1228–1236.

Rau G. H. 1978. Carbon-13 depletion in a subalpine lake: Carbon flow implications. Science 201: 901–902.

Ricketts R.D. and Anderson R.F. 1998. A direct comparison between the historical record of lake level and the $\delta^{18}O$ signal in carbonate sediments from Lake Turkana, Kenya. Limnol. Oceanogr. 43: 811–822.

Ricketts R.D. and Johnson T.C. 1996. Climate change in the Turkana basin as deduced from a 4000 year long $\delta^{18}O$ record. Earth Planet. Sc. Lett. 142: 7–17.

Roberts N., Reed J., Leng M.J., Kuzucuoglu C., Fontugne M., Bertaux J., Woldring H., Bottema S., Black S., Hunt E. and Karabiyikoglu M. 2001. The tempo of Holocene climatic change in the eastern Mediterranean region: new high-resolution crater-lake sediment data from central Turkey. Holocene 11: 721–736.

Rosenbaum J. and Sheppard S.M.F. 1986. An isotopic study of siderites, dolomites and ankerites at high temperatures. Geochim. Cosmochim. Ac. 50: 1147–1150.

Russell K.M., Galloway J.N., Macko S.A., Moody J.L. and Scudlark J.R. 1998. Sources of nitrogen in wet deposition to the Cheasapeake Bay region. Atmos. Environ. 32: 2453–2465.

Sauer P.E., Eglinton T.I., Hayes J.M., Schimmelmann A. and Sessions A.L. 2001. Compound-specific D/H ratios of lipid biomarkers from sediments as a proxy for environmental and climatic conditions. Geochim. Cosmochim. Ac. 65: 213–222.

Schwalb A. 2003. Lacustrine ostracods as stable isotope recorders of late-glacial and Holocene environmental dynamics and climate. J. Palaeolimnol. 29: 265–351.

Shemesh A., Charles C.D. and Fairbanks R.G. 1992. Oxygen isotopes in biogenic silica - global changes in ocean temperature and isotopic composition. Science 256: 1434–1436.

Shemesh A., Burckle L.H. and Hays J.D. 1995. Late Pleistocene oxygen isotope records of biogenic silica from the Atlantic sector of the Southern Ocean. Paleoceanography 10: 179–196.

Shemesh A., Rosqvist G., Rietti-Shati M., Rubensdotter L., Bigler C., Yam R. and Karlen W. 2001. Holocene climate change in Swedish Lapland inferred from an oxygen isotope record of lacustrine biogenic silica. Holocene 11: 447–454.

Siegenthaler U. and Eicher U. 1986. Stable oxygen and carbon isotope analyses. In Berglund, B.E. (ed.), Handbook of Holocene Palaeoecology and Palaeohydrology. John Wiley and Sons Ltd, London, pp. 407–422.

Smith F.A. and Walker N.A. 1980. Photosynthesis by aquatic plants: effects of unstirred layers in relation to assimilation of CO_2 and HCO_3^- and to carbon isotopic discrimination. New Phytol. 86: 245–259.

Smol J.P., Birks H.J.B. and Last W.M. 2001a. Tracking Environmental Change Using Lake Sediments. Volume 3: Terrestrial, Algal, and Siliceous Indicators. Kluwer Academic Publishers, Dordrecht, The Netherlands, 371 pp.

Smol J.P., Birks H.J.B. and Last W.M. 2001b. Tracking Environmental Change Using Lake Sediments. Volume 4: Zoological Indicators. Kluwer Academic Publishers, Dordrecht, The Netherlands, 217 pp.

Sternberg L.S.L. 1989. Oxygen and hydrogen isotope ratios in plant cellulose: Mechanisms and applications. In: Rundel P.W., Ehleringer J.R. and Nagy K.A. (eds), Stable Isotopes in Ecological Research. Springer-Verlag, New York, pp. 124–141.

Street F.A. 1980. The relative importance of climate and local hydrogeological factors in influencing lake-level fluctuations. Palaeoecol. Afr. 12: 137–158.

Street-Perrott F.A., Holmes J.A., Waller M.P., Allen M.J., Barber N.G.H., Fothergill P.A., Harkness D.D., Ivanovich M., Kroon D. and Perrott R.A. 2000. Drought and dust deposition in the West African Sahel: a 5500-year record from Kajemarum Oasis, northeastern Nigeria. Holocene 10: 293–302.

Street-Perrott F.A., Marchand D.S., Roberts N. and Harrison S.P. 1989. Global lake-level fluctuations from 18,000 to 0 years ago: a palaeoclimatic analysis. U.S. Department of Energy, Office of Energy Research, DOE/ER/60304-H1.

Street-Perrott F.A., Ficken K.J., Huang Y. and Eglinton G. 2004. Late Quaternary changes in carbon cycling on Mt. Kenya, East Africa: an overview of the $\delta^{13}C$ record in lacustrine organic matter. Quaternary Sci. Rev. 23, 861–879.

Stuiver M. 1970. Oxygen and carbon isotope ratios of fresh-water carbonates as climatic indicators. J. Geophys. Res. 75: 5247–5257.

Talbot M.R. 1990. A review of the palaeohydrological interpretation of carbon and oxygen isotopic ratios in primary lacustrine carbonates. Chem. Geol. 80: 261–79.

Talbot M.R. 2001. Nitrogen isotopes in palaeolimnology. In: Last W.M. and Smol J.P. (eds), Tracking Environmental Change Using Lake Sediments. Volume 2: Physical and Geochemical Techniques. Kluwer Academic Publishers, Dordrecht, The Netherlands, pp. 401–439.

Talbot M.R. and Johannessen T. 1992. A high resolution palaeoclimatic record for the last 27,500 years in tropical West Africa from the carbon and nitrogen isotopic composition of lacustrine organic matter. Earth Planet. Sci. Lett. 110: 23–37.

Talbot M.R. and Kelts K. 1986. Primary and diagenetic carbonates in the anoxic sediments of Lake Bosumtwi, Ghana. Geology 14: 912–916.

Talbot M.R., Livingston D.A., Palmer P.A., Maley J., Melack J.M., Delibrias G. and Gulliksen S. 1984. Preliminary results from sediment cores from Lake Bosumtwi, Ghana. Paleoecol. Afr. 16: 173–192.

Tarutani T., Clayton R.N. and Mayeda T. 1969. The effect of polymorphism and magnesium substitution on oxygen isotope fractionation between calcium carbonate and water. Geochim. Cosmochim. Ac. 33: 987–996.

Teranes J.L. and Bernasconi S.M. 2000. The record of nitrate utilization and productivity limitation provided by $\delta^{15}N$ values in lake organic matter – a study of sediment trap and core sediments from Baldeggersee, Switzerland. Limnol. Oceanog. 45: 801–803.

Teranes J.L. and McKenzie J.A. 2001. Lacustrine oxygen isotope record of 20th-century climate change in central Europe: evaluation of climatic controls on oxygen isotopes in precipitation. J. Paleolimnol. 26: 131–146.

Turney C.S.M. 1999. Lacustrine bulk organic $\delta^{13}C$ in the British Isles during the last glacial-Holocene transition (14-9 ka C-14 BP). Arct. Antarct. Alp. Res. 31: 71–81.

Urey H.C., Lowenstam H.A., Epstein S. and McKinney C.R. 1951. Measurement of palaeotemperatures and temperatures of the Upper Cretaceous of England, Denmark and Southeastern United States. Geol. Soc. Am. Bull. 62: 399–416.

von Grafenstein U., Erlenkeuser H., Muller J., Trimborn P. and Alefs J. 1996. A 200 year mid-European air temperature record preserved in lake sediments: an extension of the $\delta^{18}O_p$-air temperature relation into the past. Geochim. Cosmochim. Ac. 60: 4025–4036.

von Grafenstein U., Erlenkeuser H. and Trimborn P. 1999. Oxygen and carbon isotopes in modern fresh-water ostracod valves: assessing vital offsets and autecological effects of interest for palaeoclimate studies. Palaeogeogr. Palaeocl. 148: 133–152.

Wang C.H. and Yeh H.W. 1985. Oxygen isotopic composition of DSDP Site 480 diatoms: implications and applications. Geochim. Cosmochim. Ac. 49: 1469–1478.

Watanabe T., Naraoka H., Nishimura M. and Kawai T. 2004. Biological and environmental changes in Lake Baikal during the late Quaternary infreed from carbon, nitrogen and sulphur isotopes. Earth Planet. Sc. Lett. 222: 285–299.

Wolfe B.B., Aravena R., Abbott M.B., Seltzer G.O. and Gibson J.J. 2001a. Reconstruction of paleohydrology and paleohumidity from oxygen isotope records in the Bolivian Andes. Palaeogeogr. Palaeocl. 176: 177–192.

Wolfe B.B. and Edwards T.W.D. 1997. Hydrologic control on the oxygen-isotope relation between sediment cellulose and lake water, Taimyr Peninsula, Russia: Implications for the use of surface-sediment calibrations in paleolimnology. J. Paleolimnol. 18: 83–291.

Wolfe B.B., Edwards T.W.D., Aravena R. and MacDonald G.M. 1996. Rapid Holocene hydrologic change along boreal treeline revealed by $\delta^{13}C$ and $\delta^{18}O$ in organic lake sediments, Northwest Territories, Canada. J. Paleolimnol. 15: 171–181.

Wolfe B.B., Edwards T.W.D., Aravena R., Forman S.L., Warner B.G., Velichko A.A. and MacDonald G.M. 2000. Holocene paleohydrology and paleoclimate at treeline, north-central Russia, inferred from oxygen isotope records in lake sediment cellulose. Quaternary Res. 53: 319–329.

Wolfe B.B., Edwards T.W.D., Elgood R.J. and Beuning K.R.M. 2001b. Carbon and oxygen isotope analysis of lake sediment cellulose: methods and applications. In: Last W.M. and Smol J.P. (eds), Tracking Environmental Change Using Lake Sediments: Physical and Chemical Techniques. Kluwer Academic Publishers, Dordrecht, The Netherlands, pp. 373–400.

Wolfe B.B., Edwards T.W.D. and Hall R.I. 2002. Past and present ecohydrology of the Peace-Athabasca Delta, northern Alberta, Canada: water isotope tracers lead the way. PAGES News 10: 16–17.

Wooller M.J., Francis D., Fogel M.L., Miller G.H., Walker I.R. and Wolfe A.P. 2004. Quantitative palcotemperature estimates from $\delta^{18}O$ of chironomid head capsules preserved in Arctic lake sediments. J. Paleolimnol. 31: 267–274.

Xia J., Ito E. and Engstrom D.R. 1997. Geochemistry of ostracode calcite: Part 1: An experimental determination of oxygen isotope fractionation. Geochim. Cosmochim. Ac. 61: 377–382.

Yakir D. 1992. Potential use of O-18 and C-13 in cellulose for estimating the impact of terrestrial photosynthesis on atmospheric CO_2. Photosyn. Res. 34: 108–108.

Yu Z.C., Ito E., Engstrom D.R. and Fritz S.C. 2002. A 2100-year trace-element and stable-isotope record at decadal resolution from Rice Lake in the Northern Great Plains, USA. Holocene 12: 605–617.

5. ISOTOPES IN SPELEOTHEMS

FRANK McDERMOTT (frank.mcdermott@ucd.ie)
Department of Geology
University College Dublin
Belfield
Dublin 4, Ireland

HENRY SCHWARCZ (schwarcz@mcmaster.ca)
Department of Geography and Geology
McMaster University
Hamilton, Ontario
L8S 4M1, Canada

PETER J. ROWE (P.Rowe@uea.ac.uk)
School of Environmental Sciences
University of East Anglia,
Norwich
NR4 7TJ, UK

Key words: Speleothems, caves, palaeoclimate, oxygen isotopes, carbon isotopes, U–series, fluid inclusions.

Introduction

Interest in speleothems (secondary cave carbonates such as stalagmites) as recorders of continental palaeo–environments has increased markedly during the past decade, reflecting the need to provide reliable palaeoclimatic records from a range of continental settings. This renewed interest has been underpinned by important analytical advances that facilitate the acquisition of well–dated oxygen and carbon isotope data at a high spatial resolution, and by emerging methodologies for the extraction of stable isotope signals from speleothem fluid inclusions. Speleothems (mostly calcite, but occasionally aragonite) are deposited slowly by degassing of meteoric water–fed drips in caves. Most studies utilise stalagmites rather than stalactites or flowstones, because their simple geometry, relatively rapid growth rates and tendency to precipitate close to isotope equilibrium with the cave drip waters facilitates palaeoclimatic reconstruction. The rationale for interpreting oxygen isotopes in speleothems is broadly similar to that for other secondary carbonate archives on the continents such as lake carbonates and tufa, insofar as they reflect (i) the $\delta^{18}O$ of meteoric water (cave drips) and (ii) the temperature dependent water–calcite oxygen isotope fractionation. In common with most continental secondary carbonate deposits, the quantitative, and sometimes even the qualitative interpretation of isotopes in speleothems can be problematic because of the multiplicity of factors that potentially affect the $\delta^{18}O$ of drip waters. Importantly however, speleothems preserve several additional non–isotope climate sensitive signals (e.g.,

185

M.J. Leng (ed.), 2005. *Isotopes in Palaeoenvironmental Research.* Springer. Printed in The Netherlands.

carbonate petrography, annual band thickness, trace element ratios, luminescence and organic molecular signals) the details of which (Frisia et al. 2000; 2002; Baker and Barnes 1998; Fairchild et al. 2001) are beyond the scope of this review, but which contribute greatly to the correct interpretation of isotope–based speleothem records. Carbon isotope variations in speleothems are usually taken to reflect climate–driven palaeovegetation signals, but careful consideration of all possible fractionation effects is essential to avoid misinterpreting these data.

In this chapter we focus on the use of stable (oxygen, hydrogen, carbon) isotopes, with an emphasis on understanding the isotope systematics in the context of karst systems, and on the processes by which isotope signals are transferred from the atmosphere and near–surface environment (e.g., vegetation and shallow soil zone) through the karst system to the speleothem calcite. Key issues are those processes (climatic and non–climatic) that can affect the integrity of the recorded signal and its utility for palaeoclimate reconstruction. While the emphasis is firmly on isotope systematics and a process–based understanding, a small number of case studies are included to illustrate specific points, and to highlight some of the emerging issues in this rapidly evolving field. Also included are updates on recent methodological developments for the extraction of reliable stable isotope signals from speleothem fluid inclusions and the rationale for their interpretation.

Brief historical perspective
The potential of speleothems as palaeoclimatic recorders was first explored more than thirty–five years ago (Hendy and Wilson 1968; Thompson et al. 1974) and the emphasis in early studies was on providing palaeotemperature estimates. These pioneering researchers quickly recognised however that, in common with most continental carbonate archives, the recovery of palaeotemperatures from oxygen isotope data is inherently complex. This complexity arises because the $\delta^{18}O$ of speleothem calcite depends not only on the temperature at the time of deposition, but also on the $\delta^{18}O$ of the cave seepage waters, that in turn can reflect a multiplicity of climatic (and sometimes non–climatic) variables (e.g., McDermott 2004). These include temporal changes in the $\delta^{18}O$ of the vapour sources, changes in vapour track trajectories, rainout history and condensation, evaporative enrichment at the surface and in the epikarst zone, variable water transit times in the karst system, complex mixing histories and seasonal changes in the calcite deposition rate. As a consequence of these complexities, and because of the need to provide climate modellers with parameters against which they can test their models, the emphasis has now shifted away from attempting to reconstruct palaeotemperatures towards the provision of well–dated isotope records. The latter can, for example, provide a valuable means to detect and evaluate 'leads' and 'lags' in the climate system and to identify so–called 'teleconnections', because a major strength of speleothem isotope records is their robust chronology (see below). Another emerging application is the provision of data to test 'isotope–enabled' general circulation models (GCMs) that are currently being developed by several climate modelling groups. The latter models can predict secular changes in the $\delta^{18}O$ of precipitation at selected points on the Earth's surface, and the provision of estimates of the hydrogen and oxygen isotopic composition of palaeo–precipitation is a key priority in future research. Speleothems have a valuable role to play in providing such data, and in

particular their fluid inclusions are potentially important recorders of the H and O isotope ratios of palaeoprecipitation if, as discussed below, technical problems regarding their extraction and measurement can be overcome.

The cave environment

Caves are a product of karstification whereby relatively soluble rocks such as limestones are dissolved by downward percolating meteoric waters that have interacted with a soil zone containing elevated levels of CO_2 (Figure 1). Cave air temperatures remain essentially constant throughout the year in poorly ventilated caves (typically $\pm 1°C$), reflecting the high thermal inertia of host rocks. As a consequence, seasonal temperature variations are usually averaged out, and cave air temperatures are similar to the mean annual air temperature of the region above the cave. Cave air is characterised by high relative humidity (typically 95 to 100%) that minimises evaporation of cave drip water. Aside from areas close to cave entrances where lower humidity and air currents permit evaporation, secondary calcite deposition typically occurs by degassing of CO_2 from carbonate–saturated drip waters, and not by evaporation of water (Schwarcz 1986; Ford and Williams 1989). Details of carbonate equilibria in karst waters have been well documented elsewhere (e.g., Buhmann and Dreybrodt 1985), but a key point is that drip waters degas when they enter a cave, because cave atmospheres usually have lower CO_2 levels (0.06–0.6% vol.) compared with the overlying soil gas (0.1–3.5% vol.). Stalagmite growth rates are variable (usually 20–300 µm/year) and are influenced by the mean annual air temperature at the site, drip–water availability and the calcium concentration of drip waters (Genty et al. 2001; Polyak and Asmeron 2001; Fleitmann et al. 2004), in reasonable agreement with theoretical predictions (Dreybrodt 1988). In arid or weakly vegetated sites, growth rates are often lower than predicted by growth rate models, and growth hiatuses that compromise the reliability of age models can occur. The latter may in themselves provide valuable climatic signals, particularly in cases where subtle geochemical changes are recorded in the stalagmite related to a more arid period (e.g., McMillan et al. (in press)). At mid– to high–latitudes in the northern hemisphere, speleothem deposition typically ceases during glacials (Gordon et al. 1989; Baker et al. 1993), whereas continuous deposition throughout the glacial periods is a feature of many circum–Mediterranean and lower–latitude sub–tropical sites (McDermott 2004).

Recent advances in analytical techniques

The most important analytical developments that have occurred since the pioneering studies of the 1970's concern improvements in the provision of accurate and precise age estimates for speleothem isotope records. Unlike many other continental archives (e.g., lake sediments, peat bogs) speleothems are not amenable to dating by the radiocarbon method because a variable but difficult to quantify proportion (typically 5–20%) of their carbon is '^{14}C–dead' (Genty et al. 2001). Fortunately, speleothems are near–ideal material for dating by U–series, specifically using the 'daughter deficiency' $^{230}Th/U$ method, and the majority appear to behave as closed systems with respect to uranium and its decay products. Uranium contents in calcite speleothems are rather variable

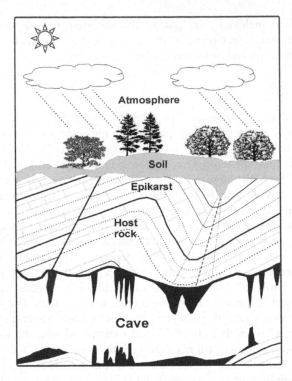

Figure 1. Schematic diagram of a cave system illustrating several of the variables that can influence the stable isotope signals recorded in speleothems. Columnar 'constant diameter' stalagmites depicted near the centre of the diagram are typically fed by slowly dripping vadose seepage water. Longer water residence times result in attenuation of seasonal $\delta^{18}O$ variations in rainwater, and drip water $\delta^{18}O$ values closely resemble those of the weighted mean $\delta^{18}O$ in precipitation. In arid regions, the O isotope ratio of precipitation may be modified by evaporation prior to infiltration, or may be biased by selective infiltration of more intense precipitation events (e.g., Bar–Matthews et al. (2003)). Judicious choice of speleothems for palaeoclimate studies using stable isotopes must take these and other factors (including cave conservation issues) into account.

(typically $0.1 - 10$ µg/g ^{238}U) and are about an order of magnitude higher in aragonite–rich speleothems (McDermott et al. 1999; McMillan et al. in press). Much of the renewed interest in the use of speleothems as palaeoclimatic recorders during the past decade can be traced to the development of Thermal Ionisation Mass–Spectrometry (TIMS) techniques for U–series measurements in the late 1980's (Edwards et al. 1988; Li et al. 1989) that quickly supplanted the established alpha–counting method. TIMS revolutionised U–series dating by allowing a very substantial (10–50 fold) reduction in sample size, and an order of magnitude improvement in analytical precision. Recent technological developments have led to a new generation of high–resolution magnetic sector Multi–Collector Inductive Coupled Plasma mass–spectrometers (MC–ICP–MS) with vastly improved ionisation efficiency for elements such as thorium and protoactinium (Shen et al. 2002; Richards et al. 2003). With the advent of these instruments and with improved within–analysis precision, new issues are beginning to emerge. These include the need for more thorough inter–laboratory comparisons to detect possible systematic errors in mixed ^{229}Th/^{236}U spike calibrations, the need for more careful correction for initial ^{230}Th in samples with modest amounts of detrital contamination, and the requirement for improved statistical methods to construct speleothem age models from the U–series data. In a minority of speleothems, the presence of annual visible or luminescent bands can provide independent verification of radiometric chronologies (e.g., Wang et al. (2001); Frisia et al. (2003); Hou et al. (2003); Fleitmann et al. (2004)), reducing many of the uncertainties outlined above.

During the past decade there have been numerous efforts to improve the spatial, and therefore the temporal resolution of O and C isotopic analyses in speleothems. A Laser–Ablation Gas–Chromatography Isotope Ratio Mass Spectrometry (LA–GC–IRMS) system developed at Royal Holloway (University of London) for example, offers rapid in–situ O and C isotope analysis of speleothem carbonate at a spatial resolution of 250 μm (McDermott et al. 2001). Comparison of LA–GC–IRMS data with conventional isotope analyses of both dental–drilled and micro–milled traverses indicates that this laser system produces accurate and precise data (McDermott et al. 2001). Replicate analyses of standards indicate that the isotope data are reproducible to better than 0.1‰ for δ^{13}C and 0.2‰ for δ^{18}O. While this system offers only a modest improvement in spatial resolution (approximately five–fold) relative to conventional dental drilling methods, data acquisition is rapid and automated.

Several laboratories have invested recently in micro–milling systems to provide samples for conventional stable isotope analysis at a much–improved spatial resolution compared with that achievable using a conventional dental drill. Frappier et al. (2002) achieved a sampling resolution of just 20 μm by micro–milling, corresponding to a weekly to monthly temporal resolution in a stalagmite from Belize. Importantly, these high–resolution data enabled the recognition of large amplitude (11‰), rapid (sub–seasonal) fluctuations in δ^{13}C, interpreted to reflect variations in the El Nino/Southern Oscillation (ENSO). Micro–sampling at such a high spatial resolution requires that the speleothem growth laminae are perfectly flat and parallel to the micro–milled tracks, and resolutions of 70–100 μm are probably routinely achievable.

Ion micro–probes offer excellent spatial resolution (c. 25 μm) for δ^{18}O analysis (Kolodny et al. 2003), but the relatively poor analytical precision and standardisation problems that characterise the current generation of instruments (c. ± 0.5‰) restricts their use to the study of high–amplitude changes in δ^{18}O and/or climate transitions. Overall, it appears that micro–milling followed by conventional stable isotope analysis currently offers the best route forward, because although labour intensive, it offers a combination of excellent spatial resolution and analytical precision.

Oxygen isotopes in speleothems

Assuming that speleothem calcite is deposited at or close to oxygen isotope equilibrium with cave dripwater (but see discussion below), the δ^{18}O of the precipitated calcite reflects both the δ^{18}O of the dripwater and the temperature at which calcite deposition occurs. Unfortunately, in common with other continental secondary carbonate archives (e.g., lacustrine carbonates), interpretation of the oxygen isotope signal in terms of palaeotemperature is not straightforward, because the temperature dependence of $\delta^{18}O$ in rainfall ($d\delta^{18}O_p/dT$) is variable and site dependent (Schwarcz 1986; McDermott et al. 1999; Serefiddin et al. 2004). In order to evaluate more critically the extent to which drip–water $\delta^{18}O$ might reflect cave air temperatures, we have compiled the available published oxygen isotope data for modern cave waters and recent carbonate precipitates for those caves whose mean annual air temperature appears to be well documented (Table 1). This compilation encompasses a wide range of cave sites and climatic regimes, with mean annual temperatures ranging from 2.8°C (Soylegrotta, Norway; Lauritzen and Lundberg (1999)) to 26.6 °C (Harrison's Cave, Barbados; Mickler et al. (2004)). Overall there is a strong correlation between drip water $\delta^{18}O$ and cave air temperature (Figure 2), reflecting the well–known spatial meteoric water $\delta^{18}O$ – air temperature relationship, although the slope is lower than that of most regional meteoric water data–sets (e.g., Rozanski et al. (1993)). Depending on the site however, $d\delta^{18}O_p/dT$ could be greater than, equal to, or less than the temperature dependence of $\delta^{18}O$ in calcite (ct) deposited in speleothems ($d\delta^{18}O_{ct}/dT$). The latter is approximately –0.24‰ °C^{-1} at 25°C (O'Neil et al. 1969). A similar number of cases have been documented where $d\delta^{18}O_{ct}/dT$ appears to be positive (Goede et al. 1990; Burns et al. 2001; Onac et al. 2002) and negative (Gascoyne 1992; Hellstrom et al. 1998; Frumkin et al. 1999a; 1999b). In principle though $\delta^{18}O_{ct}$ could fortuitously remain invariant to an increase in mean annual air temperature if $d\delta^{18}O_p/dT$ was close to 0.24‰ °C^{-1}, because $d\delta^{18}O_p/dT$ would then be cancelled out by $d\delta^{18}O_{ct}/dT$. Such cases appear to be rare in the literature but we note that the slope of the data in Figure 2 implies a spatially–determined $d\delta^{18}O_p/dT$ value close to 0.26‰ °C^{-1}, and there is no strong relationship between $\delta^{18}O_{ct}$ and cave air temperature in the compiled data (not shown). In general therefore, it remains impossible unambiguously to relate changes in $\delta^{18}O_{ct}$ to changes in mean annual temperature, especially during intervals that lack obvious first–order climate transitions (e.g., glacial to interglacial transitions). These uncertainties underline the need for additional proxy information from the same stalagmite (e.g., annual layer thickness variations, growth–rate changes) to underpin the interpretation of $\delta^{18}O$ (McDermott et al. 1999; Fleitmann et al. 2004; Fairchild et al. in press). In the Fleitmann et al. (2004) study of annually laminated speleothems from Southern Oman for example, there is a good correlation between $\delta^{18}O$ and the thickness of annual layers, providing independent support for their interpretation that $\delta^{18}O$ primarily reflects changes in the amount of precipitation. In other regions (e.g., Israel) it has been documented that changes in $\delta^{18}O_p$ reflect the so–called 'amount effect', with lower $\delta^{18}O_p$ values associated with wetter years (Bar–Matthews et al. 1996). Overall therefore, in common with other continental secondary carbonate archives, the system is usually under–determined (e.g., Leng and Marshall (2004)), in the sense that there are several inter–related factors that can influence the $\delta^{18}O$ of speleothem calcite, and derivation of absolute temperatures is generally not possible. If in the future, reliable hydrogen and possibly oxygen isotope data can be recovered from speleothem fluid inclusions (see below), this situation will improve.

It is increasingly apparent that on centennial to millennial timescales, variations in speleothem $\delta^{18}O$ probably reflect changes in the $\delta^{18}O$ of precipitation as a consequence of atmospheric circulation changes rather than local temperature changes (Burns et al.

Table 1. Compilation of published oxygen isotope data for cave drip–waters and associated modern carbonate precipitates. Values in italics are ± uncertainties (1σ around the mean) in the oxygen isotope values where these have been given in the individual studies. * denotes that winter precipitation recharge values of $\delta^{18}O$ were used in the calculation of temperatures in Figure 3, °indicates experimental precipitates on glass slides from the work of Mickler et al. (2004).

Cave site	Modern calcite $\delta^{18}O$ (PDB)	Present–day drip water mean $\delta^{18}O$ (SMOW)	Cave temperature (°C)
Calcite precipitates			
Père Noël (Belgium)[1]	−5.5 ± 0.6	−7.20 ± 0.08	9.1
Søylegrotta (Norway)[2]	−7.3	−10.0	2.8
Ernesto (Italy)[3]	−7.0 ± 0.2	−9.0 ± 0.1	6.6
Frankcombe cave (Tasmania)*[4]	−4.0	−5.7 ± 0.1	8.3
B7 cave (Germany)[5]	−6.3	−8.4 ± 0.6	9.4
Lynd's cave (Tasmania)[6]	−4.1 ± 0.1		9.5
Little Trimmer cave (Tasmania)*[7]	−3.8	−5.68	9.5
Crag cave (Ireland)[8]	−3.5	−5.3	10.4
Flint Ridge–Mammoth cave (U.S.)[9]	−5.1 ± 0.3	−5.94	13.5
Clamouse (France)[10]	−4.9 ± 0.6	−6.2 ± 0.4	14.5
Peqiin (Israel)[11]	−5.2 ± 0.3	−5.50	16.0
Victoria Fossil Cave (S. Australia)[12]	−4.8	−4.98	16.8
Cango caves (S. Africa)[13]	−5.4	−5.44	17.5
Cold Air cave (S. Africa)[14]	−4.8	−4.0	18.8
Soreq (Israel)[15]	−5.4 ± 0.1	−5.0	19.0
Hoti Cave (Oman)[16]	−1.9 ± 1	−1.0 ± 1	23.0
Harrison's Cave (Barbados)°[17]	−5.3	−3.3 ± 0.3	26.6
Harrison's Cave (Barbados)°[17]	−4.0 ± 0.3	−3.3 ± 0.3	26.6
Harrison's Cave (Barbados)°[17]	−4.0 ± 0.4	−3.3 ± 0.3	26.6
Harrison's Cave (Barbados)°[17]	−4.6 ± 0.3	−3.3 ± 0.3	26.6
Aragonite precipitates			
Cold Air Cave (S. Africa)[18]	−2.93	−3.13	18.0
Nerja Cave (S. Spain)[19]	−4.60	−5.20	20.0

[1]Verheyden et al. (2000) and McDermott unpublished data, [2]Lauritzen and Lundberg (1999), [3,8,10]McDermott et al. (1999), [4]Goede et al. (1990), [5]Niggemann et al. (2003), [6]Xia et al. (2001), [7]Desmarchelier and Goede (1996), [9]Harmon et al. (1978), [10]McDermott et al. (1999) and Plagnes et al. (2002), [11,15]Bar–Matthews et al. (2003), [12]Desmarchelier et al. (2000), [13]Talma and Vogel (1992), [14,18]Repinski et al. (1999), [16]Burns et al. (1998), [17]Mickler et al. (2004), [19]Jimenez de Cisneros et al. (2003).

1998; Bar–Matthews et al. 1999; 2000; McDermott 2001; Yuan et al., 2004). A recent study of a five year record of $\delta^{18}O$ in precipitation ($\delta^{18}O_p$) from Tasmania for example

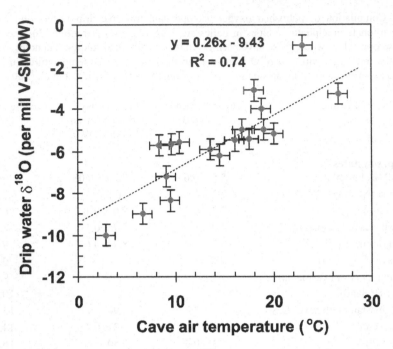

Figure 2. Relationship between cave drip water $\delta^{18}O$ and cave air temperature for data compiled in Table 1. There is a broad positive correlation between drip water $\delta^{18}O$ and cave temperature, although the slope of the data (0.26‰ $^{\circ}C^{-1}$) is lower than that of most meteoric water compilations (Rozanski et al. 1993). The slope defined by the data is approximately 0.26‰ $^{\circ}C^{-1}$, similar to, but of opposite sign to the calcite–water fractionation factor (approximately –0.24‰ $^{\circ}C^{-1}$ at 25°C, O'Neil et al. 1969). Three of the five data points that plot clearly above the regression line are from near–coastal sites (S.W. Ireland and two from Tasmania).

(Treble et al. 2005) indicates that there is little correlation between $\delta^{18}O_p$ and air temperature, and instead $\delta^{18}O_p$ is influenced strongly by synoptic patterns that influence air–mass trajectories. $\delta^{18}O$ in precipitation clearly exerts a first–order control on the δ^8O of cave drips, and the meteorological controls on this are discussed in detail elsewhere (e.g. Serefiddin et al. 2004; Treble et al. 2005). Here we note simply that the present–day spatial and seasonal variations in $\delta^{18}O_p$ arises from several so–called "effects" that include latitude, altitude, distance from the sea, amount of precipitation and surface air temperature (Rozanski et al. 1982; 1993; Gat 1996; Fricke and O'Neil 1999). On centennial to millennial timescales, additional factors such as (i) changes in the $\delta^{18}O$ of the ocean surface (ii) changes in the temperature difference between the oceanic vapour source area and the cave site (iii) shifts in moisture sources and/or storm tracks and (iv) changes in the proportion of precipitation derived from non oceanic sources, i.e., recycled from continental surface waters, must also be taken into account (see McDermott et al. (1999) for a discussion).

   One prerequisite for the successful use of $\delta^{18}O$ in speleothems as an indicator of $\delta^{18}O_p$ is that carbonate deposition occurred at or close to isotope equilibrium with the cave drip waters. It is often assumed that stalagmite calcite is deposited close to isotope equilibrium with cave drip waters in high humidity caves, but there have been relatively few critical evaluations of this assumption. The most striking result from our compilation of the available data from the literature (Table 1) is that while oxygen isotopes in speleothem calcite are typically in 'quasi–equilibrium' with their drip waters, many appear to be systematically heavier (Figure 3a) than those predicted by recently published equilibrium fractionation factors (Kim and O'Neil 1997). Some speleothem studies (e.g., Holmgren et al. (2003)) have used fractionation factors based on the early empirically–derived O isotope–temperature relationships (the Craig (1965) equation and versions thereof including the Anderson and Arthur (1983) equation). Curiously, these appear to yield temperatures that are closer to the expected (cave air) temperature for modern precipitate–drip water pairs (Figure 3b). In part this may reflect the fact that these early empirical equations were based on mixed calcite–aragonite mineralogies (Leng and Marshall 2004), because published equilibrium fractionation factors for aragonite (Grossman and Ku 1986) predict higher $\delta^{18}O$ in aragonite compared with calcite at a given temperature. It is likely that these early mixed mineralogy empirical calibrations are not appropriate for calcite speleothems. Instead, we prefer to base our assessment of present–day equilibrium deposition on the most recently published experimental calibration for calcite (Kim and O'Neil 1997). On this basis the $\delta^{18}O$ values for most modern cave calcites tend to plot above the predicted equilibrium values on a diagram of 1000 ln $\alpha_{(calcite–water)}$ vs. 1/T (Figure 3a) with typical enrichments of 0.5–1‰ relative to the predicted values. The reasons for this discrepancy are not well understood. Kinetic enrichments often accompany relatively rapid degassing of the drip waters (Mickler et al. 2004), but the relatively small and constant offset of the data from the Kim and O'Neil (1997) equilibrium fractionation line argues against kinetic effects, because in principle the latter should produce a wide range of enrichments. We note that seasonal biases in calcite precipitation rate cannot readily account for the systematic nature of the observed isotope shift in the modern cave carbonates, because given the range of climatic regimes represented, the latter should produce scatter around the equilibrium values rather than a unidirectional shift.

Two modern cave aragonite samples are included in our compilation (from Nerja Cave, southern Spain; Jimenez de Cisneros et al. (2003) and Cold Air Cave, South Africa; Repinski et al. (1999)). The sample from South Africa appears to lie on the equilibrium fractionation line for aragonite (Grossman and Ku 1986), whereas that from Nerja cave exhibits higher $\delta^{18}O$ relative to the equilibrium value predicted from its associated drip water and cave air temperature (Figure 3a). Thus, present indications are that while cave calcites are precipitated in quasi–equilibrium with their drip waters, a small (< 1‰) but systematic offset to higher $\delta^{18}O$ values appears to occur. If this effect is common, then small amplitude temporal variations in $\delta^{18}O$, in a Holocene stalagmite for example, might simply reflect non–equilibrium effects rather than a direct climate–driven rainfall $\delta^{18}O$ signal. This in turn has implications for the signal to noise ratio requirement for the identification of robust climate–related signals in time–series speleothem $\delta^{18}O$ records.

The effect of using different 'palaeotemperature equations' to calculate carbonate deposition temperatures (and to assess the extent to which equilibrium deposition has been achieved) is illustrated in Figure 3b. Clearly, the empirical Craig–derived

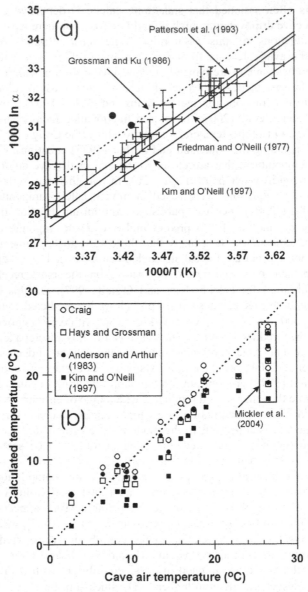

Figure 3. (a) Plot of 1000 * ln $\alpha_{(calcite-water)}$ vs. 1/T (K) for compiled modern cave calcite (crosses) and aragonite (filled circles) precipitate–drip water pairs (Table 1). Data represent a wide range of cave sites in climatic regimes with mean annual temperatures ranging from 2.8 to 26.6 °C. Error bars represent typical uncertainties on cave air temperatures and present–day $\delta^{18}O$ for compiled data (Table 1). Note that 1000 * ln $\alpha_{(calcite-water)}$ approximates to $\delta^{18}O_{calcite}-\delta^{18}O_{water}$. (b) Plot of calculated temperatures from modern cave precipitate–drip water pairs illustrating that the perception of equilibrium depends largely on the choice of 'palaeotemperature equation'. Data outlined by box in upper right part of diagram are for glass-slide precipitates (Mickler et al. 2004).

equations appear superficially to more closely recover the observed cave (air) temperature, whereas the Kim and O'Neil (1997) equation appears to yield temperatures that are too low. We interpret this to mean that most cave carbonates are not strictly in oxygen isotope equilibrium with their drip waters, as discussed above. Calculated temperatures have a range of about $4°C$, and so the choice of fractionation factor is critical in assessing the extent to which equilibrium has been achieved.

Demonstrating that modern carbonate is at or close to oxygen isotope equilibrium with cave drip waters is clearly a useful first step in evaluating the extent to which a particular cave system might be suitable for palaeoclimatic studies. More important however is the need to demonstrate that equilibrium was maintained over the growth history of a stalagmite. The criteria for recognising conditions of equilibrium deposition have been discussed previously (Hendy 1971; Schwarcz 1986). Briefly, they are (i) that $\delta^{18}O$ remains constant along a single growth layer while $\delta^{13}C$ varies irregularly, and (ii) that there is no correlation between $\delta^{18}O$ and $\delta^{13}C$ along a growth layer. In practice, consistent sampling along single growth layers is often difficult to achieve, not least because visible layers are often thinner along the flanks of stalagmites compared with their central growth axis. Nonetheless, the so–called 'Hendy criteria' are used widely by researchers as a check that carbonate was deposited at or close to isotope equilibrium with cave drip waters. In some cases it can be demonstrated that calcite deposited along the flanks of stalagmites exhibit kinetic fractionation effects, but that the material deposited close to the central growth axis may have been deposited in isotope equilibrium with the cave drip waters (Talma and Vogel 1992; Spötl and Mangini 2002). It is possible that carbon isotope ratios may be strongly affected by degassing and prior precipitation of calcite in the water flow–path, but these effects are detectable by combined stable isotope and trace element studies (see below).

Aside from issues concerning the deposition of carbonate in isotope equilibrium, it is important to evaluate the hydrogeological factors that influence the oxygen isotope ratios in the cave waters of individual drip systems in order to interpret correctly temporal changes in the oxygen isotope composition of speleothems. Selection of individual speleothems for palaeoclimatic studies depends on the goals of the research project, and the available temporal resolution in the record An understanding of the site–specific hydrological behaviour of cave drips is essential for (i) the optimal selection of speleothems that can record the isotope signals at the time resolution of interest and (ii) the correct interpretation of the isotope data in terms of climatic variables. The $\delta^{18}O$ of slow, so–called 'seepage flow' drip sites with a relatively large storage component usually reflect the mean annual $\delta^{18}O$ of precipitation (Yonge et al. 1985; Caballero et al. 1996) and speleothems deposited from such drips would be appropriate for palaeoclimate studies requiring a multi–annual resolution. Fast or 'flashy' drip sites on the other hand may exhibit large seasonal variations in $\delta^{18}O$ reflecting minimal storage and homogenisation in the epikarst. Such sites may exhibit seasonal undersaturation with respect to calcite (Baldini 2004), and are unlikely to produce the regular geometrically–simple columnar stalagmites favoured for palaeoclimatic reconstruction, but they may be of value if for example the goal is to reconstruct high–resolution records of seasonal changes in $\delta^{18}O_p$. At some arid and semi–arid circum–Mediterranean sites it has been demonstrated that recharge is dominated by heavy winter rainfall events (Cruz–San Julian et al. 1992; Bar–Matthews et al. 1996) and speleothems deposited from such sites may preserve seasonal biases in their $\delta^{18}O$. Additionally, seasonally variable increases in $\delta^{18}O$ may occur as a result of near–surface evaporative processes in arid and semi–arid sites (Bar–Matthews et al. 1996; Denniston et al. 1999a; 1999b). Another complication is that soil pCO_2 and drip water Ca contents often vary seasonally, with the result that calcite deposition rates also vary seasonally (Genty et al. 2001; Frisia et al. 2003). It is possible therefore that speleothem calcite might in some circumstances preserve seasonal biases, but this could be detected by detailed seasonal monitoring of the chosen drip sites to understand the factors controlling intra–annual variability in growth rates. These issues highlight the need for detailed site–specific present day monitoring studies to understand better the relationship between the palaeo–$\delta^{18}O$ signal preserved in speleothem calcite ($\delta^{18}O_{ct}$) and palaeoclimatic variability. Many of these uncertainties relating to the interpretation of $\delta^{18}O_{ct}$ could be reduced or eliminated if reliable fluid inclusion data are available to constrain contemporaneous changes in the $\delta^{18}O_p$ (see below).

Speleothems as recorders of teleconnections in the climate system

A major strength of speleothem isotope records is their robust chronology (Richards et al. 2003). Ages with a relative error of ± 0.5% are achievable for speleothems younger than about 100,000 years using modern mass spectrometers with high abundance–sensitivity. Perhaps the most feature of recent speleothem studies that serves to underline their remarkable potential as palaeoclimate archives is their ability to detect teleconnections in the climate system on millennial timescales (Wang et al. 2001; Spötl and Mangini 2002; Genty et al. 2003; Yuan et al. 2004). Wang et al. (2001) for example, demonstrated that oxygen isotope ratios in five partially overlapping stalagmites from Hulu Cave in China appear to be sensitive to the Dansgaard–Oeschger (DO) interstadial events documented in the Greenland ice–cores (Figure 4). $\delta^{18}O$ variations in the Hulu stalagmite were interpreted to reflect changes in the ratio of summer/winter precipitation. Importantly these results provide evidence for teleconnections in the climate system and indicate a link between the East Asian Monsoon and Greenland temperatures as predicted by recent modelling results (Chaing and Bitz in press). Similar teleconnections were noted in the Holocene record from Crag Cave (McDermott et al. 2001), although these probably reflect atmospheric circulation changes recorded contemporaneously in mid and high–latitudes.

Yuan et al. (2004) investigated the timing and duration of changes in the Asian monsoon over the past 160,000 years using oxygen isotope data for stalagmites from Dongge Cave in China, some 1200 km WSW of Hulu Cave referred to above (Figures 4 and 6). Of particular interest in the context of speleothems as recorders of teleconnections in the climate system is the remarkably good correspondence in $\delta^{18}O$ between stalagmites from Dongge and Hulu caves in times of contemporaneous growth during the last de–glaciation (e.g., Figure 7). The large amplitude of $\delta^{18}O$ variability in these Chinese speleothems indicates that they primarily record changes in $\delta^{18}O_p$, rather than temperature changes. The $\delta^{18}O$ of precipitation reaching the Hulu and Dongge sites largely reflects the progressive rainout of water vapour transported from the tropical Indo–Pacific oceanic source regions to southeastern China. Yuan et al. (2004) argue that

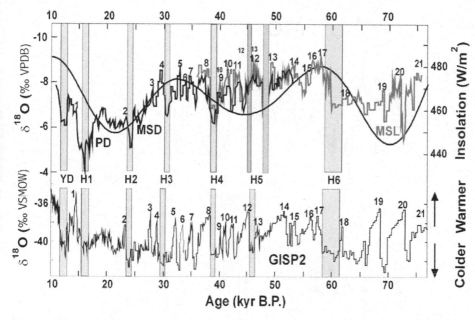

Figure 4 Oxygen isotope records for five stalagmites from Hulu Cave (32°30'N, 119°10'E), (China) exhibit good coherence between overlapping stalagmites (upper curve). The Dansgaard–Oeschger (DO) interstadial events (numbered) and the Heinrich events (lower curve) are clearly discernible in the Hulu data (upper curve), suggesting a link between Greenland temperatures and the behaviour of the East Asian Monsoon between about 75 ka and 11 ka. The large amplitude variability (about 5‰) in $\delta^{18}O$ in the Hulu stalagmites was interpreted to reflect temporal changes in the balance between precipitation associated with the summer and winter monsoon. Diagram reproduced with permission from Wang et al. (2001).

higher $\delta^{18}O$ during glacial times (Figure 7) reflect drier conditions, similar to that seen elsewhere (e.g., in Israel, Figure 5). Such millennial–scale changes are interpreted to reflect major and abrupt changes in the Asian Monsoon and appear to be contemporaneous with changes in air temperatures recorded by the high–latitude (Greenland) ice cores. By contrast, low–frequency changes (tens of millennia) recorded by the Dongge and Hulu stalagmites during the past 160,000 years appear to correlate with changes in N. hemisphere insolation (Yuan et al. (2004), not shown).

Speleothems as recorders of solar forcing of Earth's climate

One of the most remarkable results to emerge from oxygen isotope studies of speleothems in recent years has been the apparent evidence for solar forcing of Earth's climate on sub–Milankovitch timescales. Neff et al. (2001) argued that $\delta^{18}O$ in a Holocene stalagmite (H5) from Hoti Cave in northern Oman (23°05'N, 57°21'E) correlates strongly with atmospheric $\Delta^{14}C$ (derived from the tree–ring radiocarbon calibration dataset), and they interpreted this to reflect solar forcing of the tropical monsoon (Figure 8a). The high–resolution oxygen isotope record from speleothem H5 was interpreted as a proxy for variations in tropical circulation and monsoon rainfall

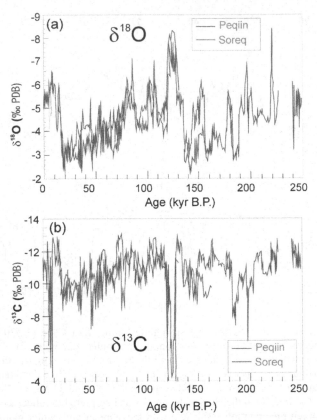

Figure 5. Speleothems from Israel have provided a remarkably coherent picture of regional climate variability over the past 250,000 years (Bar–Matthews et al. 1996; 1997; 1999; 2000; 2003). This diagram (Bar–Matthews et al. 2003) combines data from Soreq (31.45°N, 35.03°E, central Israel) and Peqiin caves (32.58°N, 35.19°E, northern Israel). Oxygen isotope ratios exhibit large amplitude variations that closely track changes in the marine $\delta^{18}O$ record (ice volume effect) for the Eastern Mediterranean reconstructed from the planktonic foraminifera *Globigerinoides ruber* (Bar–Matthews et al. 2003), with a relatively constant (5.6 ± 0.7‰) offset. Low $\delta^{18}O$ events coincide with high water stands in caves and with sapropel events in the Mediterranean, indicative of wet conditions. Sapropel event S5 (128–117 ka) is characterised by low $\delta^{18}O$ and high $\delta^{13}C$. The latter was interpreted to reflect rapid throughput of meteoric water during deluge events, with high $\delta^{13}C$ values a consequence of relatively little interaction between infiltrating water and soil CO_2. An alternative explanation for high $\delta^{13}C$ during sapropel event S5 recorded in a speleothem from a cave in Jerusalem (Frumkin et al. 2000) involved loss of vegetation and soil cover under hot–dry conditions, but this may be difficult to reconcile with the low $\delta^{18}O$ (wet conditions) recorded at Soreq and Peqiin. Reproduced with permission from Bar–Matthews et al. (2003).

over this part of Arabia. They argued that the primary control on centennial– to decadal–scale changes in tropical rainfall and monsoon intensity are variations in solar radiation, as reflected in the atmospheric (tree–ring) $\Delta^{14}C$ dataset. These results indicate that the Indian Ocean monsoon system was considerably stronger in the early to mid– Holocene than at present, indicating that the inter–tropical convergence zone (ITCZ)

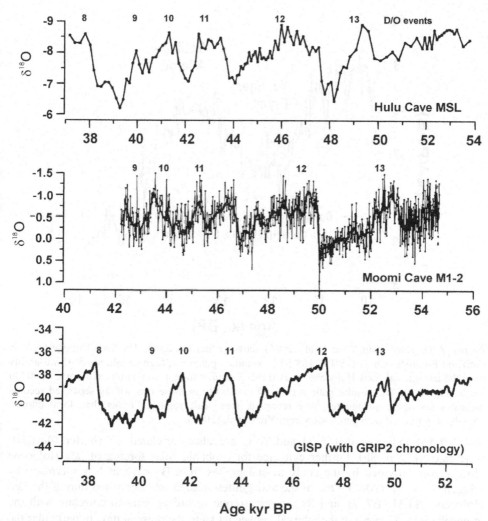

Figure 6. Stalagmite (M1–2) deposited during the last glacial at Moomi Cave (Socotra Island, Yemen, 12°30'N 54°E) exhibits relatively low–amplitude δ¹⁸O variability (middle curve) with low δ¹⁸O (wetter conditions) apparently associated with the Dansgaard/Oeschger events identified in the GISP core (plotted using GRIP 2001 chronology, lower curve). Also shown (upper panel) is the relevant part of the O isotope record from Hulu Cave in central China (Wang et al 2001). The three curves are plotted using their own independent chronologies, and that the age model for the Moomi Cave stalagmite has been revised to correct for a systematic error (Burns et al. 2003). A major finding of the Burns et al. (2003) study is that the tropics (ITCZ) vary in concert with N. hemisphere high–latitude temperature, with increased precipitation in the tropics associated with warmer conditions in Greenland. Reproduced with permission from Burns et al. (2003).

was situated distinctly further north than at present. Lower δ¹⁸O was associated with increased monsoonal precipitation because the so–called 'amount effect' is dominant in this region. The high–resolution interval of speleothem H5 (between approximately 8.3

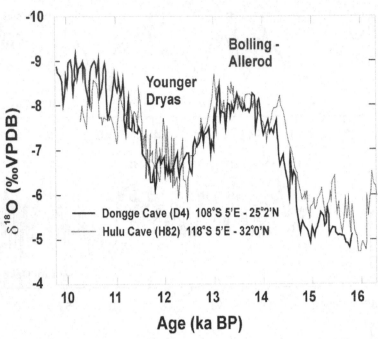

Figure 7. Diagram from Yuan et al. (2004) showing that stalagmite D4 from Dongge Cave in Guizhou Province, China (25°17', 108°5'E) exhibits a pattern of high–amplitude $\delta^{18}O$ variability over the late–glacial/early Holocene interval that is similar to that from Hulu cave, some 1200 km away (Figure 4). The similar pattern of $\delta^{18}O$ variability in these two widely separated records indicates that these stalagmites have recorded large–scale regional, rather than local climate signals. Reproduced with permission from Yuan et al. (2004).

and 7.9 ka) indicates that $\delta^{18}O$ and $\Delta^{14}C$ are also correlated on shorter (decadal) timescales (Figure 8b). Further evidence for probable solar forcing of late Holocene climate was provided by the study of stalagmites from B7 cave in NW Germany by Niggemann et al. (2003). One of the stalagmites studied was deposited during the late Holocene (STAL–B7–7), and its oxygen isotope record appears to correlate with the atmospheric $\Delta^{14}C$ curve in the interval between 4 ka to the present–day. In particular the interval around 2.8 to 2.4 ka that shows strong variability in atmospheric $\Delta^{14}C$, and for which solar forcing has been suggested previously (van Geel et al. 1998) shows a remarkably close correlation between $\Delta^{14}C$ and $\delta^{18}O$ (Figure 8c). Similar results for a longer interval of the Holocene were reported by Fleitmann et al. (2003) for a stalagmite (Q5) from a different cave in Oman (Qunf Cave, 17°N, 54°18E). Their high–resolution record spans most of the Holocene (10.3 to 2.7 ka and 1.4 to 0.4 ka), and like the record for N. Oman (Hoti Cave) discussed above, the $\delta^{18}O$ variability is interpreted to reflect changes in monsoon precipitation, reflecting the dominance of the amount effect. In the time interval after about 8.0 ka, the intensity of the monsoon system appears to have decreased (higher $\delta^{18}O$) as northern hemisphere summer insolation decreased gradually. In the early Holocene portion of the Q5 record (before c. 8 ka), changes in the monsoon intensity (inferred from speleothem $\delta^{18}O$) appear to be in phase with Greenland

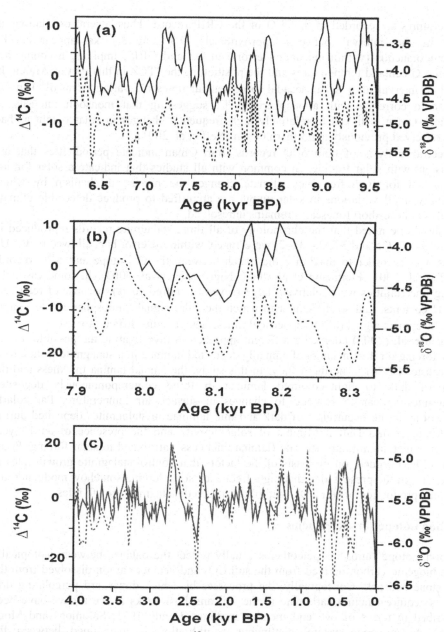

Figure 8. (a) The coherent variation of $\delta^{18}O$ and atmospheric (tree–ring) $\Delta^{14}C$ in the Holocene stalagmite H5 (Hoti Cave, Oman) over the time interval between about 9.5 and 6 ka. It was argued that variations in solar output affect the intensity of monsoonal precipitation (Neff et al. 2001). (b) Details of the correlation over a short portion of the Hoti Cave record. (c) The close correlation between $\delta^{18}O$ in the Holocene stalagmite STAL–B7-7 (B7 Cave, Sauerland, Germany) and atmospheric $\Delta^{14}C$ (Niggemann et al. 2003). Dashed curves are for $\delta^{18}O$ and solid curves represent $\Delta^{14}C$. Redrawn after Neff et al. (2001) and Niggemann et al. (2003).

temperatures as recorded in the δ^8O of the GRIP record. Thus several oscillations in $\delta^{18}O$ in the interval between approximately 9.3 and 8.7 ka appear to be contemporaneous (within the dating errors) in Q5 and GRIP, implying a connection between Greenland temperatures and the position of the ITCZ in this part of Arabia. In the time interval after 8 ka, decadal to centennial timescale variations in $\delta^{18}O$ in Q5 appear to correlate with atmospheric $\Delta^{14}C$, suggesting that monsoon intensity in southern Oman is linked to relatively high–frequency changes in solar output as had been observed previously in northern Oman (Neff et al. 2001).

Spectral analysis of both $\delta^{18}O$ records from Oman indicate periodicities that are consistent with solar forcing. In common with all studies that indicate a solar forcing mechanism for high–frequency climate change, the precise mechanism by which relatively small variations in solar output are amplified to produce detectable climate signals on these short timescales remains unresolved.

It should be noted that the chronology of all three stalagmite records reproduced in Figure 8 (H5, Q5 and STAL–B7–7) were 'tuned' within the error bars allowed by the U–series ages in order to maximise the match between the O isotope and $\Delta^{14}C$ records (Neff et al. 2001; Fleitmann et al. 2003; Niggemann et al. 2003). In some cases the dating uncertainties were relatively large (several hundred years), because of relatively low U contents, and so the extent to which the solar signal (atmospheric $\Delta^{14}C$) might lead the climate signal ($\delta^{18}O$) on decadal timescales remains difficult to assess.

Frisia et al. (2003) adopted a different approach in investigating the possible role of solar forcing on the thickness of annually deposited laminae in a stalagmite from Grotta di Ernesto (ER76) in northern Italy. In this study, the annual lamina thickness and the ratio of dark to light–coloured laminae in three contemporaneously deposited stalagmites appear to respond to changes in surface air temperature. Particularly noteworthy is the occurrence of dark and thin laminae in stalagmites deposited during the Maunder and Dalton Minima of solar activity and the presence of an 11–year cyclicity in stalagmite growth rate (lamina thickness), attributed to solar forcing. Frisia et al. (2003) argued, on the basis of the factors that control stalagmite growth rates at these sites in the present–day that high–frequency solar forcing somehow modulated the seasonal production of soil CO_2, that in turn affected stalagmite growth rates.

Carbon isotopes in speleothems

Carbon isotope ratios in speleothems usually reflect the balance between isotopically light biogenic carbon derived from the soil CO_2 and heavier carbon dissolved from the limestone bedrock. Conceptually, the processes by which downward percolating drip waters acquire calcium carbonate in the soil and host–rocks above a cave have been described in terms of two end–member systems (Hendy 1971; Salomons and Mook 1986). In an open–system, continuous equilibration is maintained between the downward percolating water and an infinite reservoir of soil CO_2 resulting in a progressive increase in the bicarbonate content of the water as it dissolves more limestone in the unsaturated zone. Under these conditions, the $\delta^{13}C$ of the dissolved species reflects the carbon isotopic ratio of the soil CO_2, with no detectable contribution to the carbon isotope signal by the carbonate host–rock. In a C3 plant system, the $\delta^{13}C$

of the dissolved inorganic carbon (DIC) in the percolating solution is typically in the range −14 to −18‰ when the solution reaches saturation with respect to $CaCO_3$, depending on soil pCO_2 and temperature (Hendy 1971; Salomons and Mook 1986; Dulinski and Rozanski 1990).

By contrast, in a closed system the downward percolating water loses contact with the soil CO_2 reservoir as soon as carbonate dissolution commences (Hendy 1971; Salomons and Mook 1986). CO_2 consumption in the carbonation reaction $H_2O + CO_2 = H_2CO_3$ limits the extent of limestone dissolution, because in this model the soil CO_2 reservoir is finite. Under closed–system conditions, the carbon isotope ratio of the host–rock exerts a strong influence on that of the DIC. For a C3 system with soil gas $\delta^{13}C$ of c. −23‰ and a host limestone with $\delta^{13}C$ of +1‰, the DIC $\delta^{13}C$ is typically c. −11‰. In reality most natural systems are likely to be partially open, with the $\delta^{13}C$ of DIC between that predicted by the end–member models (Dreybrodt 1988). Radiocarbon data on independently (e.g., U–Th) dated stalagmites can be used to calculate the mass fraction of ^{14}C–dead carbon and thereby constrain the balance between open and closed–system dissolution, although other effects such as ageing of organic matter in the overlying soil reservoir must also be taken into account at some sites (e.g., Genty et al. (2001)).

In arid regions, changes in the $\delta^{13}C$ values of speleothem calcite have been interpreted to reflect climate–driven changes in vegetation type (e.g., C3 versus C4 dominated plant assemblages (Dorale et al. (1992); (1998); Bar–Matthews et al. (1997); Frumkin et al., (2000)). In these regions, relatively large shifts in $\delta^{13}C$ can occur because soil respired CO_2 in equilibrium with a C3 dominated plant assemblage has $\delta^{13}C$ in the range −26 to −20‰, while that in equilibrium with C4 vegetation is heavier ($\delta^{13}C$ of −16 to −10‰). These differences produce distinctive ranges in $\delta^{13}C$ in secondary carbonates (typically −14 to −6‰ for carbonates deposited in equilibrium with CO_2 respired from C3 plants, and −6 to +2‰ for that from C4 plants). In stalagmites from Israel (Figure 5), sapropel event S5 (128-117 ka) is characterised by low $\delta^{18}O$ and high $\delta^{13}C$. The latter was interpreted to reflect rapid throughput of meteoric water during deluge events, with high $\delta^{13}C$ values a consequence of relatively little interaction between infiltrating water and soil CO_2.

The interpretation of carbon isotopes in moist temperate zones that lack C4 vegetation is necessarily more tentative, but occasionally a clear climate–related signal can be retrieved. As an example from the recent literature, Genty et al. (2003) noted that stalagmites deposited during the late glacial in the south of France exhibit $\delta^{13}C$ values that are much higher than in those deposited during the Holocene. These differences were attributed to changes in the relative proportions of atmospheric and biogenic (light) carbon. This interpretation implies that periods of climatic amelioration promote the production of soil biogenic CO_2, resulting in lighter carbon isotope ratios in the speleothem calcite (Figure 9). Data for the Grotta di Ernesto stalagmites (McDermott et al. 1999; Frisia et al. 2003) indicate that $\delta^{13}C$ values tend to be higher during episodes of thinner laminae in the mid–nineteenth century compared with those associated with thick laminae deposited during the last 50 years. These variations were interpreted to reflect decreases in the production rate of soil CO_2 giving rise to higher $\delta^{13}C$ values during the cooler intervals (see below). Critically, many temperate–zone speleothems exhibit $\delta^{13}C$ values > −6‰ (Figure 10); higher than predicted to be in equilibrium with

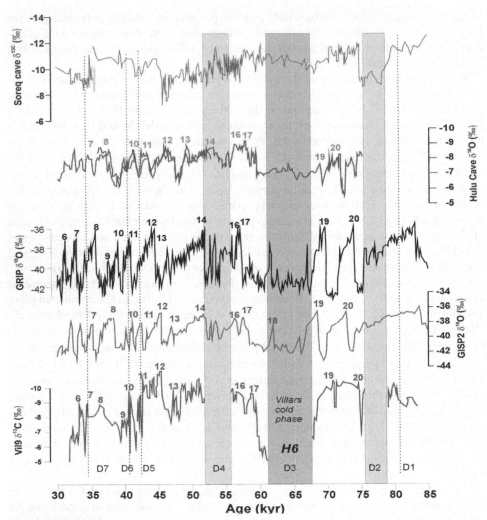

Figure 9. The lower curve shows $\delta^{13}C$ in stalagmite 'Vil9' from Villars Cave in southwest France (45°N, 0.50'E) after Genty et al. (2003). Deposition of Vil9 continued through most of the last glacial, apart from major hiatuses at 78.8–75.5 ka (D2), 67.4–61.2 ka (D3) and 55.7–51.8 ka (D4). The regular increase in $\delta^{13}C$ into event D3 ('Villars cold phase') and the decrease following re-establishment of speleothem growth points to a c. 6,000 year long cold phase with decreased vegetation intensity, leading to higher $\delta^{13}C$. Also shown is the Hulu Cave dataset (Wang et al. 2001). The poor correlation between the Vil9 and Soreq (Israel) records suggests that the latter reflects a Mediterranean (Bar–Matthews et al (1999); Vaks et al. (2003)) rather than a N.Atlantic signal. Reproduced with permission from Genty et al. (2003).

the expected C3 vegetation (Baker et al. 1997). Processes within the cave including evaporation and rapid degassing of cave drip waters may cause kinetic fractionation (Hendy 1971). Calcite precipitation in the unsaturated zone above the cave may also produce heavy carbon isotope signatures (Baker et al. 1997; Genty and Massault 1997).

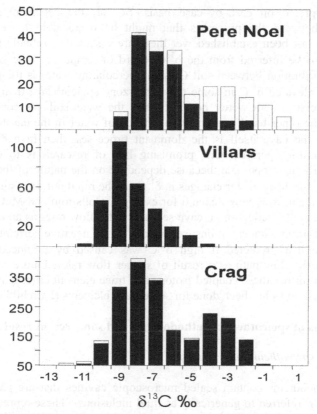

$\delta^{13}C$ ‰

Figure 10. Histograms of published carbon isotope data for three temperate–zone stalagmites from cave sites where C4 vegetation probably never existed in the late Quaternary. Filled columns are for a stalagmite deposited during the last glacial at Villars, S.W. France, (Figure 9, Genty et al. 2003) and Holocene stalagmites from Père Noël, Belgium and Crag, S.W. Ireland (Verheyden et al. 2000; McDermott et al. 1999). Unfilled segments represent modern precipitates (e.g., soda–straws, incipient stalagmites) from these caves where data are available (Père Noël and Crag Cave). Some $\delta^{13}C$ values are higher than predicted to form in equilibrium with soil CO_2 associated with the C3 vegetation at these sites (> –6 ‰). Mg/Ca ratios are available for two of these stalagmites (Crag and Père Noël), and in both, high $\delta^{13}C$ appears to be associated with elevated Mg/Ca (Figure 11). This is interpreted to reflect simultaneous degassing and prior calcite precipitation of the drip waters in the flow–paths above the stalagmites, producing a combination of high $\delta^{13}C$ and high Mg/Ca. In the absence of Mg/Ca data, high $\delta^{13}C$ could be misinterpreted as a climate–driven vegetation signal in palaeo–records.

The latter process appears may be important in samples for which the highest $\delta^{13}C$ values also exhibit high Mg/Ca (Figure 11), indicative of prior calcite precipitation from the drip waters.

In summary, the interpretation of carbon isotopes in regions where changes in the proportion of C3 and C4 plants can be verified independently (e.g., from pollen data) is relatively straightforward. In temperate regions that lack natural C4 vegetation however, the interpretation of carbon isotopes in speleothems remains more difficult, and the data are often interpreted on case–by–case basis. So far, the geochemical criteria for distinguishing between the processes that might be responsible for carbon isotope variations have not been established, yet these are essential if reliable palaeoclimatic information is to be inferred from the $\delta^{13}C$ record of temperate–zone speleothems. If incomplete equilibration between soil CO_2 and percolating water is the primary factor responsible for elevated $\delta^{13}C$ in some temperate zone speleothems, then elevated $\delta^{13}C$ should be associated with wetter periods, when the water/soil gas contact times are shorter. If, on the other hand, seasonal evaporation of water in the unsaturated zone or perhaps within the cave itself is the dominant processes, then high $\delta^{13}C$ should be associated with drier periods. One promising line of research is to combine trace element and carbon isotope data, because depending on the nature of the covariations, several possible mechanisms for changes in $\delta^{13}C$ can be ruled out. In a study of a 31,000 year old speleothem from New Zealand for example, Hellstrom and McCulloch (2000) were able to rule out a reduction in cave seepage water flow rates as an explanation for elevated $\delta^{13}C$. Barium concentrations exhibited a strong negative correlation with $\delta^{13}C$, the opposite to that predicted if high $\delta^{13}C$ was caused by enhanced prior calcite precipitation in the flow path as a result of slower flow rates. Future research should seek to develop further these coupled isotope and trace element criteria and to underpin these with modelling as has been done for some trace elements (Fairchild et al. 2000).

Fluid inclusions in speleothems: methodologies and some recent results

Significance of speleothem fluid inclusions

Almost all speleothems contain sealed microscopic cavities that are partly or wholly filled with water, referred to generically as fluid inclusions. These represent a valuable archive of palaeo–water. Individual speleothems often preserve intervals of inclusion–poor calcite that appear to alternate with depositional episodes in which abundant fluid inclusions are uniformly and densely dispersed through the calcite. Macroscopically, the presence of fluid inclusions is revealed in thick sections or cut slabs by a milky opacity, typically distributed as layers alternating with clear, apparently inclusion–free calcite. Zones of high fluid inclusion density may correspond to periods when growth rates are changing rapidly, resulting in frequent sealing off of micropores, rather than their being filled by calcite growth from their base. Fluid inclusions in stalagmites and flowstone tend to be in the form of ellipsoidal or irregular tubes, oriented parallel to the crystallographic *c* axis of the host calcite crystal (Figure 12). Water trapped as inclusions in speleothems makes up, on average, about 0.1% by weight of speleothems, ranging from c. 0.05 to 0.5 wt. %.

In detail, the mechanism of formation of inclusions is poorly understood, but is presumably the result of closing over of intercrystalline spaces on the growth surface of the speleothem (Kendall and Broughton 1978). Inclusions range in size from a few hundred nanometers to several cubic centimeters, but the majority range from 1 to 100 µm in maximum dimension. The fluid is usually a two–phase assemblage, a lower zone of water capped by a bubble of included air (Figure 12).

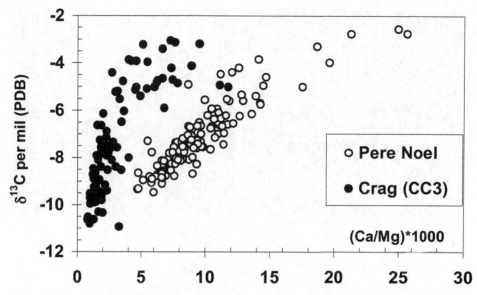

Figure 11. Diagram showing the co–variation of $\delta^{13}C$ with Mg/Ca ratios (both measurements made on the same sample aliquots from stalagmite CC3 (Crag Cave, S.W. Ireland, McDermott et al (1999)) and a stalagmite from Père Noël cave in S.E. Belgium (Verheyden et al. (2000)). Speleothems such as these from temperate zone caves that exhibit elevated $\delta^{13}C$ (> –6‰) associated with high Mg/Ca are interpreted to reflect prior degassing and calcite precipitation (increasing $\delta^{13}C$ and Mg/Ca ratios) in the water flow path above the stalagmite. Thus, high $\delta^{13}C$ values are not necessarily related to climate–driven vegetation change above the cave. For this reason caution should be exercised in interpreting carbon isotope variations, and it is recommended that Mg/Ca ratios be measured on the same sub–samples to allow detection of such degassing/prior calcite precipitation trends. Other temperate–zone speleothems exhibit a wide range in $\delta^{13}C$, with little or no accompanying change in Mg/Ca (e.g., at Grotta di Ernesto, N. Italy, McDermott et al. (1999)), and so provide more robust evidence for climate–driven vegetation change above the cave.

Preservation of the fluid inclusion isotope signals

Hendy (1971) observed that for speleothems formed at equilibrium with their drip water, their temperature of formation could be determined from the oxygen isotope fractionation between calcite and the drip water. Schwarcz et al. (1976) suggested that fluid inclusions might provide samples of that water, but noted that over the multi–millennial history of a speleothem it was likely that the $^{18}O/^{16}O$ ratio would have changed due to re–equilibration with calcite in the wall of the inclusion. This might occur for example by diffusive exchange around the perimeter of the inclusion, or by dissolution–reprecipitation at the water–calcite interface. However the extent to which re–equilibration of oxygen isotope ratios occurs remains poorly understood. Theoretical calculations suggest that diffusive exchange of oxygen isotopes in a water–calcite system should proceed very slowly at Earth surface temperatures, approaching isotope equilibrium by only one part in 10^3 on a timescale of 10^6 years for an inclusion with a

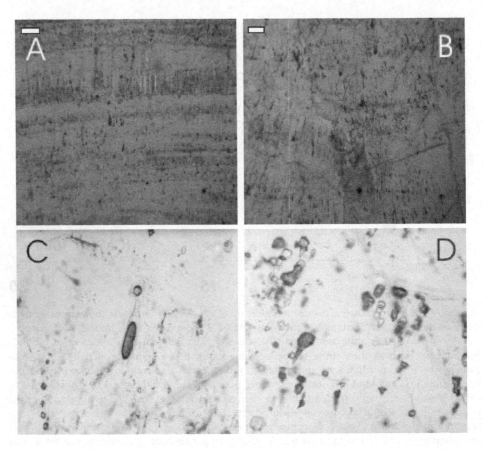

Figure 12. Photomicrographs showing fluid inclusions in speleothems. Images A and B show typical distribution of fluid inclusions in relation to (sub-horizontal) growth laminae. Inclusions are often 'thorn shape' with their long axis parallel to the crystallographic C axis of the calcite crystals. White bars in images A and B are 200 microns long. C and D are close-up images of inclusions, some of which have a trapped gas bubble. Inclusions are typically 10-30 microns in width in these images.

radius $<10^{-2}$ μm (P. Dennis, pers. comm. to PJR). For example, Rye and O'Neil (1968) noted that only 26–29% of possible oxygen isotope exchange had occurred between fluid inclusion water and the 38 Ma hydrothermal calcite host deposits at Providencia, Mexico. It appears therefore that this mechanism would produce negligible modification of the $\delta^{18}O$ of fluid inclusion waters on Pleistocene timescales. Dissolution–reprecipitation process might in principle affect the $\delta^{18}O$ of the fluid inclusion water and is likely to be concentrated in the recesses of irregular shaped inclusions (many are thorn–shaped or have triangular cross sections) as their morphology gradually modifies towards shapes that minimise surface area and energy. The extent to which this process alters the $\delta^{18}O$ of the fluid inclusion water depends on inclusion volume and age, and the relative temperatures at time of formation and alteration. However, no evidence of such alteration has been observed during petrographic examination of large numbers of speleothems of Pleistocene age (A. Kendall, pers. comm. to PJR). These observations, combined with the slow diffusive exchange of oxygen isotopes in the water–calcite system suggests that $\delta^{18}O$ re–equilibration is a less acute problem than anticipated in the early studies (see example below however), but further investigation of the issue is certainly desirable.

By contrast with the $^{18}O/^{16}O$ ratio, there is more confidence that the D/H ratio of the trapped water should remain intact after entrapment, because no opportunity exists for isotope exchange within a speleothem. In order to avoid possible problems associated with re–equilibration of the oxygen isotope ratios, Schwarcz et al. (1976) measured D/H ratios, and then calculated the initial oxygen isotope ratios of the fluid inclusion water using a meteoric water line relationship. Today, drip waters from many caves have been shown to resemble the global meteoric water line of Craig (1961) for which $\delta D = 8 \, \delta^{18}O + 10$. Schwarcz et al. (1976) assumed the same relationship for ancient waters and obtained temperatures for deposition of speleothems from various sites in North America. A subsequent study by Harmon et al. (1979) showed that, using a deuterium excess (d) of 10‰, many samples yielded impossibly low palaeotemperatures (<0°C). This problem could be avoided if $0 < d < 10‰$ during Ice Ages, as had been shown on theoretical grounds by Merlivat and Jouzel (1979), and as observed recently in high–latitude ice cores (Stenni et al. 2004).

In general, the emphasis in speleothem fluid inclusion research has now shifted from attempts to calculate absolute palaeotemperatures to the more attainable goal of reconstructing temporal changes in the $\delta^{18}O$ of palaeo–rainfall. Provided that drip waters are not evaporated, their isotope composition will, as discussed above, reflect that of the local rainwater, and the global distribution of isotopes in precipitation is amenable to modelling by isotope–enabled GCMs and so provide a useful test for such GCMs. If isotope ratio measurements of the inclusions were sufficiently precise, it should be possible to determine the deuterium excess in precipitation, which can be interpreted in terms of temperature and humidity at the moisture source areas. Documentation of this parameter through time for any given location would provide insights into vapour source changes under different climatic regimes. The noble gas content of inclusions, if determined with sufficient precision (e.g., Stuart et al., (1995)), could in principle be used as an independent thermometer recording cave temperature through time using the principles outlined by Stute and Schlosser (1993). Preliminary experiments at McMaster University, however, suggest that it may be difficult to correct for the "excess air" component.

Extracting inclusion water for isotope analysis – a historical perspective

Aside from a minority of speleothems that contain fluid inclusions large enough to permit the water to be withdrawn using a syringe (Genty et al. 2002), the recovery of inclusion water from the speleothem calcite without fractionating the oxygen and hydrogen isotopes in the process has proven to be an exceptionally difficult task. The historical development of the analytical methodologies is outlined here in order to provide a context for the more recent work. In the pioneering studies at McMaster University (1974 –1983), fluid inclusions were extracted by placing a saw–cut, c.1g block of calcite inside a stainless steel tube sealed with gaskets to a vacuum line, and crushing the block with a hydraulic press (Harmon et al. 1979). A single crushing typically yielded 10–50 μl water, adequate for D/H analyses. In later experiments Yonge (1982) progressively heated samples and continuously converted the released water to hydrogen gas using metallic U at 800°C, allowing him to obtain a thermal release history. Water was also extracted by calcining samples at c. 800°C ($CaCO_3 \rightarrow CaO + CO_2$), and by separating water from CO_2 on a vacuum line (Schwarcz and Yonge 1983). These early studies established the two primary techniques in use for extracting inclusion water for isotope measurement that are still employed today, namely thermal decrepitation and crushing. Both are carried out under high vacuum ($<10^{-5}$ torr) in order to minimise the mean free path of the water vapour molecules during recovery from the calcite and subsequent manipulation steps.

Initial studies by the McMaster group encouragingly showed that δD of fluid inclusions from small stalactites ("soda straws") agreed with that of coeval drips in caves. In addition, it was demonstrated that replicate analyses of δD from the same level in a single flowstone from Bermuda agreed within the precision of analysis (c. 1‰). In these experiments $\delta^{18}O$ of the water was analysed by reaction with BrF_5 and conversion of the resultant O_2 gas to CO_2 by reaction with graphite (M. Yamamoto, unpublished data). Replicate analyses within single growth layers of a flowstone from Bermuda agreed within analytical error (± 0.2‰), but $\delta^{18}O$ differed between layers. These analyses showed that waters trapped in Pleistocene stalagmites from N. America did not lie on the meteoric water line, but were displaced to an extent unaccountable for by shifts in 'd' alone (Schwarcz 1986). This was interpreted to reflect exchange between water and calcite as a result of change in environmental temperature between the time of entrapment and today, underlining persistent concerns about re–equilibration of the oxygen isotope ratios (but not hydrogen) of the fluid inclusion water.

Importantly, the McMaster experiments also revealed that the D/H ratios of water released by thermal decomposition of calcite were depleted by 22‰ with respect to cave drip waters. No explanation was given for this effect, and the offset was simply corrected for in calculations of palaeotemperatures. It was also observed that, with progressive decrepitation up to c. 500°C, later (higher–T) fractions of water were systematically more D–depleted. On calcining the residue after such a decrepitation sequence, Yonge (1982) found that an additional amount of water was released, comparable in volume to that liberated through all prior decrepitation steps, and which was also D–depleted by up to 80‰ with respect to drip water. The significance of this high–temperature water release is discussed further below in the context of new insights into the various forms of water that appear to be present in speleothem calcite.

In their initial paper, Schwarcz et al. (1976) used the relationship between fluid inclusion–derived temperatures and inferred $\delta^{18}O$ of drip waters to determine the local, long–term relation between these two quantities. They observed that the T–dependence of $\delta^{18}O$ of meteoric precipitation $d(\delta^{18}O_p)/dT$, varied between 0 and 0.3‰/°C and was always less than the short–term average of c. 0.7‰/°C observed today in North America and other temperate regions (Rozanski et al. 1993). Harmon et al. (1983) observed a significant drop in average δD over North America during glacial periods. In a study of a speleothem from Coldwater Cave, Iowa, although $\delta^{18}O$ of calcite was almost constant throughout the growth history of this 83 cm long stalagmite, δD varied by 12‰, indicating that the temperature of deposition had varied by 5°C (Harmon et al. 1979).

Following the renaissance in interest in speleothems as palaeoclimate recorders in the 1990s, the group at the University of East Anglia (UEA) have been active in developing techniques for the reliable extraction of isotope data from speleothem fluid inclusions (Dennis et al. 2001). They focussed on crushing procedures in order to investigate the $\delta^{18}O$ signal. Investigation of the latter is precluded by decrepitation techniques since the calcite is strongly heated, typically to ~500°C to rupture the inclusions by vapour pressure, or to ~900°C to disintegrate the sample by conversion of $CaCO_3$ to CaO. At such elevated temperatures, rapid exchange of oxygen isotopes occurs between the released water and the large quantities of CO_2 produced. After the trapped water is released and transferred to the vapour phase, fractionation can occur by preferential adsorption of the heavier isotopes (^{18}O, D) onto the reactive surfaces of freshly crushed calcite (see below) or onto the surfaces of the extraction apparatus. Conversely, there may be preferential escape of the lighter isotopes (^{16}O, ^{1}H) during sample manipulation in the high–vacuum handling lines. Fractionation must be minimised in order to maximise the signal–to–noise ratio since the amplitude of changes in the oxygen isotope ratio of groundwater is generally very low on other than glacial–interglacial timescales. Uncertainties of $>\pm1$‰ for oxygen and $>\pm8$‰ for hydrogen are of limited use for validating GCMs, or for calculating cave temperatures to a precision of better than ±5°C.

Crushing can be achieved either with the sample attached to a vacuum line by screw crushers, hammer crushers or electro–magnetically operated pistons, or offline by crushing sample capsules in a vice or agitating calcite in a ball mill. It is necessary to heat samples moderately (~150°C) to desorb the water from the freshly activated calcite surfaces and to drive the water out of those inclusions that have been accessed by narrow cracks only. Dennis et al. (2001) devised an improved method of crushing speleothem samples that quantitatively removed inclusion water in a controlled fashion. In a series of experiments in which anhydrous Iceland Spar was crushed with water of known mass and isotope composition sealed in capillary tubes, Dennis et al. (2001) showed that when released, the water adsorbed strongly onto the fresh calcite grain surfaces, probably in a series of monolayers. Attempts to desorb the water cryogenically at room temperature resulted in recoveries of only 50–80% for $H_2O/CaCO_3$ mass ratios of 1–2 mg.g^{-1}, typical of observed speleothem values. The recovered water was invariably isotopically depleted by up to 25‰ for hydrogen, and 10‰ for oxygen. Following Gammage and Gregg (1972), Dennis et al. (2001) found that heating the sample to 150°C completely desorbed the water, and allowed recovery of the correct isotope ratios. These results contrast with those of some inclusion studies on hydrothermal deposits, e.g., Yang et al. (1995), who successfully recovered water from crushed dolomite without sample heating.

Crushing and decrepitation techniques both require that the water is recovered cryogenically to a liquid nitrogen trap at −196° C and the temperature raised to boil off any trapped CO_2. The water is then transferred to an appropriate vessel for analysis, generally a pyrex tube that can be flame sealed or closed with a tap. The recovered water (typically 0.1 to 5 μL for calcite samples of 0.1–several grams) may be subjected to both oxygen and hydrogen isotope measurements, or to hydrogen isotope measurements only. In the former case, a metered amount of CO_2 of known isotope composition is transferred into the sample vessel with the water. It is sealed and the oxygen isotopes allowed to exchange between H_2O and CO_2 for several days at a carefully controlled known temperature, usually 25°C, until equilibration is completed (Kishima and Sakai 1980; Socki et al. 1999). The CO_2 and water are cryogenically separated at −75°C and the water transferred for high temperature reduction to hydrogen using zinc (Coleman et al., 1982, Tanweer et al. 1988), chromium (Donnelly et al. 2001), manganese (Shouakar–Stash et al. 2000) or uranium (Hartley 1980). The isotope composition of the CO_2 and the hydrogen are then measured directly by mass spectrometry.

Applying the crushing and heating technique to a speleothem from SW England, Dennis et al. (2001) were able to produce an oxygen and hydrogen isotope record of mid–late Holocene drip water. The data plotted on, or very close to, the Global Meteoric Water Line, with the youngest samples plotting within the field of modern drip water compositions. This result would not be expected if post–trapping exchange of oxygen with calcite had occurred, but it should be noted that (i) the samples analysed were all of Holocene age and little temperature change is expected to have occurred over this interval; and ii) the duration of storage is relatively short. On balance, caution is urged in inferring $\delta^{18}O$ values of drip water directly from corresponding analyses of fluid inclusion waters, but the alternative approach of inferring drip water $\delta^{18}O$ from measured D/H ratios is also problematic because it hinges on the correct choice of d.

The latter issue was highlighted in the Matthews et al. (2000) decrepitation study of fluid inclusions from Soreq Cave in Israel where today $d = 22‰$. Matthews et al. (2000) showed that drip water was enriched in D by 30‰ compared to trapped inclusions, comparable to the 22‰ offset observed by Schwarcz and Yonge (1983). When waters trapped during ice ages (MIS 3 and 4) were used to calculate palaeotemperatures, agreement with independent estimates of temperature was best if the authors assumed a deuterium excess of 10‰ (Craig's (1961) modern global meteoric water line). The implication is that d changes significantly between glacial and interglacial climate states, and it underlines the difficulty in objectively choosing the appropriate value for the deuterium excess.

Evidence for multiple forms of water in speleothems

An improved understanding of the nature of trapped waters in speleothems may provide useful insights into how best to improve fluid–inclusion extraction techniques in the future. As alluded to above, Yonge (1982) observed that even clear, apparently inclusion–free speleothems yielded, on decomposition, significant amounts of water, confirming that water was present in some form other than microscopically resolvable fluid inclusions. In recent years the nature of this 'residual' water became the subject of further investigation at McMaster University. Schwarcz (unpublished data) prepared portions of inclusion–rich stalagmites by grinding them to particles <10 μm, and heating them to 500 °C in a furnace prior to analysis using Fourier Transform Infrared Spectrophotometry (FTIR). In all cases, a broad absorption peak at c. 3400 cm^{-1} corresponding to liquid water was observed in these heated samples (Figure 13). Importantly, this did not appear to be due to the presence of water of hydration (lattice–bound H_2O). For example, gypsum ($CaSO_4 \cdot 2\ H_2O$) displays a number of discrete peaks in this spectral region, but no peak due to liquid water. Also, no hydroxyl (OH) peak was observed, except where they had been inadvertently overheated leading to calcination and subsequent formation of CaOH. Attempts to freeze the water by carrying out FTIR measurements at −30°C showed no change in the shape of the peak, whereas the IR spectrum of liquid water attached to calcite was seen to transform to that of ice. Although considerable care was taken not to expose the samples to atmospheric water vapour after heating, it was possible that this broad water peak was due to water inadvertently adsorbed prior to FTIR analysis. To test this, freshly heated calcite was exposed to a saturated vapour of D_2O. No absorption peak for liquid D_2O (at 2470 cm^{-1}) was observed, nor was there any change in the intensity of the 3400 cm^{-1} peak, leading to the conclusion that liquid water was still present inside these strongly decrepitated samples.

G. Rossman and C. Verdel at the Califoria Institute of Technology (pers. comm. to HPS, 2004) suggested recently that this water might be trapped in these samples as inclusions ≤ 100 nm in maximum dimension, which would not be liberated by crushing. (pers. comm. to HPS 2004). These sub–μm–sized inclusions might also be present in optically clear calcite since their small size would not lead to detectible scattering of light. During heating, the vapour pressure of steam in a nanoinclusion would be insufficient to fracture the surrounding calcite. Also, at sub–zero temperatures they might remain supercooled due to lack of nuclei for freezing. Water in these nanoinclusions might be isotopically fractionated with respect to water in larger inclusions, due to larger surface/volume ratio of the inclusions, as a result of surface/water interactions during trapping which would tend to preferentially trap H_2O vs D_2O. This proposal is currently under experimental investigation by the McMaster group.

It is noteworthy that ion–probe analyses of the hydrogen contents of speleothems (Fairchild et al. 2001) provide an independent line of evidence for the nature of trapped water. Hydrogen is typically present at concentrations of 100–1000 ppm, with a continuous distribution on the scale of a few microns. These observations are consistent with the presence of molecular or nano–inclusion hosted water.

Recent applications of fluid inclusion studies

Fleitmann et al. (2003) used a 'Dennis–type' crusher to extract fluid inclusions from speleothems from Oman. They observed large excursions in δD, the highest values approaching those of modern precipitation in the area of the cave, while mid–glacial stage values were up to 60‰ lower. From these data they inferred changes in atmospheric circulation, but did not attempt to calculate palaeotemperatures. They were able to identify the Indian Ocean as the dominant moisture source for groundwater recharge during Mid–Pleistocene interglacial periods based upon the δD values of the fluid inclusions.

Construction of a 'Dennis–type' crusher at the McMaster laboratory allowed Serefiddin et al. (2002) to obtain a series of palaeotemperatures on stalagmites from a cave in South Dakota, USA. Here, coeval portions of stalagmites deposited a few metres apart in a single chamber of the cave were found to differ in $\delta^{18}O$ by up to 2‰

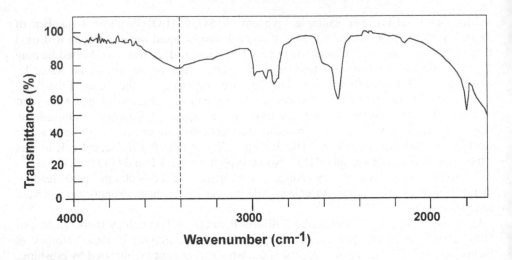

Figure 13. Fourier transform infra–red absorption spectrum of a powdered sample of a stalagmite from Coldwater Cave, Iowa. Superimposed spectra at –11 ° C and –36 °C are indistinguishable from spectrum at 25°C (not shown). The broad peak centered at 3400 cm⁻¹ indicates the presence of liquid (but not solid) H_2O (H. Schwarcz, unpublished data).

(Serefiddin et al. 2004). However, agreement was found between fluid inclusion based palaeotemperatures for the coeval deposits, showing that the difference in $\delta^{18}O$ was due to systematic difference in $\delta^{18}O$ of drip waters feeding the two sites. This was ascribed to differential delivery of winter *vs* summer precipitation to the recharge areas feeding these two stalagmites. As in earlier studies, it was necessary to assume $d < 10$ to avoid obtaining palaeotemperatures <0°C.

McGarry et al. (2003) employed a modified thermal decrepitation technique to provide δD measurements of fluid inclusions for speleothems from three caves in Israel. As in previous studies (Matthews et al. 2000), a large (30‰) fractionation in δD was observed between the recovered and parent fluid inclusion water. After correcting for this effect, a range of $\delta^{18}O$ values were inferred for the included water using both the global and Mediterranean meteoric water lines (MWL and MMWL). On the basis of these inferred $\delta^{18}O$ data, a range of depositional temperatures were calculated using the associated carbonate $\delta^{18}O$ values and a carbonate–water temperature equation. This approach yielded a trend of progressive cooling during the interval between about 130–20 ka, broadly in line with those inferred from independent marine (alkenone) based palaeotemperature estimates from the Eastern Mediterranean and Crete (Emeis et al. 1998; 2000; Kallel et al. 2000).

Current and future developments

Despite the very significant technical and analytical difficulties that surround the extraction of reliable fluid inclusion stable isotope data, such studies are worth pursuing because of their potential to unlock important insights into several issues in palaeoclimatology including:

(i) Reconstruction of secular variation in δD of meteoric water during shifts from glacial to interglacial climates in temperate and subtropical regions. Studies of fluid inclusions in dated speleothems could provide evidence of climate–controlled shifts in circulation patterns and sources of water vapour, and provide valuable tests for isotope–enabled GCMs;

(ii) Comparison of δD in nanoinclusions vs macroinclusions (>1 μm) may reveal a source of 'noise' in fluid inclusion isotope data, and may lead to strategies for selecting a population of inclusions more representative of the drip waters;

(iii) Comparison of δD of fluid inclusions along single growth layers (or even more grossly, between axial and distal parts of the same stalagmite) may provide a more unequivocal test of whether evaporation occurred during deposition, possibly leading to disequilibrium deposition of $CaCO_3$;

(iv) High resolution measurement of palaeotemperature variations on a scale of a few decades to hundreds of years will require development of techniques for analysis of fluid inclusions in smaller amounts of calcite and sub–μl amounts of water. Advances in continuous flow isotope ratio mass spectrometer (CF–IRMS) technology now allow hydrogen isotopes ($^1H^1H^+$ and $^1H^2H^+$) to be measured without interference from the He carrier gas ($^4He^+$) (e.g., Prosser and Scrimgour (1995); Begley and Scrimgour (1997)). This provides the opportunity for using the high temperature pyrolysis furnaces that are routinely used in CF work to reduce H_2O samples online to CO and H_2 (Eiler and Kitchen 2001; Sharp et al. 2001), thus avoiding the time consuming and difficult offline equilibration and liquid–gas separation steps necessary for dual inlet mass spectrometry. The omission of these sample–handling steps is likely to compensate for any slight loss of precision using the CF technique. Water samples ≤ 0.1 μl can have both oxygen and hydrogen isotopic ratios determined in ~10 minutes. This compares with minimum sample size of about 0.3 μl and measurement times of 20 minutes each for O and H_2 using dual inlet mass spectrometry. It may be possible, therefore, to improve the resolution of fluid inclusion studies by up to an order of magnitude from the 300–1000 year range normally attempted at present. Such advances in measurement procedures do not, however, reduce the problems inherent in extracting the inclusion water from the calcite matrix, except insofar as they may permit the extraction device to be coupled to a mass spectrometer sample introduction system, thus streamlining sample handling.

(iv) A remaining question is the cause of the inhomogeneous distribution of fluid inclusions in speleothems? Without knowing the mechanism of entrapment, we cannot fully comprehend why, during extended growth periods, the abundance of fluid inclusions can drop nearly to zero, resulting in the precipitation of optically clear calcite (which may still of course contain nanoinclusions). A related issue is that the temporal resolution for inclusion studies may ultimately be dictated by their closure rate, which may be on the order of decades, depending on the growth rate of the speleothem.

Summary

In summary, despite the apparently widespread lack of precise isotope equilibrium between cave drip–waters and their carbonate precipitates, speleothems are usually deposited sufficiently close to isotope equilibrium to retain useful information about climate–driven changes in the $\delta^{18}O$ of meteoric water that falls on a cave site. Several studies have demonstrated that speleothems can record details of past climates such as the Dansgaard/Oeschger events that were previously accessible only via ice cores. The apparently synchronous changes in the position of the ITCZ and Greenland temperatures recorded by several mid–latitude northern hemisphere speleothems and ice cores respectively, yields important new insights into the operation of the climate system during the last glacial and explanations of these phenomena are currently the subject of modelling work. Speleothem O isotope data have also provided unexpected and challenging new insights into the probable influence of solar forcing on monsoonal systems on centennial to millennial time–scales in the Holocene, and underline the need to improve our understanding of the amplification mechanisms of solar forcing. Despite these undoubted successes it remains essential to check, on a sample by sample basis, that isotope equilibrium was achieved and maintained during the growth history of individual stalagmites using Hendy–type tests. Present–day monitoring studies of cave drip waters are also essential to understand fully the site–specific relationships between speleothem $\delta^{18}O$ and meteorological variables at the surface.

Temporal changes in speleothem $\delta^{13}C$ offer considerable potential to reconstruct climate–driven changes in both the nature of vegetation (e.g., C3 vs. C4 type) and the intensity of vegetation above a cave site (e.g., Dorale et al. (1998); Genty et al. (2003)). However carbon isotope data should ideally be accompanied by appropriate trace element data (e.g., Ca/Mg ratios) in order to evaluate other possible causes for temporal changes in $\delta^{13}C$. Strongly correlated carbon isotope and Mg/Ca ratios may point to a role for partial degassing of cave drip–waters and calcite precipitation in the hydrological flow–path above the cave (so–called 'prior calcite precipitation') that could produce elevated $\delta^{13}C$ values unrelated to changes in vegetation type or intensity.

By virtue of their strong chronological control and the long–term integrity of their stable isotope signals, speleothems offer considerable potential for the identification of teleconnections that in turn can provide new tests for both internal and external forcing mechanisms that drive Earth's climate on a variety of time scales. Thus, the emphasis in recent studies has been on the identification of relatively large shifts in the $\delta^{18}O$ recorded by speleothems, and the evaluation of the extent to which these changes are recorded on a hemispheric scale. A remarkable result to emerge during the past few years has been evidence for strongly coherent in–phase climate signals on millennial timescales during the last glacial (e.g., Dansgaard–Oeschger events), between the Greenland ice core records and several northern hemisphere speleothem records such as caves in China (Wang et al. 2001; Yuan et al. 2004), Kleegruben Cave in the Austrian Central Alps (Spötl and Mangini 2002), Moomi Cave in Socotra Island, Yemen (Burns et al. 2003) and Villars Cave in south-west France (Genty et al. 2003). At some northern tropical sites (e.g., Burns et al. (2003)) the oxygen isotope signals have been interpreted to reflect changes in the meridional position of the ITCZ, that in turn may be linked to changes in high latitude (e.g., Greenland) air temperatures. These observations can provide tests for emerging models for teleconnections between high latitude ice sheets and the ITCZ (e.g., Chiang and Bitz in press), but a key issue is that chronological control on all records must be improved to test the exact phase relationships between these effects. It is likely that the speleothem records will continue to refine the chronology for the high–latitude ice core records. There is still a dearth of long well–dated speleothem records from equivalent low to mid–latitudes in the southern hemisphere, and so it remains difficult to delineate the southern limit of the apparent coherence between Greenland temperatures and the ITCZ.

On shorter timescales, in the Holocene for example, there has been considerable progress in providing high–resolution speleothem records. One of the most striking results to emerge from some of these datasets is the surprisingly strong coherence between oxygen isotope ratios in speleothems and atmospheric $\Delta^{14}C$, pointing to a solar influence on centennial to millennial timescales (Neff et al. 2001; Niggemann et al. 2003; Fleitmann et al. 2003). These correlations with $\Delta^{14}C$ are best seen in speleothem records in which the 'amount effect' appears to dominate the speleothem oxygen isotope record, suggesting that on these timescales solar forcing may cause changes in atmospheric circulation that in turn lead to changes in precipitation amount at a given site. In the case of the sub–tropical sites, reduced solar irradiance (high $\Delta^{14}C$) appears to cause a reduction in the intensity of monsoonal rainfall intensity (inferred from heavier $\delta^{18}O$), possibly related to southward shifts in the position of the ITCZ (Neff et al 2001; Fleitmann et al. 2003). Similarly, in the case of the central European site (Germany), periods of reduced insolation during the late Holocene were associated with drier conditions. If similar data sets can be established for a range of latitudes it may be possible to test the competing models for the amplification of weak changes in solar irradiance (Shindell et al. 2001; Bond et al. 2001). In the early Holocene there is evidence from a variety of northern hemisphere proxy records including speleothems (McDermott et al. 2001; Fleitmann et al. 2003) and lake sediments (Andresen et al. 2004) that subtle variability in climatic conditions were contemporaneous with changes in Greenland air temperatures.

Few available continental archives record climate–driven changes in, for example the oxygen isotope ratio of meteoric water, with the resolution, faithfulness and chronological control offered by speleothems. During the past decade, stable isotope studies of speleothem carbonate have provided important new insights into the operation of the Earth's past climate, arguably rivalling those from the high–latitude ice cores. As with the ice cores, the quantitative interpretation of stable isotope shifts in speleothems is seldom straightforward, and a detailed knowledge of the meteorological variables that control these in modern cave systems is essential to provide a sound interpretation of past climates. Demonstration that speleothem carbonate was deposited at or close to isotope equilibrium with cave drip waters and an understanding of the site–specific relationships between the $\delta^{18}O$ of the latter and that of local meteoric water are also important requirements. Replication of records both within and between caves in the same region (e.g., Bar–Matthews et al. (2003); Wang et al. (2001); Yuan et al. (2004)), would do much to improve confidence in the climate signals and clearly should be a goal of future studies. In the future it seems likely that data from speleothem fluid inclusions will provide a more robust basis for palaeoclimate interpretations, in particular enabling the relative effects of air temperature and atmospheric circulation changes to be unravelled.

References

Anderson T.F. and Arthur M.A. 1983. Stable isotopes of oxygen and carbon and their application to sedimentological and palaeoenvironmental problems. In: Arthur M.A., Anderson T.F., Kaplan I.R., Veizer J. and Land L.S. (eds), Stable Isotopes in Sedimentary Geochemistry: Society of Economic Palaeontologists and Mineralogists Short Course, 10: pp. 1.1–1–151.

Andresen C.S., Bjorck S., Bennike O. and Bond G. 2004. Holocene climate changes in southern Greenland; evidence from lake sediments. J. Quaternary Sci. 19: 783–795.

Baker A., Smart P.L. and Ford D.C. 1993. Northwest European paleoclimate as indicated by growth frequency variations of secondary calcite deposits. Palaeogeogr. Palaeocl. 100: 291–301.

Baker A., Ito E., Smart P.L. and McEwan R.F. 1997. Elevated and variable values of $\delta^{13}C$ in speleothems in a British cave system. Chem. Geol. 136: 263–270.

Baker A. and Barnes W.L. 1998. Comparison of the luminescence properties of waters depositing flowstone and stalagmites at Lower Cave, Bristol. Hydrol. Processes 9: 1447–1459.

Baldini J.U.L. 2004. Geochemical and hydrological characteristics of stalagmites as palaeoclimate proxies, with an emphasis on trace element variability. Ph.D. thesis, University College Dublin, 295 pp.

Bar–Matthews M., Ayalon A., Matthews A., Sass E. and Halicz L. 1996. Carbon and oxygen isotope study of the active water–carbonate system in a karstic Mediterranean cave: implication for palaeoclimate research in semiarid regions. Geochim. Cosmochim. Ac. 60: 337–347.

Bar–Matthews M., Ayalon A. and Kaufman A. 1997. Late Quaternary palaeoclimate in the Eastern Mediterranean region from stable isotope analysis of speleothems at Soreq cave, Israel. Quaternary Res. 47: 155–168.

Bar–Matthews M., Ayalon, A., Kaufman A. and Wasserburg G.J. 1999. The Eastern Mediterranean palaeoclimate as a reflection of regional events: Soreq cave, Israel. Earth Planet Sc. Lett. 166: 85–95.

Bar–Matthews M., Ayalon A. and Kaufman A. 2000. Timing and hydrological conditions of Sapropel events in the Eastern Mediterranean, as evident from speleothems, Soreq cave, Israel. Chem. Geol. 169: 145–156.

Bar–Matthews M., Ayalon A., Gilmour M., Matthews A. and Hawkesworth C.J. 2003. Sea–land oxygen isotopic relationships from planktonic foraminifera and speleothems in the Eastern Mediterranean region and their implications for paleorainfall during interglacial intervals. Geochim. Cosmochim. Ac. 67: 3181–3199.

Begley I.S. and Scrimgour C.M. 1997. High–Precision δ^2H and $\delta^{18}O$ Measurement for water and volatile organic compounds by continuous–flow pyrolysis isotope ratio mass spectrometry. Anal. Chem., 69: 1530–1535.

Bond G., Kromer B., Beer J., Muscheler R., Evans M.N., Showers W. Hoffmann S., Lotti–Bond R., Hajdas I. and Bonani G. 2001. Persistent solar influence on North Atlantic climate during the Holocene. Science 294: 2130–2136.

Buhmann D and Dreybrodt W. 1985. The kinetics of calcite solution and precipitation in geologically relevant situations in karst areas. 1. Open system. Chem. Geol. 48: 189–211.

Burns S.J., Fleitmann D., Matter A., Neff U. and Mangini A. 2001. Speleothem evidence from Oman for continental pluvial events during interglacial periods. Geology 29: 623–626.

Burns S.J., Fleitmann D., Mattey A., Kramers J. and Al–Subbary A.A. 2003. Indian Ocean Climate and absolute chronology over Dansgaard/Oeschger events 9 to 13. Science 301: 1365–1367.

Burns S.J., Matter A., Frank N. and Mangini A. 1998. Speleothem–based palaeoclimate record from northern Oman. Geology 26: 499–502.

Caballero E., Jimenez De Cisneroc C. and Reyes E. 1996. A stable isotope study of cave seepage waters. Applied Geochemistry 11: 583–587.

Chiang J.C. and Bitz C.M. In press. The influence of high latitude ice on the position of the marine Intertropical Convergence Zone. Clim. Dynam.

Coleman M.L., Shepherd T.J., Durham J.J., Rouse J.E. and Moore G.R. 1982. Reduction of water with zinc for hydrogen isotope analysis. Anal. Chem., 54: 993–995.

Craig H. 1961. Isotopic variations in meteoric waters. Science 133: 1702–1703.

Craig H. 1965. The measurement of oxygen isotope palaeotemperatures. In: Tongiorgi E. (ed.), Stable isotopes in oceanographic studies and palaeotemperatures. Pisa, Consiglio Nazionale delle Ricerche Laboratorio di Geologia Nucleare, pp. 161–182.

Cruz–San Julian J., Araguas L., Rozanski K., Benavente J., Cardenal J., Hidalgo M.C., Garcia–Lopez S., Martinesz–Garrido J.C., Moral F. and Olias M. 1992. Sources of precipitation over southeast eastern Spain and groundwater recharge. An isotopic study. Tellus 44B: 226–236.

Dennis P.F., Rowe P.J. and Atkinson T.C. 2001. The recovery and isotopic measurement of water from fluid inclusions in speleothems. Geochim. Cosmochim. Ac. 65: 871–884.

Denniston R.F., González L.A., Asmerom Y., Baker R.G., Reagan M.K. and Bettis III E.A. 1999a. Evidence for increased cool season moisture during the middle–Holocene. Geology 27: 815–818.

Denniston R.F., González L.A., Baker R.G., Amersom Y., Reagan M.K., Edwards R.L. and Alexander E.C. 1999b. Speleothem evidence for Holocene fluctuations of the prairie–forest ecotone, north–central USA. Holocene 9: 671–676.

Desmarchelier J.M. and Goede A. 1996. High resolution stable isotope analysis of a Tasmanian speleothem. Pap. And Proc. R. Soc. Tasmania 130: 7–13.

Desmarchelier J.M., Goede A., Ayliffe L., McCulloch M.T. and Moriarty K. 2000. Stable isotope record and its palaeoenvrionmental interpretation for a late Middle Pleistocene speleothem from Victoria Fossil Cave, Naracoorte, South Australia. Quaternary Sci. Rev. 19: 763–774.

Donnelly T., Waldron S., Tait A., Dougans J. and Bearhop S. 2001. Hydrogen isotope analysis of natural abundance and deuterium–enriched waters by reduction over chromium on–line to a dynamic dual inlet isotope–ratio mass spectrometer. Rapid Commun. Mass Sp. 15: 1297–1303.

Dorale J.A., González L.A, Reagan M.K., Pickett D.A., Murrell M.T. and Baker R.G. 1992. A high–resolution record of Holocene climate change in speleothem calcite from Cold Water Cave, northeast Iowa. Science 258: 1626–1630.

Dorale J.A., Edwards R.L., Ito E. and Gonzalez L.A. 1998. Climate and vegetation history of the mid–continent from 75 to 25 ka: A speleothem record from Crevice Cave, Missouri, USA. Science 282: 1871–1874.

Dreybrodt W. 1988. Processes in Karst Systems. Springer Series in Physical Environment. Springer, Heidelberg, 282 pp.

Dulinski M. and Rozanski K. 1990. Formation of $^{13}C/^{12}C$ isotope ratios in speleothems: a semi–dynamic model. Radiocarbon 32: 7–16.

Edwards R.L., Chen J.H. and Wasserburg G.J. 1988. $^{238}U-^{234}U-^{230}Th$ systematics and the precise measurement of time over the last 500,000 years. Earth Planet. Sc. Lett. 81: 175–192.

Eiler J.M. and Kitchen N. 2001. Hydrogen–isotope analysis of nanomole (picolitre) quantities of H_2O. Geochim. Cosmochim. Ac. 65: 4467–4479.

Emeis K.C., Schulz H.M., Struck U., Sakamoto T., Dosse H., Erlenkeuser H., Howell M., Kroon D. and Paterne M. 1998. Stable isotope and alkenone temperature records of sapropels from sites 964 and 967: constraining the physical environment of sapropel formation in the eastern Mediterranean Sea. In: Robertson A.H.F., Emeis K.-C., Richter C. and Camerlenghi A. (eds), Proceedings of the Ocean Drilling Program 160, pp 309–331.

Emeis K.C., Schulz H.M., Struck U., Rosenberg R., Bernasconi S., Erlenkeuser H., Sakamoto T. and Martininez–Ruiz F. 2000. Temperature and salinity variations of Mediterranean Sea surface waters over the last 16,000 years from records of planktoic stable oxygen isotopes and alkenone unsaturation ratios. Palaeogeogr. Palaeocl. 158: 259–280.

Fairchild I.J., Borsato A., Tooth A.F., McDermott F., Frisia S., Hawkesworth C.J., Huang Y. and Spiro B. 2000. Controls on trace element (Sr–Mg) compositions of carbonate cave waters: implications for speleothem climatic records. Chemical Geology 166: 255–269.

Fairchild I.J., Baker A., Borsato A., Frisia S., Hinton R.W., McDermott F. and Tooth A.F. 2001. Annual to sub–annual resolution of multiple trace–element trends in speleothems. J. Geol. Soc. London 158: 831–841.

Fairchild I.J., Frisia S., Borsato A. and Tooth A.F. In press. Speleothems in their geomorphic, hydrological and climatological context. In: Nash D.J. and McLaren S.J. (eds), Geochemical Sediments and Landscapes. Blackwell Publishers.

Fleitmann D., Burns S.J., Neff U., Mangini A. and Matter A. 2003. Changing moisture sources over the last 330,000 years in Northern Oman from fluid–inclusion evidence in speleothems. Quaternary Res. 60: 223–232.

Fleitmann D., Burns S.J., Mudelsee M., Neff U., Kramers J., Mangini A. and Matter A. 2003. Holocene forcing of the Indian Monsoon recorded in a stalagmite from southern Oman. Science 300: 1737–1739.

Fleitmann D., Burns S.J., Neff U., Mudelsee M., Mangini A. and Matter A. 2004. Palaeoclimatic interpretation of high–resolution oxygen isotope profiles derived from annually laminated speleothems from Southern Oman. Quat. Sci. Rev. 23: 935–945.

Ford D.C. and Williams P.W. 1989. Karst Geomorphology and Hydrology. Chapman and Hall, London, 601 pp.

Frappier A., Sahagian D., González L.A. and Carpenter S.J. 2002. El Niño events recorded by stalagmite carbon isotopes. Science 298: 565.

Fricke H.C. and O'Neil J.R. 1999. The correlation between $^{18}O/^{16}O$ ratios of meteoric water and surface temperature: its use in investigating terrestrial climate change over geologic time. Earth Planet. Sc. Lett. 170: 181–196.

Frisia S., Borsato A., Fairchild I.J. and McDermott F. 2000. Calcite fabrics, growth mechanisms, and environments of formation in speleothems from the Italian Alps and southwestern Ireland. J. Sediment. Res. 70: 1183–1196.

Frisia S., Borsato A., Fairchild I.J., McDermott F. and Selmo E.M. 2002. Aragonite–calcite relationships in speleothems (Grotte de Clamouse, France): Environment, fabrics and carbonate geochemistry. J. Sediment. Res. 72: 687–699.

Frisia S., Borsato A., Preto N. and McDermott F. 2003. Late Holocene annual growth in three Alpine stalagmites records the influence of solar activity and the North Atlantic Oscillation on winter climate. Earth Planet. Sc. Lett. 216: 411–424.

Frumkin A., Carmi I., Gopher A., Ford D.C., Schwarcz H.P. and Tsuk T. 1999a. A Holocene millennial–scale climatic cycle from a speleothem in Nahal Qanah Cave, Israel. Holocene 9: 677–682.

Frumkin A., Ford D.C. and Schwarcz H.P. 1999b. Continental palaeoclimatic record of the last 170,000 years in Jerusalem. Quaternary Res. 51: 317–327.

Frumkin A., Ford D.C. and Schwarcz H.P. 2000. Palaeoclimate and vegetation of the last glacial cycles in Jerusalem from a speleothem record. Global Biogeochem. Cy. 14: 863–870.

Gammage R.B. and Gregg S.J. 1972. The sorbtion of water vapour by ball–milled calcite. Journal Colloid Interf. Sci. 38: 118–124.

Gascoyne M. 1992. Palaeoclimate determination from cave calcite deposits. Quaternary Sci. Rev. 11: 609–632.

Gat J.R. 1996. Oxygen and hydrogen isotopes in the hydrologic cycle. Annu. Rev. Earth and Pl. Sc. 24: 225–262.

Genty D. and Massault M. 1997. Carbon transfer dynamics from bomb–C–14 and delta C–13 time series of a laminated stalagmite from SW France – Modelling and comparison with other stalagmite records. Geochim. Cosmochim. Ac. 63: 1537–1548.

Genty D., Baker A. and Vokal B. 2001. Intra- and inter-annual growth rate of modern stalagmites. Chem. Geol. 176: 191–212.

Genty D., Baker A., Massault M., Procter C., Gilmour M., Pons–Banchu E. and Hamelin B. 2001. Dead carbon in stalagmites: Carbonate bedrock paleodissolution vs. ageing of soil organic matter. Implications for ^{13}C variations in speleothems. Geochim. Cosmochim. Ac. 65: 3443–3457.

Genty D., Plagnes V., Causse C., Cattani O., Stievenard M., Falourd S., Blamart D., Ouahdi R. and Van–Exter S. 2002. Fossil water in large stalagmite voids as a tool for palaeoprecipitation reconstruction and palaeotemperature calculation. Chem. Geol. 184: 83–95.

Genty D., Blamart D., Ouahdi R., Gilmour M., Baker A., Jouzel J. and Van–Exter S. 2003. Precise dating of Dansgaard–Oeschger climatic oscillations in western Europe from stalagmite data. Nature 421: 833–837.

Goede A., Veeh H.H. and Ayliffe L.A. 1990. Late Quaternary palaeotemperature records for two Tasmanian speleothems. Aust. J. Earth Sci. 37: 267–278.

Gordon D., Smart P.L., Ford D.C., Andrews J.N., Atkinson T.C., Rowe P.J. and Christopher N.S.J. 1989. Dating of late Pleistocene interglacial and interstadial periods in the United Kingdom from speleothem growth frequency. Quaternary Res. 31: 14–26.

Grossman E.L. and Ku T.L. 1986. Oxygen and carbon isotope fractionation in biogenic aragonite–temperature effects. Chem. Geol. 59: 59–74.

Harmon R.S., Schwarcz H.P. and Ford D.C. 1978. Stable isotope geochemistry of speleothems and cave waters from the Flint Ridge-Mammoth Cave System, Kentucky: implications for terrestrial climatic change during the period 230,000 to 100,000 years BP. Journal of Geology 86: 373–384.

Harmon R.S., Schwarcz H.P., Ford D.C. and Koch D.L. 1979. An isotopic paleotemperature record for late Wisconsinan time. Geology 7: 430–433.

Harmon R., Schwarcz H.P. and O'Neil J.R. 1979. D/H ratios in speleothem fluid inclusions: A guide to variations in the isotopic composition of meteoric precipitation? Earth Planet. Sc. Lett. 42: 254–266.

Hartley P.E. 1980. Mass spectrometric determination of Deuterium in water at natural levels. Anal. Chem. 52: 2232–2234.

Hellstrom J., McCulloch M.T. and Stone J. 1998. A detailed 31,000 year record of climate and vegetation change, from the isotope geochemistry of two New Zealand speleothems. Quaternary Res. 50: 167–178.

Hellstrom J.C. and McCulloch M.T. 2000. Multi–proxy constraints on the climatic significance of trace element records from a New Zealand speleothem. Earth Planet. Sc. Lett. 179: 287–297.

Hendy C.H. 1971. The isotopic geochemistry of speleothems 1. The calculation of the effects of the different modes of formation on the isotopic composition of speleothems and their applicability as palaeoclimatic indicators. Geochim. Cosmochim. Ac. 35: 801–824.

Hendy C.H. and Wilson A.T. 1968. Palaeoclimatic data from speleothem. Nature 216: 48–51.

Holmgren K., Lee–Thorp J.A., Cooper G.R.J., Lundblad K., Partridge T.C., Scott L., Sithaldeen R., Talma A.S. and Tyson P.D. 2003. Persistent millennial–scale climatic variability over the past 25,000 years in Southern Africa. Quaternary Sci. Rev. 22: 2311–2326.

Hou J.Z., Tan M., Cheng H. and Liu T.S. 2003. Stable isotope records of plant cover change and monsoon variation in the past 2200 years: evidence from laminated stalagmites in Bejing, China. Boreas 32: 304–313.

Jimenez de Cisneros C., Caballero E., Vera J.A., Duran J.J. and Julia R. 2003. A record of Pleistocene climate from a stalactite, Nerja Cave, southern Spain. Palaeogeogr. Palaeocl. 189: 1–10.

Kallel N., Duplessey J.–C., Labeyrie L., Fontugne M., Paterne M. and Montacer M. 2000. Mediterranean and pluvial periods and sapropel formation during last 200,000 years. Palaeogeogr., Palaeocl. 157: 45–58.

Kendall A.C. and Broughton P.L. 1978. Origin of fabrics in speleothems composed of columnar calcite crystals. J. Sediment Petrol. 48: 519–538.

Kim S.T. and O'Neil J.R. 1997. Equilibrium and nonequilibrium oxygen isotope effects in synthetic carbonates. Geochim. Cosmochim. Ac. 61: 3461–3475.

Kishima N. and Sakai H. 1980. Oxygen–18 and deuterium determination on a single water sample of a few milligrams. Anal. Chem. 52: 356–358.

Kolodny Y., Bar–Matthews M., Ayalon A. and McKeegan K.D. 2003. A high spatial resolution $\delta^{18}O$ profile of a speleothem using an ion–microprobe. Chem. Geol. 197: 21–28.

Lauritzen S.E. and Lundberg J. 1999. Calibration of the speleothem delta function: an absolute temperature record for the Holocene in northern Norway. Holocene 9: 659–669.

Leng M.J. and Marshall J.D. 2004. Palaeoclimate interpretation of stable isotope data from lake sediment archives. Quaternary Sci. Rev. 23: 811–831.

Li W.X., Lundberg J., Dickin A.P., Ford D.C., Schwarcz H.P., McNutt R. and Williams D. 1989. High–precision mass–spectrometric uranium–series dating of calcite deposits and implications for palaeoclimatic studies. Nature 339: 534–536.

Liu Z.H. and Dreybrodt W. 1997. Dissolution kinetics of calcium carbonate minerals in H_2O – CO_2 solutions in turbulent flow. The role of the diffusion boundary layer and the slow reaction $H_2O + CO_2 = H^+ + HCO_3^-$. Geochim. Cosmochim. Ac. 61: 2879–2889.

Matthews A., Ayalon A. and Bar–Matthews M. 2000. D/H ratios of fluid inclusions of Soreq Cave (Israel) speleothems as a guide to the Eastern Mediterranean Meteoric Water Line relationships in the last 120 Ky. Chem. Geol. 166: 183–191.

McDermott F., Frisia S,, Huang Y., Longinelli A., Spiro B., Heaton T.H.E., Hawkesworth C.J., Borsato A., Keppens E., Fairchild I.J., van der Borg K., Verheyden S. and Selmo E. 1999. Holocene climate variability in Europe: evidence from $\delta^{18}O$ and textural variations in speleothems. Quaternary Sci. Rev. 18: 1021–1038.

McDermott F., Mattey D.P. and Hawkesworth C.J. 2001. Centennial–scale Holocene climate variability revealed by a high–resolution speleothem $\delta^{18}O$ record from S.W. Ireland. Science 294: 1328–1331.

McDermott F. 2004. Palaeo–climate reconstruction from stable isotope variations in speleothems: A review. Quaternary Sci. Rev. 23: 901–918.

McGarry S., Bar–Matthews M., Matthews A., Vaks A. and Ayalon A. 2003. Constraints on hydrological and paleotemperature variations in the Eastern Mediterranean region in the last 140 ka given by the δD values of speleothem fluid inclusions. Quaternary Sci. Rev. 23: 919–934.

McMillan E., Fairchild I.J.F., Frisia S., Borsato A. and McDermott F. In press. Annual trace element cycles in calcite–aragonite speleothems: evidence of drought in the western Mediterranean 1200–1100 years BP. J. Quaternary Sci.

Merlivat L. and Jouzel J. 1979. Global climatic interpretation of the deuterium – ^{18}O relationship for precipitation. J. Geophys. Res. 84: 5029–5033.

Mickler P.J., Banner J.L., Stern L., Asmerom Y., Edwards R.L. and Ito E. 2004. Stable isotope variations in modern tropical speleothems: Evaluating equilibrium vs. kinetic isotope effects. Geochim. Cosmochim. Ac. 68: 4381–4393.

Neff U., Burns S.J., Mangini A., Mudelsee M., Fleitmann D. and Matter A. 2001. Strong coherence between solar variability and the monsoon in Oman between 9 and 6 kyr ago. Nature 411: 290–293.

Niggemann S., Mangini A., Richter D.K. and Wurth G. 2003. A paleoclimate record of the last 17,600 years in stalagmites from the B7 cave, Sauerland, Germany. Quaternary Sci. Rev. 22: 555–567.

Onac B.P., Constantin S., Lundberg J. and Lauritzen S.E. 2002. Isotopic climate record in a Holocene stalagmite from Ulisor Cave (Romania). J. Quaternary Sci. 17: 319–327.

O'Neil J.R., Clayton R.N. and Mayeda T.K. 1969. Oxygen isotope fractionation in divalent metal carbonates. J. Chem. Phys. 51: 5547–5558.

Plagnes V., Causse C., Genty D., Paterne M. and Blamart D. 2002. A discontinuous climatic record from 187 to 74 ka from a speleothem of the Clamouse Cave (south of France). Earth Planet. Sc. Lett. 201: 87–103.

Polyak V.J. and Asmeron Y. 2001. Late Holocene climate and cultural changes in the Southwestern United States. Science 294: 148–151.

Prosser S.J. and Scrimgour C.M. 1995. High–Precision Determination of $^2H/^1H$ in H_2 and H_2O by continuous flow isotope ratio mass spectrometry. Anal. Chem. 67: 1992–1997.

Repinski P., Holmgren K., Lauritzen S.E. and Lee–Thorp J.A. 1999. A late–Holocene climate record from a stalagmite, Cold Air Cave, Northern Province, South Africa. Palaeogeogr. Palaeocl. 150: 269–277.

Richards D.A. and Dorale J.A. 2003. Uranium–series chronology and environmental applications of speleothems. In: Bourdon B., Henderson G.M., Lundstrom C.C. and Turner S.P. (eds), Uranium Series Geochemistry. Rev. Mineral. Geochem. 52 : pp. 407–460.

Rozanski K., Sonntag C. and Munnich K.O. 1982. Factors controlling stable isotope composition of European precipitation. Tellus 34: 142–150.

Rozanski K., Araguás–Araguás L. and Gonfiantini R. 1993. Isotopic patterns in modern precipitation. In: Swart P.K., Lohmann K.C., McKenzie J. and Savin S. (eds), Climate change in continental isotopic records. Geophysical Monograph 78, American Geophysical Union, Washington, D.C. pp. 1–36.

Rye R.O. and O'Neil J.R. 1968. The ^{18}O content of water in primary fluid inclusions from Providencia, North–Central Mexico. Econ. Geol. 63: 232–238.

Salomons W. and Mook W.G. 1986. Isotope geochemistry of carbonates in the weathering zone. In: Fritz P. and Fontes J.C. (eds). Handbook of Environmental Isotope Geochemistry, Vol. 2, The Terrestrial Environment, B. Elsevier, Amserdam, pp. 239–270.

Schwarcz H.P., Harmon R.S., Thompson P. and Ford D.C. 1976. Stable isotope studies of fluid inclusions in speleothems and their paleoclimate significance. Geochim. Cosmochim. Ac. 40: 657–665.

Schwarcz H.P. and Yonge C. 1983. Isotopic composition of paleowaters as inferred from speleothem and its fluid inclusions. In: Gonfiantini R. (ed.), Paleoclimates and Paleowaters: A Collection of Environmental Isotope Studies. Int. Atomic Energy Agency, Proceedings of Advisory Group Meeting, Vienna STI/PUB/621: pp. 115–133.

Schwarcz H.P. 1986. Geochronology and Isotope Geochemistry of Speleothem. In: Fontes J. C. and Fritz P. (eds), Handbook of Environmental Isotope Geochemistry. The Terrestrial Environment, B. Elsevier, p. 271–303.

Serefiddin F., Schwarcz H.P. and Ford D.C. 2002. Palaeotemperature reconstruction using isotopic variations in speleothem fluid inclusion water. Geochimica Cosmochimica Acta 66: Supl. 1: p. A697.

Serefiddin F., Schwarcz H.P., Ford D.C. and Baldwin S. 2004. Late Pleistocene paleoclimate in the Black Hills of South Dakota from oxygen isotope records in speleothems. Palaeogeog. Palaeocl. 203: 1–17.

Sharp Z.D., Atudorei V. and Durakiewicz T. 2001. A rapid method for determination of hydrogen and oxygen isotope ratios from water and hydrous minerals. Chem. Geol. 178: 197–210.

Shen C., Edwards R.L., Cheng H., Dorale J., Thomas R.B., Moran S.B., Weinstein S.E. and Edmonds H.N. 2002. Uranium and thorium isotopic and concentration measurements by magnetic sector inductively coupled plasma mass spectrometry. Chem. Geol. 185: 165–178.

Shindell D., Rind D., Balachandran N., Lean J. and Lonergan P. 2001. Solar cycle variability, ozone and climate. Science 284: 305–308.

Shouakar–Stash O., Drimmie R., Morrison J., Frape S.K., Heemskerk A.R. and Mark W.A. 2000. On–line D/H analysis for water, natural gas, and organic solvents by manganese reduction. Anal. Chem. 72: 2664–2666.

Socki R.A, Romanek C.S. and Gibson E.K. 1999. On-line technique for measuring stable oxygen and hydrogen isotopes from microlitre quantities of water. Anal. Chem. 71: 2250–2253.

Spötl C. and Mangini A. 2002. Stalagmite from the Austrian Alps reveals Dansgaard–Oeschger events during isotope stage 3: Implications for the absolute chronology of Greenland ice cores. Earth Planet. Sc. Lett. 203: 507–518.

Stenni B., Jouzel J., Masson–Delmotte V., Röthlisberger R., Castellano E., Cattani O., Falourd S., Johnsen S.J., Longinelli A., Sachs J.P., Selmo E., Souchez R, Steffensen J.P. and Udisti R. 2004. A late–glacial high–resolution site and source temperature record derived from the EPICA Dome C isotope records (East Antarctica). Earth Planet. Sc. Lett. 217: 183–195.

Stuart F.M., Burnard P.G., Taylor R.P. and Turner G. 1995. Resolving mantle and crustal contributions to ancient hydrothermal fluids: He–Ar isotopes in fluid inclusions from Dae Hwa W–Mo mineralisation, South Korea. Geochim. Cosmochim. Ac. 59: 4663–4673.

Stute M. and Schlosser P. 1993. Principles and applications of the noble gas paleothermometer. In: Swart P.K., Lohmann K.C., McKenzie J. and Savin S. (eds), Climate Change in Continental Isotopic Records, Geophysical Monograph 78, American Geophysical Union, pp. 89–100.

Talma A.S. and Vogel J.C. 1992. Late Quaternary paleotemperature derived from a speleothem from Cango Caves, Cape Province, South Africa. Quaternary Res. 37: 203–213.

Tanweer A., Hut G. and Burgman J.O. 1988. Optimal conditions for the reduction of water to hydrogen by zinc for mass spectrometric analysis of deuterium content. Chem. Geol. 73: 199–203.

Thompson P., Schwarcz H.P. and Ford D.C. 1974. Continental Pleistocene climatic variations from speleothem age and isotopic data. Science 184: 893–894.

Treble P.C., Budd W.F., Hope P.K. and Rustomji P.K. 2005. Synoptic–scale climate patterns associated with rainfall $\delta^{18}O$ in southern Australia. J. Hydrol. 302: 270–282.

Verheyden S., Keppens E., Fairchild I.J., McDermott F. and Weiss D. 2000. Mg, Sr and Sr isotope geochemistry of a Belgian Holocene speleothem: implications for palaeoclimate reconstructions. Chem. Geol. 169: 131–144.

Vaks A., Bar–Matthews M., Ayalon A., Schilman B., Gilmour M., Hawkesworth C.J., Frumkin A., Kaufman A. and Matthews A. 2003. Paleoclimate reconstruction based on the timing of speleothem growth and oxygen and carbon isotope composition in a cave located in the rain shadow in Israel. Quaternary Res. 59: 182–193.

Van Geel B., van der Plicht J., Kilian M.R., Klaver E.R., Kouwenberg J.H.M., Ressen H., Reynaud–Farrera I. and Waterbolk H.T. 1998. The sharp rise in delta C–14 ca. 800 cal BC: Possible causes, related climatic teleconnections and the impact on human environments. Radiocarbon 4: 535–550.

Wang Y.J., Cheng H., Edwards R.L., An Z.S., Wu J.Y., Shen C.C. and Dorale J.A. 2001. A high–resolution absolute–dated late Pleistocene monsoon record from Hulu Cave, China. Science 294: 2345–2348.

Xia Q., Zhao J. and Collerson K.D. 2001. Early–Mid Holocene climatic variations in Tasmania, Australia: Multi–proxy records in a stalagmite from Lynds Cave. Earth Planet Sc. Lett. 194: 177–187.

Yang W., Spencer R.J. and Krouse H.R. 1995. Stable isotope and major element composition of fluid inclusions in Devonian and Cambrian dolomite cements, western Canada. Geochim. Cosmochim. Ac. 59: 3159–3172.

Yonge C.J. 1982. Stable isotope studies of water extracted from speleothems. Ph.D. thesis, McMaster University. 298 pp.

Yonge C., Ford D.C., Gray J. and Schwarcz H.P. 1985. Stable isotope studies of cave seepage water. Chem. Geol. 58: 97–105.

Yuan D., Cheng H., Edwards R.L., Dykoski C.A., Kelly M.J., Zhang M., Qing J., Lin Y., Wang Y., Wu J., Dorale J., An Z. and Cai Y. 2004. Timing, duration and transitions of the last interglacial Asian monsoon. Science 304: 575–578.

6. ISOTOPES IN MARINE SEDIMENTS

MARK A. MASLIN (mmaslin@geog.ucl.ac.uk)
GEORGE E. A. SWANN (g.swann@ucl.ac.uk)
Environmental Change Research Centre
Department of Geography
University College London
26 Bedford Way
London
WC1H 0AP, UK

Key words: Oxygen isotopes, carbon isotopes, nitrogen isotopes, silicon isotopes, boron isotopes, oceans, marine sediments, palaeoceanography.

Introduction

Marine sediments provide long continuous records of past climate changes at intra-annual, annual to centennial scale resolutions enabling insights into past changes within both oceanic and continental environments. Stable isotopes provide palaeoceanographers with the means to reconstruct a range of variables including surface and deep ocean circulation patterns, sea surface and bottom water temperature, sea surface salinity, iceberg activity and origin, upwelling intensity, productivity, nutrient utilisation, surface-water dissolved carbon dioxide content and water-column oxygen content in addition to inferences on global ice volume, ice sheet failure, river discharge, aridity, vegetation composition, and continental erosion rates. The fundamental source of stable isotope information derived from marine sediment archives primarily originates from oxygen and carbon. Comprehensive introductions on oxygen and carbon isotopes and their physico-chemical behaviour/systematics are given in Craig and Gordon (1965), Garlick (1974), Hoefs (1997), Criss (1999), Rohling and Cooke (1999) and Zeebe and Wolf-Gladrow (2001). This chapter describes the current state of palaeoceanographic and palaeoclimatic uses of both oxygen and carbon isotopes, alongside less frequently applied stable isotopes such as nitrogen, silicon and boron.

Composition and distribution of marine sediments

Marine sediments contain two components: biogenic and lithogenic (Figure 1). The biogenic component primarily originates in surface waters, although further contributions originate from the ocean bottom dwelling community together with material transported by rivers, wind and icebergs. The biogenic component can be divided into three sub-components: organic material, calcium carbonate and opal-A (biogenic silica). Organic matter can range from individual molecules to pollen grains, organic walled microfossils, opaque organic matter and wood plant fragments. The major contribution to the calcium

227

M.J. Leng (ed.), 2005. *Isotopes in Palaeoenvironmental Research.* Springer. Printed in The Netherlands.

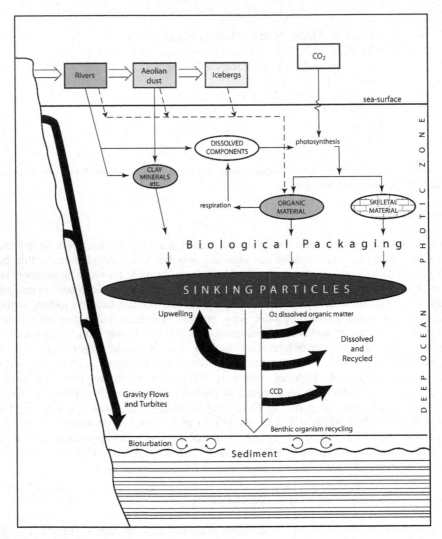

Figure 1. Formation of marine sediments, sources and depositional pathways (adapted from Maslin et al. (2003)).

carbonate sediment load originates from coccolithophores (nannofossils) and planktonic foraminifera, with a smaller contribution from benthic foraminifera and ostracods. The major contributor to the siliceous or opal sediment load are diatoms, with a small contribution from silicoflagellates, radiolarians and siliceous sponge spicules and occasional additional input from phytoliths and freshwater diatoms, transported to the ocean by either aeolian and/or riverine processes. The lithogenic component in marine sediment cores is primarily composed of clays, though larger material up to boulder size can be deposited as icebergs melt. The majority of the lithogenic components originates from continental rocks and soils which become eroded and deposited into the ocean via riverine or aeolian transportation or through icebergs.

The global distribution of marine sediment can be classified into five categories based on its primary composition: 1) carbonate, 2) siliceous, 3) ice-rafted, 4) red clay and 5) terrigenous (Figure 2a). The primary control on sediment type relates to the proximity of the site to the continent. This controls whether ice rafted debris and/or terrigenous (mainly riverine) input dominates the sediment and the magnitude and type of marine productivity. Whether carbonate or siliceous material dominates, the biogenic component in the marine sedimentary record is primarily controlled by the surface productivity regime which in turn is influenced by a plethora of factors including light, temperature, salinity, and nutrient concentrations. Hence, both the latitude and continental configurations are important due to their role in bringing about nutrient availability through seasonal changes and upwelling. Highly productive areas such as the centres of the six main upwelling zones in the world (California margin, Peru margin, Angola-Benguela, North African-Portuguese margin, Arabian Sea and Antarctica, see Figure 2b) and most coastal areas are dominated by diatom based productivity, while less productive areas such as the North East Atlantic are dominated by both coccolithophores and foraminifera. As highlighted below, stable isotopes of carbon, nitrogen and silicon are all excellent tools for reconstructing past productivity and nutrient utilisation. In contrast in areas remote from continental influences, where productivity is low, the dominant input to the marine archive is wind-blown clay, which results in the formation of red clay sediments with extremely low sedimentation rates. An additional source of chemical and stable isotope information within marine sediments lies in sediment pore water, sea water trapped and entrained within the marine archive by subsequent sediment accumulation. As demonstrated below, analysis of the oxygen isotope composition of trapped pore waters can provide a valuable insight into past variations in both ice volume and the chemical composition of ocean water.

One of the main considerations in reconstructing past climates from marine sediments is the temporal resolution. Sedimentation rates in the deep-ocean are on average between 1-5 cm kyr^{-1} with highly productive areas producing a maximum of 20 cm kyr^{-1}. This limits the temporal resolution of any study to about 200 yr cm^{-1} (50 yr cm^{-1} for productive areas). Additionally though, in marine sediments below oxygenated bottom waters, an active biological benthic community can mix the top 20 cm of sediment. This 'bioturbation' can reduce the temporal resolution in deep ocean marine studies to as low as 1000 years. In contrast, on continental shelves, marginal seas and other specialised natural sediment traps, such as anoxic basins and fjords, sedimentation rates can exceed 10 m kyr^{-1} (e.g., Jennings and Weiner (1996); Pike and Kemp (1997); Maslin et al. (1998)) providing a temporal resolution greater than 1 yr cm^{-1}. A potential draw-back to such high resolution locations, however, stems from the highly localised environmental and climatic signal which may be recorded by the stable isotope data in contrast to the global climatic signal which is more readily obtained from deep ocean sites. Additional problems associated with continental margins revolve around possible reworking, erosion and redistribution of the sediment by mass density flows such as turbidities and slumps which can prevent or significantly hinder any palaeoenvironmental reconstruction (Figure 2c). Consequently, while sediments from continental margins hold may notable advantages, caution and consideration is needed towards possible artefacts which may erode the validity of the environmental signal present within the isotope data.

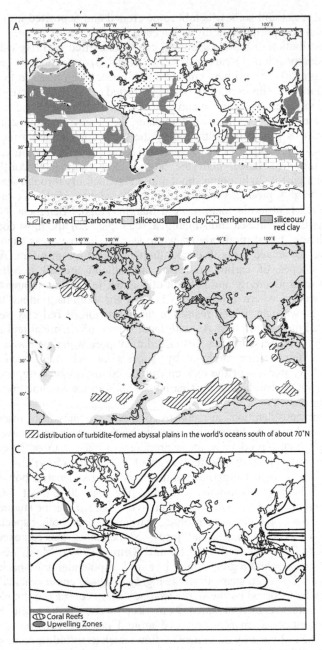

Figure 2. A) Global distribution of marine sediment types. B) Global distribution of continental shelf and abyssal plains, both of which can be highly unstable depositional environments with significant erosion and redistribution of marine sediments. C) Surface ocean circulation and the location of the major high productivity upwelling zones and coral reefs around the world (adapted from Maslin et al. (2003)).

Oxygen isotopes in marine sediments

Three major influences affect the marine oxygen isotope record (Equation 1; Table 1). The first is the water temperature at which the mineral, for example calcite or opal, is precipitated (δ_T). As ocean water warms, $\delta^{18}O$ in the inorganic precipitates is lowered. Second is the actual $\delta^{18}O$ of the water ($\delta^{18}O_w$) in which the inorganic precipitate are produced. $\delta^{18}O_w$ is controlled by both global influences, for example the amount of isotopically light freshwater that is stored in ice sheets (δ_{GIV}) and the local influences of evaporation versus freshwater inputs (δ_{local}) i.e., precipitation, river and iceberg discharge. From this, the $\delta^{18}O$ value of each of these different parameters can be reconstructed and used to infer palaeoenvironmental conditions over a range of time-scales given:

$$\Delta\delta^{18}O_M = \Delta\delta^{18}O_w + \Delta\delta_T = \Delta\delta_{GIV} + \Delta\delta_{local} + \Delta\delta_T \text{ + vital effects} \qquad (1)$$

where:
$\Delta\delta^{18}O_M$ = oxygen isotopes of the mineral precipitated (e.g., calcite or opal).
$\Delta\delta^{18}O_w$ = oxygen isotope composition of the water which is a combined affect of variations in global ice volume ($\Delta\delta_{GIV}$) and local influences ($\Delta\delta_{local}$).
$\Delta\delta_T$ = water temperature.
Vital effects = deviations from equilibrium due to the incorporation of oxygen into biogenically produced calcite or opal. These effects include ontogeny, symbiont photosynthesis, respiration and life history (see Rohling and Cooke (1999) for further details).

Table 1. Environmental influences on marine oxygen isotopes.

Environmental Factor	Increase	Decrease
Temperature	$\delta^{18}O$ decrease	$\delta^{18}O$ increase
Global Ice Volume	$\delta^{18}O$ increase	$\delta^{18}O$ decrease
Salinity*	$\delta^{18}O$ increase	$\delta^{18}O$ decrease
Density*	$\delta^{18}O$ increase	$\delta^{18}O$ decrease

* an exception is sea ice formation which increases surface water salinity and density but lowers the surface water $\delta^{18}O$ (O'Neil 1968) as $\delta^{18}O$ always increases in the formation of ice from water. This affect though is minimal and is not believed to influence past climatic records.

Palaeotemperature ($\Delta\delta_T$)

The historic development and use of oxygen isotopes in marine studies (e.g., Urey (1947); (1948)) resulted in most initial $\delta^{18}O$ records derived from foraminiferal being interpreted in terms of temperature. Emiliani (1955; 1971) assumed that 80% of the planktonic foraminifera oxygen isotope record was attributable to changes in ocean temperatures with only 20% related to changes in the global ice volume. Such assumptions regarding the ^{18}O content of the continental ice sheets during glacial periods, however, have been shown to be untenable (Dansgaard and Tauber 1969) with Antarctica ice glacial $\delta^{18}O$ values approximately four time lower than originally envisaged (see Duplessy et al.

(2002)). Shackleton (1967) utilised $\delta^{18}O$ in benthic foraminifera to look at changes in the ice volume signal. Assuming that there were only minimal changes in deep ocean water temperatures over glacial-interglacial cycles, the ice volume signal observed in planktonic and benthic foraminifera were found to be considerably greater than previously indicated. A number of equations have since been produced to combine the changes in water temperature and the calcite-water fractionation of $\delta^{18}O$. While others have been proposed (e.g., Epstein et al. (1953)), the two most commonly used and/or quoted equations for foraminiferal palaeotemperature studies are:

Shackleton (1974) based on O'Neil et al. (1969)

$$T = 16.9 - 4.38(\delta_c - \delta_w) + 0.1(\delta_c - \delta_w)^2 \tag{2}$$

Hays and Grossman (1991) based on O'Neil et al. (1969)

$$T = 15.7 - 4.36(\delta_c - \delta_w) + 0.12(\delta_c - \delta_w)^2 \tag{3}$$

where T is the isotopic temperature estimate; δ_c, $\delta^{18}O$ of the foraminiferal calcite test; δ_w, $\delta^{18}O$ of the surface waters. From this, changes in sea surface temperature (SST) recorded by foraminifera can be calculated for:

$$\Delta\delta_c = \Delta\delta_w - 0.23\Delta T \tag{4}$$

Consequently, for approximately every 1‰ change in foraminiferal $\delta^{18}O$, assuming no change in ice volume or local influences, a 4°C change in SST can be inferred. The above equations apply only to foraminifera with others equations developed for the temperature relationship of $\delta^{18}O_{(diatom)}$ and $\delta^{18}O_{(coral)}$ (see below). The existence of past changes in the ocean $\delta^{18}O$ caused by local influence and global ice volume does not, however, discredit or prevent the use of oxygen isotopes for palaeotemperature reconstructions. In order to interpret $\delta^{18}O$ as a palaeotemperature signal, either there should be little or no change in global ice volume and/or local conditions or independent estimates of both these influences are required (see below).

    For the period prior to major changes in global ice volume, however, oxygen isotope records are a highly effective palaeo-thermometer, provided local $\delta^{18}O$ influences can be accurately accounted for. Zachos et al. (2001) produced a composite benthic foraminiferal $\delta^{18}O$ record for the last 70 Ma from 40 Deep Sea Drilling Project (DSDP) and Ocean Drilling Program (ODP) sites (Figure 3). The global distribution of these sites minimises the role of local influences on the foraminifera $\delta^{18}O$ record. Moreover the sites selected are in relatively deep water (>1000 m) which, due to their very nature, are more conservative in terms of temperature variation relative to both surface and intermediate depth waters.

    Prior to the onset of Antarctic glaciation (c. 33 Ma) it is possible to interpret the stacked global benthic foraminifera $\delta^{18}O$ record purely in terms of changes in bottom water temperature. The oxygen isotopes suggest the deep oceans of the Cretaceous and Paleocene periods may have been as warm as 10-15°C which is radically different to the present day when deep water temperatures vary from between 4 °C and −1 °C (see papers in Barrera and Johnson (1999)). The results also indicate a marked bottom water

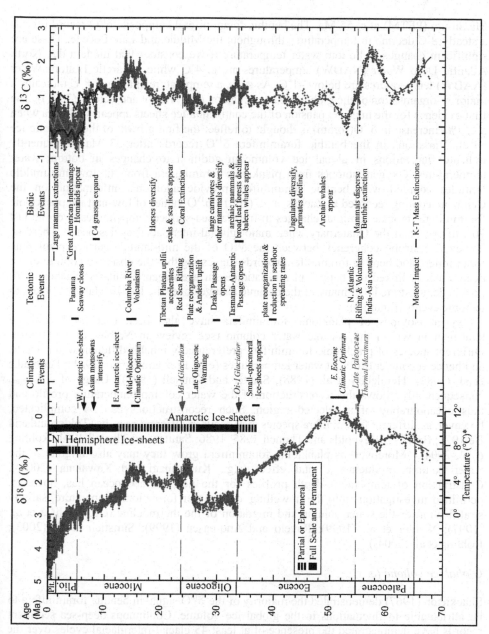

Figure 3. Global deep-sea oxygen and carbon isotopes record for the last 70 Ma based on benthic foraminifera data from 40 DSDP and ODP Sites compared with climatic, tectonic and biotic events (reproduced with permission from Zachos et al. (2001)).

temperature increase of c. 5°C over five million years during the Paleocene to the Early Eocene climatic optimum (Zachos et al. 2001) with an additional abrupt 5°C warming event and subsequent 100 kyr cooling period occurring at the Paleocene-Eocene Thermal

Maximum (PETM) at c. 55 Ma. The benthic foraminiferal composite $\delta^{18}O$ record reveals a steady 8°C decline in temperatures throughout the Middle and Late Eocene. These are significant changes in bottom water temperature if we consider that modern day North Atlantic Deep Water (NADW) temperatures are c. 4°C while Antarctic Bottom Water (AABW) temperatures are below −1°C. As such, a temperature change of 8°C indicates a major re-organisation of the global ocean circulation (e.g., Haupt and Seidov (2001)). The first evidence for the major expansion of the continental ice sheets appears at 33 Ma when a c. 1‰ increase in $\delta^{18}O$(foram) is thought to reflect the first growth of the Antarctic ice sheet. Variations in the benthic foraminifera $\delta^{18}O$ records after 33 Ma consequently indicate fluctuations in global ice volume in addition to changes in bottom water temperatures. Oxygen isotopes from planktonic foraminifera from the low and middle latitudes coupled with benthic foraminifera provide additional information on the Cenozoic cooling recorded in Zachos et al. (2001). $\delta^{18}O$ values of low-latitude planktonic foraminifera indicate minimal changes in tropical sea-surface temperatures over the last 70 Ma except in the Quaternary when major glacial-interglacial cycles occur. There is, however, a strong coherency between the $\delta^{18}O$ of the mid-latitude surface planktonic foraminifera and benthic foraminifera records suggesting that the observed cooling in the deep water is linked to changes in mid-latitude surface ocean. Similarly, Zachos et al. (2001) demonstrated that many of the oceanic changes over the last 70 Ma were linked to tectonic events (Figure 3).

Oxygen isotopes of planktonic foraminifera have also been used to investigate stratification within the oceanic water column (see review in Niebler et al. (1999)). Different species of planktonic foraminifera preferentially inhabit different water depths and hence encounter different water temperatures (e.g., Erez and Honjo (1981); Fairbanks et al. (1982); Hemleben et al. (1988); Sautter and Thunell (1991)). Usage of $\delta^{18}O$(foram) consequentially allows the reconstruction of the water column temperature profile and thus oceanography of a selected region. Such reconstructions can be complicated, however, as different foraminifera species may occur in different seasons (Tolderlund and Bé 1971; Bé 1977; Reynolds and Thunell 1985; 1986; Sautter and Thunell 1991; Rohling et al. 2004). Moreover, as planktonic foraminifera grow they may also migrate in the water column producing a vital effect (e.g., Kuroyanagi and Kawahata (2004)). Construction of temperature-depth profiles for the near surface ocean has, however, permitted investigations into the upwelling, deep water formation, temperature, salinity stratification of the water column and migration of the thermocline (e.g., Mulitza et al. (1997); Niebler et al. (1999); Ravelo and Andreasen (1999); Simstich et al. (2003); Rohling et al. (2004)).

Global ice volume (δ_{GIV})

Shackleton (1967) demonstrated the majority of the $\delta^{18}O$ signal in benthic foraminifera to be attributable to fluctuations in the global ice volume. Continuous deep-sea sediment records have demonstrated the presence of at least 45 glacial-interglacial cycles over the last 2.5 Ma (e.g., Tiedemann et al. (1994); Shackleton et al. (1995)). Marine oxygen-isotope records thus provide an excellent stratigraphic tool permitting the correlation of cores collected around the globe and the creation of the Marine Isotope Stage (MIS) record (Emiliani 1955; Shackleton and Opdyke 1973; Imbrie et al. 1984; Prell et al. 1986; Tiedemann et al. 1994) (see Figure 4).

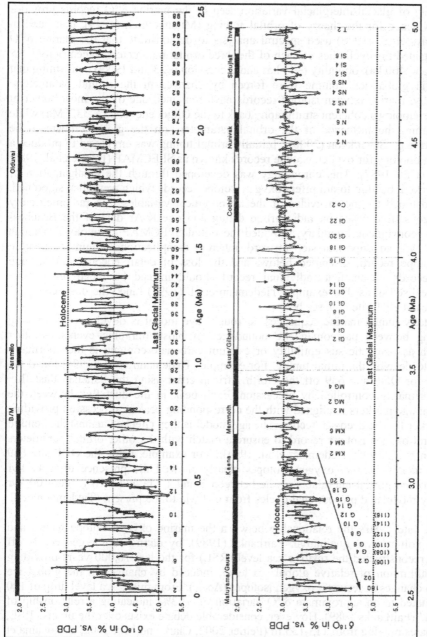

Figure 4. Benthic foraminifera δ¹⁸O record from ODP site 659 in the tropical North East Atlantic over the last 5 Ma with palaeomagnetic reversal boundaries and Marine Isotope Stages (MIS) defined. Prior to MIS 104 (c. 2.6 Ma) the nomenclature of Tiedemann et al. (1994) is used to reflect the palaeomagnetic period and the position of the interval within the palaeomagnetic period. For example, stages within the Gauss are described as G1, G2, G3 etc., stages within the Nunivak as N1, N2, N3 etc. (adapted from Tiedemann et al. (1994)).

The cyclicity of glacial-interglacial variation seen in the marine $\delta^{18}O$ records provides essential evidence for the theory of orbital forcing (Milankovitch (1949), translated in (1969)). Hays et al. (1976) used spectral analysis to demonstrate that the marine $\delta^{18}O$ record contained the cyclicities of each of the three orbital parameters, eccentricity (~95 ka, 125 ka and 400 ka), obliquity (~41 ka) and precession (~23 and 19 ka) confirming that variations in global ice volume were forced by changes in the orbital parameters. Consequently, marine oxygen isotope records have remained one of the most powerful tools for developing a coherent stratigraphy back to the Oligocene (c. 23.8-33.7 Ma) while also permitting the influence of the orbital parameters on the global climate to be investigated (e.g., Shackleton (2000)). Initially orbital tuning was employed to produce a standard chronology for oxygen isotope records known as SPECMAP (Imbrie et al. 1984; Martinson et al. 1987). This chronology was developed through the amalgamation of several, stacked, benthic foraminifera oxygen isotope records which were then smoothed, filtered and tuned to known cycles of the astronomical variables. Six radiometrically dated horizons, five based on radiocarbon dating and one K-Ar date of the Brunhes-Matuyama geomagnetic boundary, provided the initial SPECMAP chronology. Though SPECMAP was an important step forward, inherent complications within the benthic oxygen isotope record, discussed below, and the loss of detail encountered through stacking records means that individual record should instead be tuned on a more regional/localised basis. For example Tiedemann et al. (1994) tuned the dust record off North Africa at ODP site 659, see below.

For orbitally tuning marine sequences, a 'rough' age model is developed by linearly interpolating between palaeomagnetic boundaries or biostratigraphic events. Another proxy, such as magnetic susceptibility or calcium carbonate content, is then normally employed to develop the initial tuning. For example, Tiedemann et al. (1994) tuned the age model for ODP site 659 off the North African coast using the Sahelian dust flux which is primarily controlled by precession. Each peak in the dust flux between the independent age markers is aligned with the corresponding peak in precession providing an age marker for the deep-sea record. The age model is then checked against the benthic foraminiferal oxygen isotope record to ensure a match with the other orbital parameters (Shackleton et al. 1995; Tiedemann et al. 1994). For example, using the ODP site 659 tuned age model, benthic oxygen isotopes clearly showed the presence of a 41 kyr obliquity forced glacial-interglacial cycle between 5 and 1 Ma and then longer 100 kyr eccentricity modulated precessional cycles from c. 1 Ma to the present (Tiedemann et al. 1994).

Until the late 1980's the relationship between the marine oxygen isotope record and global ice volume was qualitative. Fairbanks (1989), by dating submerged corals off Barbados, reconstructed the relative sea level (RSL) for the last 18,000 years providing the first calibration of relative global sea level, induced by changes in the global ice volume, to changes in seawater oxygen isotopes ($\Delta\delta_{GIV}$). From this, a sea level drop of 120 m was equated to an approximate $\delta^{18}O$ variation of ~1.3‰ indicating a relationship of 0.011‰ m^{-1} (Fairbanks 1989). However, considerable debate exists over the precise RSL, with estimates ranging from 113-135 m (Peltier 2002; Clark and Mix 2002; Yokoyama et al. 2000), and the subsequent proportion of the glacial-interglacial variation in benthic foraminifera oxygen isotopes that can be attributed to ice volume fluctuations. Estimates for this have ranged from 1.7‰ (Broecker and Denton 1989) to 0.8‰ (Schrag et al. 1996; Burns and Maslin 1999) with the current best estimate 1‰ $\pm$ 0.1‰ (Schrag et al. 2002; Waelbroeck et al. 2002; Duplessy et al. 2002). This uncertainty surrounding the calibration arises due to the unknown glacial-interglacial temperature variations of different deep water masses (see below section on carbon isotopes). Whereas the total benthic foraminifera oxygen isotope variation in the North Atlantic from the LGM to modern day is 2‰, in the equatorial Pacific this decreases to 1.5‰ (Waelbroeck et al. 2002). Detailed work by Waelbroeck et al. (2002) comparing $\delta^{18}O$ records from the deep North Atlantic, Southern Indian Ocean and deep Equatorial Pacific Ocean suggests the lowest possible increase of $\delta^{18}O$ during the LGM was 0.95‰ enabling the production of a detailed RSL, or ice-volume Equivalent Sea Level (ESL), curve for the last four climatic cycles.

The size of δ_{GIV} can also be estimated from pore waters (Schrag et al. 1996; 2002; Burns and Maslin 1999), trapped glacial waters which are not influenced by temperature but can be affected by advection and mixing. Consequently, advection models are needed to accurately calculate the full glacial shift in oxygen isotopes requiring knowledge of the core sedimentation rate and type of core used (e.g., Schrag et al. (1996)). Detailed studies of pore waters have produced a glacial-Holocene variation in $\Delta\delta_{GIV}$ of 1.0‰ ± 0.1‰ (Schrag et al. 1996; 2002; Adkins and Schrag 2001). If a simple linear relationship between RSL or ESL and marine oxygen isotopes existed then using the best current estimate for both variables (130 m and 1‰) produces a relationship of 0.0077‰/m. However, it is known that the relationship between RSL and benthic foraminifera $\delta^{18}O$ is non-linear and varies for both deglaciation and glaciation. Even after removal of the temperature affect originating from the different deep water mass temperature histories (Waelbroeck et al. 2002), a non-linear relationship between RSL and $\Delta\delta_{GIV}$ remains. First there is clear evidence of large differences between the average isotope composition of each of the major ice sheets (see summary in Duplessy et al. (2002)). The Laurentide ice sheet probably had a $\delta^{18}O$ value of between –28‰ to –34‰, whereas the Eurasian ice sheets may have had a value anywhere between –16‰ and   –40‰. In addition, the glacial aged ice on Greenland was between –41 and –42‰ while on Antarctic similar ice may have varied between –40‰ and –60‰. As such, the timing of the growth and melting of each individual ice sheet influences the given RSL-$\Delta\delta_{GIV}$ relationship. Secondly, the extent of isotope fractionation during snowfall changes as the ice sheet expands, typically increasing as the ice sheet matures through a combination of continentality and altitude effects. Consequently, making reliable estimates over the proportion of the $\delta^{18}O_{(foram)}$ signal related to changes in global ice volume is extremely complex and subject to many uncertainties.

Burns and Maslin (1999) attempted to resolve the problems of advection and advection modelling when analysing sediment pore-water by sampling glacial aged waters within the Amazon fan. Sites drilled on the fan were found to have glacial sedimentation rates of over 10 m kyr⁻¹ ensuring that there were no interglacial pore waters close enough to mix and dilute the glacial aged waters. However, only a 0.3-0.4‰ glacial $\delta^{18}O$ increase was observed relative to modern values due to the origin of bottom water varying from NADW at present to the lower $\delta^{18}O$ waters of the AABW during the last glacial period. Taking this difference into consideration suggests a minimum glacial AABW $\delta^{18}O$ increase of 0.8‰ while comparison of the pore water and benthic foraminifera $\delta^{18}O$ suggests a bottom water temperature decrease of at least 4°C during the last glacial period compared to interglacial periods.

The latest novel approach to estimating global sea level involves using planktonic foraminifera $\delta^{18}O$ to reconstruct RSL in the Red Sea (Siddall et al. 2003). The Red Sea is unique in possessing a narrow (18 km) and shallow (137 m) connection with the open ocean, the Strait of Bab el Mandab, which makes the site highly sensitive to changes in global RSL. Lowering of sea level reduces the exchange of water between the Red Sea and open ocean, thus increasing the residence time of water. As evaporation is extremely high in this region, any changes in the residence time are reflected in surface water salinity and therefore oxygen isotopes. Siddall et al. (2003) modelled the relationship between sea level, Red Sea water residence times, surface water salinity and oxygen isotopes. Importantly, sea level and surface salinity can be independently derived and thus the global sea level estimate is not dependent on the modelled Sea Surface Salinity (SSS). From this work, the planktonic foraminiferal oxygen isotope records over the last 470 kyrs were converted into past sea level changes demonstrating that sea levels during Dansgaard-Oeschger cycles varied by 35 m ± 12 m at rates of 2 cm yr^{-1} (Siddal et al. 2003) with equal contributions originating from the Antarctica and Northern Hemisphere ice sheets (Rohling et al. 2004). This represents rates of change equivalent to the last deglaciation, indicating the importance and magnitude of these glacial millennial-scale events.

Palaeo-salinity ($\Delta\delta_{local}$)

Calculation of Sea Surface Salinity (SSS) is possible in locations where the $\Delta\delta_{local}$ signal is primarily controlled either by the balance of evaporation verses precipitation (E-P) or by the input of lower $\delta^{18}O$ freshwater, such as from melting icebergs or rivers. The local influence on the marine oxygen isotope record is calculated by subtracting the temperature influence and the global ice volume variations from the mineral precipitate record:

$$\Delta\delta_{local} = \Delta\delta^{18}O_M - \Delta\delta_T - \Delta\delta_{GIV} \qquad (5)$$

where $\Delta\delta_{local}$, $\delta^{18}O$, $\delta_T - \Delta\delta_{GIV}$ are defined in Equation (1).

Sea-surface temperature can be reconstructed in three ways; planktonic foraminifera assemblage data (e.g., Imbrie and Kipp (1971); Pflaumann et al. (2003)), coccolith *n*-alkenone U^K_{37} biomarkers (e.g., Haug et al. (2005)) and Mg/Ca ratios in planktonic foraminifera tests (e.g., Elderfield and Ganssen (2000)). These estimates of temperature can be converted to $\Delta\delta_T$ using either Equation 2 or 3. The isotope affect of changing global ice volume can be estimated by converting RSL or ESL to oxygen isotope variation using a simple linear relationship ~0.008-0.01 ‰ m^{-1}. Depending on the level of detail and length of record being examined, either the detailed RSL records for the last 20 ka (Fairbanks 1989, Bard et al. 1990; 1996) or the longer RSL record (e.g., Waelbroeck et al. (2002)) can be used to reconstruct δ_{GIV}. Once the δ_{local} has been separated then the relationship between it and SSS can be calculated. This relationship depends on the oxygen isotope value of the end-member inputs i.e., freshwater (δ_{fresh}) and mean marine sea water (δ_{ocean}). These values vary with locality due to the inputs of meltwater, precipitation and river runoff and the affects of evaporation and ocean circulation resulting in an oxygen isotope-salinity relationship:

$$\Delta\delta_{local} = (\delta_{ocean} - \delta_{fresh} / S_{ocean} - S_{fresh})S_{local} + \delta_{fresh} \qquad (6)$$

where $\Delta\delta_{local}$, δ_{ocean} and δ_{fresh} are defined above and/or in Equation (1); S_{ocean} is mean ocean salinity; S_{fresh}, freshwater salinity; S_{local}, mean local salinity. When salinity of freshwater

(S_{fresh}) is virtually equal to zero it can be omit and the equation can be solved for local salinity:

$$S_{local} = (\delta_{fresh} - \Delta\delta_{local})(S_{ocean}/\delta_{fresh} - \delta_{ocean}) \qquad (7)$$

Duplessy et al. (1991; 1992) first used this approach to reconstruct SSS for the North Atlantic Ocean during the LGM. Because this was a time-slice reconstruction they were able to ignore the ice volume influence, calculating a relationship of $\Delta\delta_{local}$ to surface water salinity of 0.5‰ psu^{-1} (psu = practical salinity unit in ‰). This method has subsequently been adapted to reconstruct down core variations in surface water salinity (Fillion and Williams 1984; Rostek et al. 1993; Maslin et al. 1995a; Chapman and Maslin 1999; Pahnke et al. 2003). For example Maslin et al. (1995a) demonstrated that large quantities of meltwater were released into the North Atlantic during each of the last four Heinrich events suggesting the resultant sea water became too light to sink regardless of any cooling, thereby preventing the normal formation of glacial deep water. Calculation of SSS, however, is fraught with difficulties due to the propagation of errors (Schmidt 1999; Rohling 2000). These errors include:

1) uncertainties in the palaeo-SST estimates, which are ±1-1.5°C for mid to high latitude sites (e.g., Maslin et al. (1995a); Pflaumann et al. (2003)) and at least ±2°C in the tropics (Kucera et al. in press a; b);

2) estimation of $\Delta\delta_{GIV}$ which contains inherent errors that are usually quoted at ±0.001‰ m^{-1};

3) calculation of the mean ocean salinity (S_{ocean}) due to the concentration effect caused by changes in global ice-volume. From estimates of RSL, the later can be calculated using Equation (8) but significant uncertainties remain causing the confidence in estimates of reconstructed RSL to be no more than ±5 m (Fairbanks 1989):

$$S_{ocean} = 34.74 \times RSL / (3900 - RSL) \qquad (8)$$

4) estimates of the oxygen isotope freshwater end-member (δ_{fresh}), which also contains inherent uncertainties. In the North Atlantic calibration of Duplessy et al. (1991) the GEOSECS (1987) data provides a modern end-member of c. −20‰ producing a relationship of:

$$\Delta S_{local} = 1.76\Delta\delta_{local} \qquad (9)$$

This relationship changes significantly when estimating the freshwater end member for the North Atlantic during the LGM. First, a large range of estimates exists for the average isotopic value of the Laurentide ice sheet (−28‰ to −35‰, Duplessy et al. (2002)). Second, there is uncertainty over whether meltwater from the adjacent ice sheet or melting icebergs is represented by this value or by a less negative coastal values of −20‰, as suggested by Mix and Ruddiman (1984) and Maslin et al. (1995a). If a value of −35‰ is used, however, the relationship changes to:

$$\Delta S_{local} = 1.12\Delta\delta_{local} \qquad (10)$$

In other locations the isotope composition of the freshwater end-member can also be markedly different. Core sites influenced by Antarctic glacial meltwater could have values of between −40 and −60‰, while in the tropical Atlantic Ocean the modern freshwater input from the Amazon River lies around −5‰ (Grootes 1993). Not only are there large errors associated with each of the above assumptions but these are accumulative within palaeosalinity calculations. The propagation of these errors has been examined in detail by Schmidt (1999) and Rohling (2000). For example, if a full error propagation analysis is carried out on the Heinrich event data presented in Maslin et al. (1995a) then the maximum errors during the glacial are ±0.9‰ for the $\Delta\delta_{local}$, representing a ΔS_{local} error of ±1.6 psu. Although an extreme example, this demonstrates that any salinity reconstruction using the subtraction approach requires full awareness of the errors involved. However, since it is still one of the few methods that is available to provide some indication of past salinity, we can not yet dismiss the technique until a better method is developed. There are, however, situations where the errors involved in calculating SSS are much lower. One example is the calculation of SSS in the South China Sea during the past 3,000 years (Wang et al. 1999a; b). Here SSS errors are considerably lower due to negligible change in ice volumes over the last 3 ka, minimal SST variations over the same period and the presence of a good understanding on the controls and thus isotopic end members of the South China Sea.

The SSS record of Wang et al. (1999a; b) provides a unique insight into the Chinese monsoons with the major control on SSS the evaporation-precipitation regime, i.e., salinity falls during heavy monsoon periods and rises during dry periods. As detailed records have been kept by the Chinese over the last 3,000 years, Wang et al. (1999a,b) successfully related changes in the reconstructed SSS to the occurrence of droughts, floods, changes in the Chinese Dynasties, periods of peasant uprising and construction phases of the Great Wall of China.

It is also possible to circumvent the direct calculation or modelling of the surface salinity-$\Delta\delta_{local}$ relationship by comparing the relative inputs of the δ_{ocean} verses δ_{fresh}. For example Maslin and Burns (2000) used planktonic foraminiferal oxygen isotopes recovered from the Amazon Fan as a proxy for the relative mixing of fresh Amazon River water and the Atlantic Ocean. To circumvent the problem of isolating the $\Delta\delta_{local}$ signal they subtracted a planktonic isotope record of a core south of the Amazon River, which is not affected by the freshwater discharge, from the ODP core north of the River which is. A $\delta^{18}O_{(foram)}$ increase of over 1‰ was detected during the Younger Dryas suggesting a marked reduction in the amount of negative $\delta^{18}O$ being received from river outflow. Using isotopic end-members of δ_{fresh} = −5‰ and δ_{ocean} = +1‰, Maslin and Burns (2000) estimated the percentage change in the outflow of the Amazon River and hence changes in rainfall over the Amazon Basin. Extension of this record suggests that during both the Younger Dryas and LGM Amazon River discharge was reduced by c. 50% ± 15% (Maslin and Burns 2000; Ettwein 2005). A similar study subtracting oxygen isotopic records either side of the Congo river discharge plume revealed periodic freshwater peaks and enhancement of river discharge during the last 200 ka (Uliana et al. 2002). The limitation to this approach is the quality of the age models for each core which determines the accuracy of the $\delta^{18}O$ subtraction.

Water masses

Benthic foraminifera $\delta^{18}O$ enable the tracing of water masses and consequently changes in ocean circulation to be observed. This was briefly discussed above with regards to pore water $\delta^{18}O$ from the Amazon Fan (Burns and Maslin 1999) showing a glacial-interglacial shift between the AABW and NADW. Lynch-Stieglitz et al. (1999a; b) also employed oxygen isotope records to reconstruct water masses, in this case the LGM water density and thus circulation through the Florida Straits. Benthic foraminifera $\delta^{18}O$ values were used as a proxy for water density with increases in $\delta^{18}O$ indicating either increases salinity and/or decreasing temperatures which both result in increased density (Lynch-Stieglitz et al. 1999a). Taking a time-slice approach, the affect of global ice volume on $\delta^{18}O$ is standardised across all sites in the Florida Straits with their results only limited by the relative accuracy of the age model for each core. Comparison of the LGM results with modern data demonstrates that the LGM density difference across the Florida Strait was reduced by geostrophic flow, being only two-thirds of the modern value, suggesting the Gulf Stream, which originates at the Florida Strait, was much weaker (Lynch-Stieglitz et al. 1999b).

Skinner et al. (2003) also used benthic foraminifera $\delta^{18}O$ to trace different past water masses in the North East Atlantic Ocean using estimates of bottom water temperatures from Mg/Ca ratios in benthic foraminifera to remove the $\Delta\delta_T$ component and the RSL data (Lambeck and Chappell 2001; Siddall et al. 2003) to remove the $\Delta\delta_{GIV}$. By comparing the residual benthic foraminifera $\delta^{18}O$ and the Mg/Ca bottom water temperature estimate, millennial scale variations in dominant bottom water sources were demonstrated at this site. As relatively high values of $\delta^{18}O$ and warmer temperatures are indicative of northern-sourced NADW, with opposite conditions indicative of southern-sourced AABW, Skinner et al. (2003) linked the occurrence of Greenland ice core millennial stadial events with reduced NADW and increased incursion of AABW into the North East Atlantic.

Biogenic silica oxygen isotopes

In addition to foraminifera based oxygen isotopes, analysis of isotopes within sources of biogenic silica (BSi), such as diatoms and radiolarians, represents an additional technique that holds significant potential as a palaeoclimatic tool. More importantly, analysis of BSi oxygen isotopes enables the provision and extension of existing isotope records into oceanic regions depleted in $CaCO_3$. This is most prominent in high latitude regions, such as the Southern Ocean, where foraminifera are not generally well preserved within the sedimentary record (Figure 1). By contrast, sources of BSi (SiO_2), in particular diatoms, flourish in these regions (Figure 3a).

A variety of techniques have been developed for the extraction of BSi oxygen isotope data leading to measurement on diatoms, phytoliths and radiolarians. To date, however, most success within marine archives has been achieved by analysing the oxygen isotope ratios within diatoms ($\delta^{18}O_{(diatom)}$). A key pre-requisite for BSi work is the necessity of ensuring samples are free from contamination. Existing work has demonstrated the $\delta^{18}O_{(diatom)}$ signal to be highly sensitive to the level of non-BSi contamination within the sample, particularly when sample purity falls below 90% (Morley et al. 2004). Obtaining such material is physically, though not technically, difficult with sources of BSi commonly intermixed with similar sized clay and/or silt particles necessitating the need for a series of chemical and/or physical preparation stages (e.g., Shemesh et al. (1995); Morley et al. (2004); Rings et al. (2004); Lamb et al. (in press)).

A notable problem associated with the isotope analysis of diatoms is the presence of a loosely bonded outer hydrous layer, Si-OH, which results in oxygen freely exchanging with water used in the laboratory during sample preparation. As such, early studies were thwarted by poor precision and sample contamination (e.g., Mopper and Garlick (1971); Labeyrie (1974); Mikkelsen et al. (1978)). For reliable $\delta^{18}O_{(diatom)}$ data to be generated, the isotope signal from the tetrahedrally bonded (-Si-O-Si) layer needs to be analysed, which is not subjected to isotope exchange during sample preparation. Instead the signal is perceived to reflect the $\delta^{18}O$ of the ambient seawater in which the BSi was formed, thereby providing a reliable palaeo-$\delta^{18}O$ signal. Extraction of the inner -Si-O-Si layer, however, is technically challenging, preventing to date widespread adoption of $\delta^{18}O_{(diatom)}$ measurements as a standard palaeoceanographic technique. At present, three techniques for the analysis of $\delta^{18}O_{(diatom)}$ exist: Controlled Isotope Exchange (CIE) (Labeyrie and Juillet 1982; Juillet-Leclerc and Labeyrie 1987), stepwise fluorination using a fluorine based reagent such as ClF_3 or BrF_5 (Haimson and Knauth 1983; Matheney and Knauth 1989) and inductive High-Temperature carbon Reduction (iHTR) (Lücke et al. 2005). Analysis of SST, $\delta^{18}O_{(seawater)}$ and $\delta^{18}O$ values of the -Si-O-Si layer of surface sediment diatom assemblages, has resulted in the production of a palaeotemperature calibration equation between 1.5°C and 24°C (Juillet-Leclerc and Labeyrie 1987):

$$t = 17.2 - 2.4(\delta^{18}O_{(diatom)} - \delta^{18}O_{(water)} - 40) - 0.2(\delta^{18}O_{(diatom)} - \delta^{18}O_{(water)} - 40)^2 \qquad (11)$$

where t is the ocean surface water temperature; $\delta^{18}O_{(diatom)}$, $\delta^{18}O$ of the non-exchangeable silica within the diatom frustule; $\delta^{18}O_{(water)}$, $\delta^{18}O$ of the ambient water. Shemesh et al. (1992) suggested the calibration of Juillet-Leclerc and Labeyrie (1987) to be in-accurate in high latitude waters where localised upwelling imparts a significant control on the diatom-water oxygen isotope equilibrium. Although controversial, support for this originates from evidence that core-top $\delta^{18}O_{(diatom)}$ samples from the Southern Ocean would provide a reconstructed SST of 8.8°C under the equation of Juillet-Leclerc and Labeyrie (1987), which is incompatible with observed SST in the region that peak below 4°C (Shemesh et al. 1992). Consequently, Shemesh et al. (1992) proposed a high latitude palaeotemperature equation:

$$t = 11.03 - 2.03(\delta^{18}O_{(diatom)} - \delta^{18}O_{(water)} - 40) \qquad (12)$$

A key factor of both the Juillet-Leclerc and Labeyrie (1987) and Shemesh et al. (1992) temperature calibrations for $\delta^{18}O_{(diatom)}$ are the markedly different gradients they display relative to those calculated for the calcite-water curve (Epstein et al. 1953). Using Equation (12), Shemesh et al. (1992) combined $\delta^{18}O_{(diatom)}$ with $\delta^{18}O_{(foram)}$ data, assuming the incorporation of oxygen into the foraminifera calcite shell and diatom frustule occurred in the same season and water depth, to calculate changes over time of palaeo-$\delta^{18}O_{(water)}$ and palaeotemperature in Southern Ocean surface waters. A major benefit of this approach is the ability to avoid using benthic $\delta^{18}O_{(foram)}$ records and their associated limitations to estimate the $\delta^{18}O_{(GIV)}$ component present in both the $\delta^{18}O_{(diatom)}$ and planktonic $\delta^{18}O_{(foram)}$ signal. Using this, a c. 2.0°C SST increase was observed in the shift from glacial to Holocene conditions with a concordant 1.2 ± 0.2‰ decrease in $\delta^{18}O_{(water)}$

(Shemesh et al. 1992). These interpretations are entirely dependent upon the accuracy of the $\delta^{18}O_{(diatom)}$-temperature calibration. The precise calibration of $\delta^{18}O_{(diatom)}$ and temperature, however, is at present uncertain with estimates ranging from 0.2‰ to 0.5‰ per °C (Juillet-Leclerc and Labeyrie 1987; Shemesh et al. 1992; Brandriss et al. 1998) leading to considerable debate on the exact temperature calibration. Consequently, as with measurements of $\delta^{30}Si_{(diatom)}$ (see section below on silicon isotopes), making detailed quantitative palaeoclimatic interpretations of down-core $\delta^{18}O_{(diatom)}$ data, especially in terms of temperature, is at present problematic.

Investigation of a series of cores across the Southern Ocean has revealed periodic glacial decreases in $\delta^{18}O_{(diatom)}$ over MIS 2 and the last c. 430,000 years (Shemesh et al. 1994; 1995; 2002). Relating these changes to an increasing temperature signal contradicts the numerous other proxy records which indicate the last glacial to have been much colder than present. The marked, c. 2‰, decreases in $\delta^{18}O_{(diatom)}$ are consequently instead connected to meltwater influxes originating from Antarctica, releasing water which contain an inherently lower values of $\delta^{18}O$ than that for seawater (Shemesh et al. 1994; 1995).

An additional problem when using $\delta^{18}O_{(diatom)}$ is the possibility for inter-species fractionation effects (vital effects) within diatoms (see Equation 1). Unlike foraminifera, diatoms can not generally be picked out to create mono-specific samples due to their significantly smaller sizes. Limited analysis of different size fractions which contain different relative proportions of individual diatom species, however, suggests no species related vital effect within the errors (±0.2-0.3) of the technique (Juillet-Leclerc and Labeyrie 1987; Shemesh et al. 1995). Connected to this are the possible problems associated with samples containing a mixture of diatom species which bloom at different times in the year. As such, $\delta^{18}O_{(diatom)}$ studies can potentially be hampered by bloom-signal dilution resulting in abnormal $\delta^{18}O_{(diatom)}$ results (Raubitschek 1999; Leng et al. 2001). Recent work from the North West Pacific across the onset of major Northern Hemisphere Glaciation (NHG), c. 2.75-2.73 Ma has illustrated the importance of extracting season-specific samples for diatom isotope analysis. Whereas the $\delta^{18}O$ record from spring calcifying planktonic foraminifera indicate a significant, 7.5°C, cooling of the surface waters at the ONHG (Maslin et al. 1996), an autumnal blooming $\delta^{18}O_{(diatom)}$ species record consisting of *Cosinodiscus radiatus* and *Cosinodiscus marginatus* reveals a 4.6‰ decrease, equating to a significant freshening and warming of the surface waters (Haug et al. 2005). Consequently, by analysing $\delta^{18}O_{(diatom)}$ the provision exists not only for existing isotope records from foraminifera to be extended further into the Southern Ocean, where diatoms often represent the solitary source of palaeoclimatic information, but to also extract further information from existing study regions where interpretations solely based on foraminiferal isotopes records often provide single season-specific information (e.g., Haug et al. (2005)).

A further issue concerning the use of $\delta^{18}O_{(diatom)}$ as a palaeoenvironmental proxy stems from the culture analysis of modern marine and lacustrine diatom populations which demonstrate $\delta^{18}O_{(diatom)}$ values significantly below those modelled by the fossil derived silica-water fractionation curve of Juillet-Leclerc and Labeyrie (1987) (Schmidt et al. 1997; Brandriss et al. 1998). While the reasoning behind this remains unclear, evidence suggests that the $\delta^{18}O_{(diatom)}$ record may reflect a function of successive isotopic exchanges between the diatom silica and deep ocean/surface sediment pore water (Schmidt et al. 1997; 2001). Consequently, in some situations, $\delta^{18}O_{(diatom)}$ may simply reflect the $\delta^{18}O$ in the surface sediment pore-water rather than condition in the surface waters as the final isotope signal may only become permanently incorporated into the inner layer of the diatom silica following burial. As such further laboratory and in-field studies are required to replicate the experiments of Schmidt et al. (1997; 2001) and to investigate the presence of so called secondary isotope exchanges in the -Si-O-Si layer outside the surface waters. If replicated, these results cast a significant level of uncertainty upon the reliability of $\delta^{18}O_{(diatom)}$ based reconstruction. Additional uncertainties also surrounds the role of diatom growth rates in causing non-equilibrium oxygen isotope fractionation into diatom frustules with, for example, analysis of $\delta^{18}O_{(diatom)}$ from culture experiments indicating less oxygen isotope fractionation in faster growing diatoms (Schmidt et al 2001). If shown to produce significant fractionation differences, increased interpretation and attention to the data is needed where diatom growth rates, for example through changes in iron fertilisation in the Southern Ocean, may have altered through time.

At present, despite significant successes in developing $\delta^{18}O_{(diatom)}$ as a palaeoenvironmental proxy, comparatively few studies have taken full advantage of the new technique to provide detailed palaeoceanographic reconstructions when compared to its increasing widespread use in lacustrine sediments (see Leng et al. this volume). As highlighted here, a number of issues remain to be resolved through culture experiments and laboratory work. Despite this, the potential for $\delta^{18}O_{(diatom)}$ to answer many questions about changes within the ocean over glacial and inter-glacial cycles is considerable and deserving of future attention. This is emphasised if season specific diatom species can be regularly extracted and separated to provide palaeoceanographic records for both the spring and autumn time intervals.

Corals

An additional source of marine stable isotope data lies in the record obtained from living and fossil corals. Two main types of corals exist; near-surface-dwelling photosynthesising corals and non-photosynthesising deep ocean corals. Near-surface-dwelling photosynthesising corals form reefs distributed along tropical coastlines wherever relatively clear, warm water is present in depths ranging from low tide level down to tens of metres (see Figure 2c). In contrast, non-photosynthesising deep ocean corals are present throughout the deep ocean providing an insight into the bottom water condition in which they live. One species, *Desmophyllum cristagalli*, grows between 0.2 mm y^{-1} and 1.0 mm yr^{-1} enabling one specimen to potentially provide a record of bottom water conditions over a few hundred years (Adkins et al. 1998). Corals construct their skeletons from aragonite ($CaCO_3$) and incorporate multiple geochemical tracers; including stable isotopes of carbon and oxygen and other trace contaminants retained within the skeletal lattice (e.g., Sr, U, Mg, Ba, Cd, Mn).

Corals can be dated by annual banding, radiocarbon or U-series methods (e.g., Edwards et al. (1987); Bard et al. (1990); Adkins et al. (1998); Cole (2001)). However, care is necessary when analysing coral in palaeoceanographic studies since each species' geochemistry is intimately connected with its own biology. Consequently, individual taxa possess significantly different relationships between individual geochemical proxies and the target climatic parameters. In addition, the structural complexity of coral skeletons can complicate the development and interpretation of it's climate records. As with foraminifera, coralline aragonite $\delta^{18}O$ reflects seawater temperatures, seawater $\delta^{18}O$ and a biological offset. Coral studies that investigate mid-Holocene to recent climatic change, however, do not have to contend with the ice volume $\delta^{18}O$ signal, which for the last 5,000 years is almost negligible. Coral calibration studies have obtained a temperature-$\delta^{18}O$ relationship of 0.18-0.26‰°C^{-1} (Cole 2003), similar to the foraminiferal calibration of 0.23‰°C^{-1}. However, as many of the most important near surface coral records are from the tropics, rainfall imparts a significant influence onto the $\delta^{18}O$ records. There is also a significant biological offset or vital effect of –2‰ to   –5‰ which, although large when compared to the isotope climate signal in corals, appears constant for individual taxa through time.

Coral $\delta^{18}O$ records have been used to place the 20th century climate of the tropics into the context of the last three to four centuries. If the $\delta^{18}O_{(coral)}$ record is interpreted purely in terms of sea surface temperature then a 1-3°C warming must have occurred at most locations since the 18th century (Cole 2003). The only exception to this is the Galapagos where upwelling maintains relatively cool surface waters. However, Hendy et al. (2002) compared records of $\delta^{18}O_{(coral)}$ with U/Ca and Sr/Ca to suggest a significant salinity reduction throughout the 15th to 19th Century. If accurate, this freshening of the tropical ocean during the Little Ice Age would have lowered the $\delta^{18}O$ of the ocean and hence $\delta^{18}O_{(coral)}$ affecting any comparison of temperature records through time. Large $\delta^{18}O_{(coral)}$ differences over the past 4-5 centuries may also be related to El-Niño Southern Oscillation (ENSO) variability, thereby preventing accurate quantitative interpretation of coral oxygen isotopes as a SST record.

Coral $\delta^{18}O$ records have, however, been central to the study of past ENSO variability since both salinity and temperature influence $\delta^{18}O$ in the same direction. Comparisons between the instrumental records of ENSO and $\delta^{18}O$ records reconstructed from the western, central and eastern Pacific demonstrates an excellent correlation. During an El Niño event in the western Pacific conditions are drier (increased evaporation and higher $\delta^{18}O$ rainfall) and colder, both of which increase surface water $\delta^{18}O$. Conversely, in the eastern and central Pacific conditions are wetter (less evaporation and lower $\delta^{18}O$ rainfall) and warmer which lead to lower $\delta^{18}O$ surface waters. Holocene coral $\delta^{18}O$ records indicate that $\delta^{18}O$ variability associated with ENSO only began in the Mid-Holocene, at c. 5,000 yrs BP (Gagan et al. 1998; Tudhope et al. 2001). Coral $\delta^{18}O$ records indicate a shift in the ENSO cycle at the beginning of the twentieth century from a 10-15 year to a 5-7 year cycle (Cole et al. 2001) with an abrupt shift to more intense and frequent El Niño events (3-5 years) from 1976.

Carbon isotopes in marine sediments

Carbon isotope records from marine carbonates, organic matter and intrinsic carbon within biogenic silica enable an understanding of the global carbon system documenting the complex interactions within the atmosphere-ocean-biosphere system (e.g., Rosenthal et al. (2000); Shemesh et al. (2002); Houghton (2004); Sundquist and Vissr (2004)). Carbon isotope ratios, $\delta^{13}C$, in carbonate are measured alongside oxygen isotopes. However, $\delta^{13}C$ ratios have only routinely been published with $\delta^{18}O$ data since the early 1970s. In many of the early papers limited interpretation was made of the $\delta^{13}C$ records as they are inherently complicated containing large spatial variability. Unlike $\delta^{18}O$, only a small direct relationship exists between carbonate carbon isotopes and temperature (c. 0.0035‰ °C^{-1}) although temperature affects the solubility of CO_2 into sea water and thus indirectly the $\delta^{13}C$ composition of phytoplankton and marine carbonates. Instead, $\delta^{13}C$ in marine carbonates reflects a combination of oceanic productivity, ocean circulation and the relative storage of carbon in organic matter and carbonates while isotopes in organic matter reflect ocean productivity, surface water CO_2, input of continental derived material and the occurrence of diagenesis.

The primary fractionation process in the carbon cycle is photosynthesis, thereby enabling carbon isotopes to provide an insight into past changes in ocean productivity. Nearly all carbon within organic compounds in the biosphere is ultimately derived from photosynthetically produced material. Photosynthetic fixation of carbon involves a very large fractionation with ^{12}C preferentially incorporated into the organic compound causing very low $\delta^{13}C$ in organic matter with shifts of up to 25‰. Average $\delta^{13}C$ values of organic carbon in land plants (higher plants) vary according to the chemical pathway of photosynthesis (Smith and Epstein 1971): plants using the C3 (e.g., trees) (Calvin-Benson or non-Kranz) photosynthesis pathway have a fractionation of between −22‰ and −30‰ (average −27‰) (Anderson and Arthur 1983; Ehleringer et al. 1997), compared to an atmosphere isotope composition of −6‰ to −7‰. Plants using the C4 (Hatch-Slack or Kranz) pathway, including tropical and marsh grasses, have a $\delta^{13}C$ range of between −9‰ to −15‰ (average −13‰). C4 plants re-evolved about 7 Ma (Cerling et al. 1997) and generally out-compete C3 plants at low humidity and low pCO_2 conditions, such as during the LGM, due to their internal 'carbon pump' system in addition to their primary carbon fixing enzyme (PEP-carboxylase) which does not react with oxygen. Consequently, increased photorespiration under low carbon dioxide conditions does not inhibit photosynthesis (e.g., Ehleringer et al. (1997)). This different photosynthetic pathway also explains why C4 photosynthesis does not fractionate CO_2 as much as C3 plants. An additional third type of plant, the CAM (Crassulacean Acid Metabolism), uses either C3 or C4 photosynthesis depending on water availability. In terms of occurrence though, CAM is negligible compared with other pathways (see review in Anderson and Arthur (1983)).

The photosynthetic reaction pathways of marine phytoplankton (protista, which use dissolved CO_2 in photosynthesis) are much less well understood than that those for land plants. However, organic matter in marine plankton indicate a range of values from between −10‰ and −31‰ with most results constrained between −17‰ and −22‰. As such, Descdas-Gros and Fontugue (1985) have speculated that phytoplankton utilise a spectrum of pathways ranging from the C3 to the C4 dependent on the environmental stress (e.g., nutrient and light availability). There is also a difference in fractionation during the formation of organic matter dependent on the temperature of surface waters. At high temperatures the solubility of carbon dioxide is low, thus the availability of CO_2 to phytoplankton is limited. Under these conditions the fractionation is reduced resulting in $\delta^{13}C_{org}$ being as high as −13‰. By contrast at low temperatures the solubility of CO_2 increases resulting in a much larger fractionation of up to −32‰. Experimental work has demonstrated this relationship between $\delta^{13}C_{org}$ and temperature to be non-linear with no significant decrease in the $\delta^{13}C$ of plankton between 14°C and 30°C but a rapid decrease in $\delta^{13}C$ between −2°C and 10°C. Consequently, the precise relationship may vary between 0.5‰ °C^{-1} and 1.5‰ °C^{-1}. (Sackett et al. 1965; Rau et al. 1989). Hence, palaeoceanographic reconstructions using tropical cores can minimise the possible indirect influence of SST on organic and calcite carbon isotopes as the relationship is minimal at high temperatures (e.g., Fontugne and Duplessy (1986); Sarnthein et al. (1988); Jasper and Gagosian (1990)).

Palaeoproductivity

Once organic matter leaves the photic zone, particulate organic carbon (POC) rain has three possible fates. Firstly, before the POC reaches the intermediate or deep waters, oxidisation and respiration can occur forming Dissolved Inorganic Carbon (DIC), which is recycled to the surface; secondly, oxidisation and respiration and subsequent upwelling to the surface from the sub-surface waters or thirdly, incorporation and preservation within the sediment. Speculation and debate prevails about the exact relative proportions of the POC which enters each of these fates. Despite the complexities listed above, photosynthetic activity in the photic zone of the oceans results in a strong depletion of the surface water dissolved CO_2. The carbon isotopes of organic matter ($\delta^{13}C_{org}$) incorporated in marine sediments may therefore often reflect primarily surface water productivity. For this interpretation to be valid, other influences such as ocean circulation, inputs of terrestrial organic matter (e.g., Sackett (1964); Fontugue and Duplessy (1986); Jasper and Gagosian (1990)) and diagenesis of organic matter within the sediment must be taken into account.

Sarnthein et al. (1988) conducted a detailed study in an attempt to overcome these problems with bulk organic carbon isotopes ($\delta^{13}C_{org}$). Using data from a variety of locations, including coastal and deep water cores, $\delta^{13}C_{(org)}$ results were interpreted purely in terms of changes in oceanic productivity. High late glacial/Holocene $\delta^{13}C_{org}$ values observed in the equatorial eastern Atlantic by Sarnthein et al. (1988) were linked to similar, $\delta^{13}C_{org}$ values of −18.7‰ found in the modern Peru upwelling region. This would suggest that export productivity upwelling cells that existed in the low and mid-latitudes during the last glacial period declined at the onset of deglaciation. Based on two lines of evidence, changes in the terrestrial biosphere storage and level of terrestrial sourced organic matter in the oceans were argued to be small and thus insignificant (see section below section on carbon storage and release). Firstly, average $\delta^{13}C$ value for C3 plants is − 26‰. Consequentially, if C3 plants were a major contributor to bulk organic matter then higher $\delta^{13}C_{org}$ values would have been expected than the observed −18.7‰. Secondly, if C4 plants were the dominant terrestrial source, Carbon:Nitrogen (C/N) ratios of about 20-50 would have been expected whereas core C/N ratios were limited to c. 8-9, levels characteristic of marine plankton (Müller et al. 1983).

Other planktonic organisms living in the surface water that form calcareous tests from DIC also record the surface water productivity isotope effect. Common values recorded in modern planktonic foraminifera are about +2 to +3‰ while the average modern ocean $\delta^{13}C$ of the total DIC is around 0‰, reflecting the reduced dissolved CO_2 available after phytoplankton usage for photosynthesis. The interpretation that planktonic foraminifera and coccolith (nannofossil) carbonate carbon isotopes both solely represent changes in surface water productivity requires attention as nannoplankton are autotrophic and therefore use photosynthesis. Consequently, the dissolved carbonate used in test formation may be influenced by metabolic processes in the cell itself. Measurements made with microprobes around actively photosynthesising individuals have demonstrated local deviations in the isotope composition of dissolved carbonate close to the cells. In addition many species of planktonic foraminifera, especially many spinose species, contain algal symbionts (Hemleben et al. 1988) causing the DIC used in test calcification to not be entirely representative of ocean water with potential influences originating from metabolic processes. Examples of the use of planktonic foraminifera carbon isotopes to reconstruct surface water productivity can be found in Berger et al. (1978), Ganssen and Sarnthein (1983), Mortlock et al. (1991) and Ruhlemann et al. (1999). Additional information concerning surface water productivity can also be obtained by using two or more species that live at different water depths and/or in different seasons (e.g., Kroon and Ganssen (1989)). Investigations have shown that $\delta^{13}C$ values of *G. bulloides* increase with higher productivity whereas $\delta^{13}C$ values in *G. ruber* decrease with increased upwelling of ^{12}C-rich subsurface water (e.g., Ganssen (1983); Kroon and Ganssen (1989); Curry et al. (1992)). Schneider et al. (1994) used the difference in *G. ruber* (pink) and *G. bulloides* $\delta^{13}C$ as a proxy for changes in upwelling and biological productivity off the Congo Basin to support other proxy data which reveals temporal variations in surface water productivity over the last 200 kyr.

It has been argued though that planktonic foraminiferal carbon isotopes do not represent surface water productivity, but nutrient availability and usage (e.g., Wefer et al. (1999)). Berger et al. (1997) argue there is a clear difference between productivity which controls the organic flux out of the photic zone and the availability of nutrients which controls the amount of biomass in the photic zone. Expansion of low $\delta^{13}C$ organic matter in the photic zone and thus export from the surface waters depends on the amount of nitrate and phosphate available to produce organic matter. Moreover the nutrients supplied to the photic zone usually originate from subsurface water which may contain an excess of ^{12}C from CO_2 liberated from oxidised recycled organic matter. Hence planktonic foraminifera carbon isotopes contain a relationship between both nutrient supply and carbon isotopes (e.g., Broecker and Peng (1982)) with $\delta^{13}C$ a mixed signal based on CO_2 exchange with the atmosphere, removal of carbon in solid form by export productivity and re-supply of dissolved carbon from subsurface water.

Wefer et al. (1999) consequentially argue that organic and carbonate surface water carbon isotopes reflect surface water fertility rather than just productivity. As such, shallow-living and deep-living planktonic foraminifera can instead be utilised to reconstruct thermocline fertility, thickness of the mixed layer in addition to general changes in surface water fertility. Carbon isotopes of benthic foraminifera may also reflect surface water productivity since as organic matter falls from the surface zone into deeper water it interacts with oxygen in the water column. Oxidation of the organic matter releases carbon dioxide lower in $\delta^{13}C$ resulting in a decrease in the $\delta^{13}C$ DIC pool. Benthic foraminifera, therefore, usually have low $\delta^{13}C$ values compared with the surface water. So, theoretically, the difference between the planktonic and benthic foraminifera $\delta^{13}C$ at a particular location should reflect the efficiency of organic carbon export from the surface to the deeper ocean (Figure 5): referred to as the carbon pump (Broecker 1982). Benthic-planktonic foraminifera $\Delta\delta^{13}C$ may, however, reflect other factors including the amount of surface water productivity, the amount of recycling of organic matter in the surface ocean and the amount of oxygen in the water column which is connected to both the total amount of organic matter falling and the age of the deep water (see below section on palaeo-pCO_2 reconstructions).

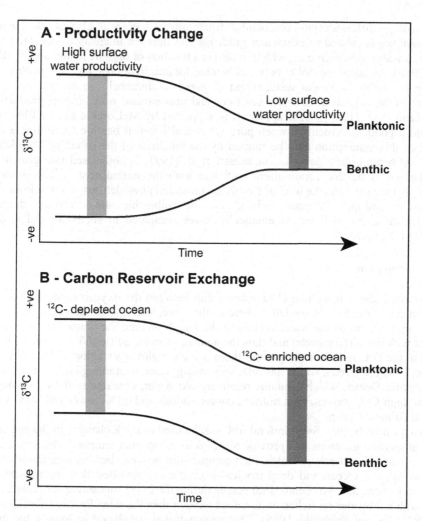

Figure 5. Theoretical planktonic and benthic carbon isotope records illustrating: A) the affects of surface water productivity changes on the planktonic to benthic carbon isotopic ratio and B) changes in the whole ocean carbon isotopic value through exchanges with other reservoirs such as the land biomass.

One example of the use of the difference between planktonic and benthic foraminiferal $\delta^{13}C$ is at the K/T boundary (~65 Ma) (e.g., Ivany and Salawitch (1993)). The carbon isotope record has been used to suggest the presence of both reduced surface water productivity and anoxic/hypoxic bottom water conditions after the K/T impact/extinction event (Coccioni et al. 1993; Kaiho et al. 1999; Norris et al. 2001; Culver 2003; Alegret et al. 2001; 2002; 2003). However, the interpretation may be more complicated as both the surface and deep ocean became progressively lower in $\delta^{13}C$ over the subsequent 0.5 Myrs, suggesting a possible release of highly negative $\delta^{13}C$ carbon from gas hydrates or the terrestrial biosphere (Melosh et al. 1990; Day and Maslin in press) (see below section on carbon storage and release).

The isotope difference between benthic foraminiferal species is also of interest since this difference is related to porewater gradients and thus the intensity of organic matter oxidation within the sediments, which in turn is a function of organic matter supply (Zahn et al. 1986). As such, the $\Delta\delta^{13}C$ between benthic foraminifera taxa, e.g., *C. wuellerstorfi* (epifaunal – on the sediment surface) and *U. peregina* (infaunal – in the sediment), is a measure of the organic flux to the sea floor and thus surface water export productivity (e.g., Zahn et al. (1986)). This approach is supported by McCorckle et al. (1990) who revealed a good agreement between pore water and infaunal benthic foraminifera $\delta^{13}C$. However, this assumption may be limited by the influence of the so called "Mackensen effect" on benthic foraminifera. Mackensen et al. (1993) hypothesised that growth and reproduction of benthic foraminifera coincides with the intermittent, usually seasonal, flux of phytodetritus. As the $\delta^{13}C$ of freshly accumulated phytodetritus is 3-4‰ lower than the surface sediment organic carbon, $\delta^{13}C$ of epibenthic shells formed during a phytodetritus event, will be systematically lower compared to shells formed at other periods of the year.

Ocean circulation

As discussed above there is a clear relationship between the oxygen content of oceanic deep water and the $\delta^{13}C$ of the DIC. There is, therefore, a shift to lower values in the DIC $\delta^{13}C$ with the 'age' of the water related to the length of time the water has been out of contact with the surface ocean and thus the amount of oxidised (low $\delta^{13}C$) organic matter added to the DIC reservoir. At present there is a clear global variation of DIC $\delta^{13}C$ with the deep Atlantic Ocean, which has relatively young water, containing high $\delta^{13}C$ while the deep Pacific Ocean, which contains relatively old water, characterised by low oxygen content, high CO_2 content, high nutrient concentrations and c.1‰ lower DIC $\delta^{13}C$ values than the Atlantic Ocean.

Consequently, benthic foraminiferal $\delta^{13}C$ can be used to track changes in the age of the water at specific locations and provide a means to reconstruct temporal changes in deep ocean circulation patterns and rates. First comparisons between benthic foraminiferal $\delta^{13}C$ in deep Atlantic Ocean and deep tropical Pacific cores revealed that the present $\delta^{13}C$ gradient between the two oceans disappeared during the LGM suggesting the Atlantic was filled with cold water no richer in dissolved oxygen than the deep Pacific (Shackleton et al. 1983; Mix and Fairbanks 1985). This approach was developed to look at the north-south cross section of the Atlantic and has demonstrated significant changes in ocean circulation during the LGM (e.g., Duplessy et al. (1988); Curry et al. (1988); Sarnthein et al. (1994)). Comparison of modern and LGM benthic foraminiferal $\delta^{13}C$ also demonstrates a significant reduction in the amount of NADW formed during the LGM with the reduced NADW not penetrating as far south as the modern day (e.g., Duplessy et al. (1988); Curry et al. (1988); Sarnthein et al. (1994), see Figure 6). This evidence, combined with bottom water temperature estimates based on oxygen isotopes, modelling studies (see Seidov et al. (2001)) and foraminifera Cd/Ca, Ba/Ca, Zn/Ca, Mg/Ca, radio-carbon and $^{231}Pa/^{230}Th$ ratios are critical in highlighting that the circulation of the deep ocean can be highly variable and a key component of the global climate system (Lynch-Stieglitz 2004). In addition, benthic foraminiferal $\delta^{13}C$ has been employed to track variations in the strength of the NADW during the onset of major NHG (e.g., Haug and

Tiedemann (1998)), the Mid-Pleistocene Revolution (e.g., Raymo et al. (1997)) and glacial terminations (e.g., Sarnthein and Tiedemann (1990)) including the last deglaciation (e.g., Charles and Fairbanks (1992); Boyle (1995)).

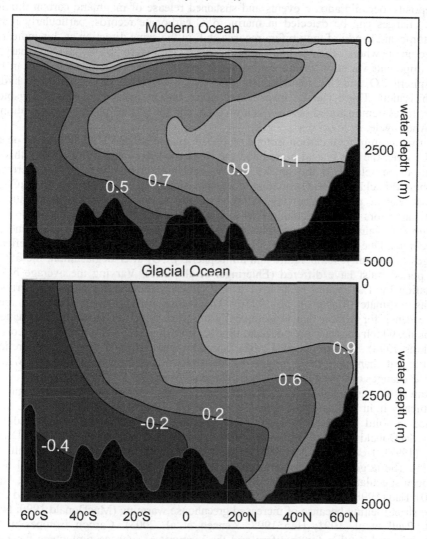

Figure 6. Modern and LGM distributions of $\delta^{13}C$ in the Atlantic Ocean inferred from benthic foraminifera. Figure amended and adapted from Labeyrie et al. (1992) and www.ngdc.noaa.gov

Carbon storage and release

The global carbon cycle can alter on a variety of time-scales up to tens of millions of years (Sundquist and Visser 2004). Each carbon reservoir, whether the deep mantle or the atmosphere, has a carbon isotope signature which can be used to trace exchanges between these reservoirs. On longer geological timescales these exchanges include the carbonate weathering-sedimentation cycles and the organic carbon production-consumption-oxidation cycles (Sundquist and Visser 2004) (e.g., the formation and erosion of coal and oil deposits, oceanic anoxic events and sustained release of magmatic carbon dioxide). These changes can be detected in marine $\delta^{13}C$ carbonate records; particularly if both planktonic and benthic foraminifera records shift in the same direction as this indicates a change on the whole ocean carbon isotopic content (see Figure 5b). In the Quaternary, the most important variation in the carbon cycle are the glacial-interglacial shifts in both atmospheric CO_2 and CH_4, in addition to short-term releases of carbon such as gas hydrate destabilisation. These can be investigated using carbon isotope records, and isotopic budgeting as demonstrated below when investigating the glacial-interglacial variations in the carbon cycle.

Benthic foraminiferal carbon isotope records display a c.0.4‰ $\delta^{13}C$ increase in oceanic DIC between the last glacial period and the Holocene (see Table 2). Originally this was though to be related to the uptake of low $\delta^{13}C$ carbon by the expanding terrestrial biosphere (Shackleton 1977) equating to an expansion of the biosphere by 500 GtC since the last glacial period (Crowley 1995). Two assumptions, however, may introduce significant errors in this calculation (Maslin and Thomas 2003). First, an estimate of the average $\delta^{13}C$ value for the terrestrial biosphere is required for both the LGM and after deglaciation. Due to differences in the ratio of C3 to C4 plants at both intervals, reflecting changes in humidity, temperature and atmospheric pCO_2 after deglaciation, the $\delta^{13}C$ for each period must have differed (Ehleringer et al. 1997). Varying the average $\delta^{13}C$ of vegetation by 1‰ results in at least a ±50 GtC variation in the modelled terrestrial biomass estimate (Maslin et al. 1995b). The second assumption is that there are no sources/sinks for carbon that is isotopically different from terrestrial organic matter or carbonate, which can vary on a decadal time scale (Kump and Arthur 1999; Francois and Godderis 1998). Brovkin et al. (2002), however, demonstrated that it is theoretically possible that changes in the average $\delta^{13}C$ of marine biota during deglaciation account for the $\delta^{13}C$ changes observed in atmospheric carbon dioxide.

It can also be argued that the most important source of carbon are the large volumes of isotopically light methane gas, with $\delta^{13}C$ values of –40‰ to –100‰ (average c.–60‰), trapped in solid hydrates (clathrates) in permafrost and continental margin sediments (e.g., MacDonald (1990); Kvenvolden (1993); (1998); Matsumoto et al. (1996); Dickens et al. (1997); Henriet and Mienert (1998); Buffett (2000); Dickens (2001); Paull et al. (2003)). The large increase in atmospheric methane at the end of the last glacial period has been speculated to originate from methane release from gas hydrates (e.g., Nisbet (1990); Haq (1998); Kennett et al. (2003)) which may have accelerated the rate and extent of ice sheet melting because of increased greenhouse warming (MacDonald 1990; Nisbet 1990; Paull et al. 1991; Haq 1998; Kennett et al. 2003). Carbon isotopes in both planktonic and benthic foraminifera and the biomarkers such as diploptene have been used to trace rapid releases of highly negative $\delta^{13}C$ methane from gas hydrate dissociation. During the last glacial period these negative carbon isotope spikes have been observed in the Coral Sea off the coast of Papua New-Guinea (de Garidel-Thoron et al. 2004), North West Pacific Ocean (Uchida et al. 2004), Okhotsk Sea (Lembke. pers comm.), Santa Barbara Basin (Kennett et al. 2000; Kennett and Fackler-Adams 2000; Kennett et al. 2003) and the Amazon Fan (Maslin et al. in press). By adapting carbon isotope budgeting method to include methane release from hydrates it becomes possible to calculate the amount of methane hydrate potentially released during deglaciation. Maslin and Thomas

(2003) adapted the Shackleton (1977) method to (see also Figure 7):

$$(C_O \times \Delta\delta^{13}C_O) + (C_A \times \Delta\delta^{13}C_A) + (\Delta C_L \times \delta^{13}C_L) + (\Delta C_{GH} \times \delta^{13}C_{GH}) = 0 \qquad (13)$$

where C_O is the total carbon storage of the ocean (~38,000 GtC); $\Delta\delta^{13}C_O$, deglacial ocean carbon isotopic shift (~0.4‰, see text below); C_A, atmosphere carbon storage at the end of the last deglaciation (~ 560 GtC); $\Delta\delta^{13}C_A$, deglacial atmospheric carbon isotopic shift (in the range of 0.3-0.6‰, average 0.45‰ used); ΔC_L, shift in carbon storage of the land biomass (vegetation and soils) in GtC; $\delta^{13}C_L$, average carbon isotope composition of land biomass expansion (–15 to –25‰, see text below); ΔC_{GH}, gas hydrate release during the last deglaciation in GtC; $\delta^{13}C_{GH}$, average carbon isotope composition of gas hydrates released (–40‰ to –100‰).

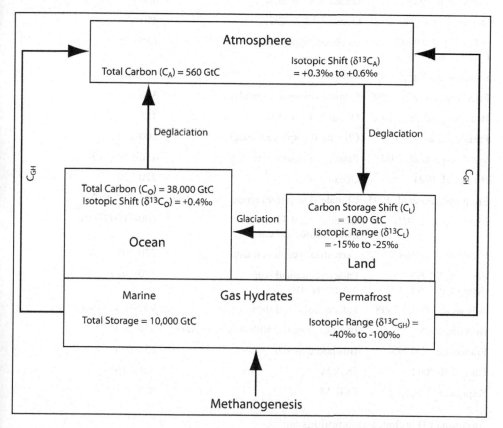

Figure 7. Adapted carbon isotope budget method for estimating glacial–interglacial shifts in land biomass and the release of gas hydrate carbon (adapted from Maslin and Thomas (2003)).

Table 2. Estimates of the expansion of carbon stored in the terrestrial biosphere since the LGM using different methods illustrating a range of glacial-Holocene $\delta^{13}C$ shifts in oceanic carbon isotope studies.

Source	Method	Terrestrial biomass increase (GtC) range (average)
Shackleton 1977	Ocean $\delta^{13}C$, 0.70‰	1000
Duplessy et al. 1984	Ocean $\delta^{13}C$, 0.15‰	220
Berger and Vincent 1986	Ocean $\delta^{13}C$, 0.40‰	570
Curry et al. 1988	Ocean $\delta^{13}C$, 0.46‰	650
Duplessy et al. 1988	Ocean $\delta^{13}C$, 0.32‰	450
Adams et al. 1990 Faure et al. 1996	Palaeoecological data	1350
Prentice and Fung 1990	Climate model	−30±50
Friedlingstein et al. 1992	Climate (biosphere) model	300
Broecker and Peng 1993	Ocean $\delta^{13}C$, 0.35‰	425
Prentice et al. 1993	Climate (biosphere) model	300-700
van Campo et al. 1993	Palaeoecological data	430-930 (715)
Bird et al. 1994	Ocean $\delta^{13}C$	270-720
Friedlingstein et al. 1995	Climate (biosphere) model	507-717 (612)
Maslin et al. 1995b	Ocean $\delta^{13}C$, 0.40‰, and palaeoecological data	400-1100 (700)
Crowley et al. 1995	Terrestrial vegetation data	750-1050
Peng et al. 1995 Peng et al. 1998	Palaeoecological data biosphere data	470-1014
Adams and Faure 1998	Palaeoecological data	900-1900 (1500)
Beerling 1998	Biosphere model with $\delta^{13}C$ inventory	550-680
Francois et al. 1998	Biosphere model	134-606
Otto et al. 2002	DGVM	828-1106
Kaplan et al. 2002	DGVM	821

Equation (13) includes assumptions on:

1) The average shift in the global oceanic carbon $\delta^{13}C$ values ($\Delta\delta^{13}C_O$). The average of the best estimates is c. 0.40±0.05‰ (Berger and Vincent (1986); Curry et al. (1988); Duplessy et al. (1988); Broecker and Peng (1993), see Table 1). However, these estimates do not include changes in the carbon isotope composition of DIC in the top 1 km of the marine water column.

2) The expansion of the terrestrial biomass (ΔC_L). Maslin and Thomas (2003) reviewed

the palaeovegetation to estimate that the maximum shift in terrestrial biomass expansion between 18 and 8 ka is c. 1,000 GtC. This is in agreement with the conservative palaeovegetation mapping approach (Adams et al. 1990; Maslin et al. 1995b) and recent Dynamic Global Vegetation Models (DGVMs) which suggest a shift of between 820 and 1100 GtC (Otto et al. 2002; Kaplan et al. 2002). However, neither DGVMs results consider peat accumulation during the late Holocene, which could contribute several hundred GtC (e.g., Gajewski et al. 2001).

3. The average isotope value of the deglacial vegetation expansion ($\delta^{13}C_L$). This requires assumptions over the relative expansion of C3 and C4 plants since the last deglaciation. From palaeovegetation maps it can be estimated that C3-dominated vegetation types expanded approximately four times more than C4-dominated types (Maslin et al. 1995b; Adams et al. 1990; Adams and Faure 1998; Beerling 1999) during the last deglaciation, which produces an estimated mean $\delta^{13}C$ value of the terrestrial biomass of c. –22‰.

4. The average isotope value of the gas hydrate methane released ($\delta^{13}C_{GH}$). The isotope composition of gas hydrates can vary between –40‰ or –100‰ (MacDonald 1990; Matsumoto et al. 1996; Dickens et al. 1997; Kvenvolden 1998; Buffett 2000) with a value of –60‰ used as an approximation (Dickens 2001).

Using these constraints, it is possible to balance the global carbon isotope inventory with palaeovegetation estimates of deglacial expansion of the terrestrial biomass with a gas hydrate release in the range of 70 to 180 GtC. This represents 1% to 2% of the present gas hydrate reservoir (MacDonald 1990; Kvenvolden 1998) and less than half of the present reservoir in the Arctic permafrost (MacDonald 1990). If gas hydrate $\delta^{13}C$ composition was close to the average observed value of –60‰ and the average biosphere $\delta^{13}C$ value close to –22‰, a release of 120 GtC would be sufficient which represents only 10% of all the methane produced between 18 kyr BP and 8 kyr BP (Brook et al. 1999; 2000; Severinghaus and Brook 1999). Consequentially, for the carbon budget model to release sufficient gas hydrates to account for all the additional methane released during deglaciation, extreme estimates of each variable need to be used, e.g., $\delta^{13}C_L = -25‰$, $\Delta C_L = 1300$ GtC, $\Delta\delta^{13}C_O = 0.3‰$, and $\delta^{13}C_{GH} = -40‰$, suggesting that arguments that clathrate methane is the dominant source of the atmosphere rise in pCH_4 (Kennett et al. 2003), may not be correct. While gas hydrates provide a significant contribution to the carbon budget, other possible sources must also reside in tropical wetlands and temperate wetlands-peatlands to account for the other 70% of the enhanced deglaciation atmospheric methane increase.

Palaeo-pCO₂ reconstruction

It has been recognised that photoautotrophs (e.g., photosynthesising algae) have widely varying carbon isotope compositions, in part due to carbon fractionation during photosynthesis. Degens et al. (1968), Calder and Parker (1973) and Morris (1980), however, observed that all algae grown in cultures have increasingly $\delta^{13}C$ as the supply of CO_2 increased, suggesting that anomalously lower $\delta^{13}C$ values in geological samples could be interpreted as a result of enhanced pCO_2 (Popp et al. 1989; Freeman and Hayes 1992). While subsequent developments have refined this approach, the principle remains the same with organic matter $\delta^{13}C$ values being interpreted as a palaeo-pCO_2 proxy (e.g., Jasper and Hayes (1990); Pagani et al. (1999)). In this section the basic principles of this approach and some secondary influences are discussed. In addition, recent developments have enabled palaeo-pCO$_2$ to be reconstructed using boron isotopes in foraminiferal carbonate, as discussed below.

To avoid the problems related to multiple sources of carbon which affect bulk organic matter δ^{13}C (see above section on palaeoproductivity) the carbon isotope composition of the 'biomarker' C$_{37}$ di-unsaturated alkenones is commonly used to estimate the amount of isotope fractionation that occurred during photosynthesis, thereby providing an estimate of the dissolved CO$_2$ concentration in surface waters. Popp et al. (1989) suggested an empirical relationship linking isotope fractionation during photosynthesis and CO$_2$ concentration:

$$\varepsilon_p = a \log CO_{2aq} + b \qquad (14)$$

where ε_p is carbon isotope discrimination during photosynthesis; CO$_{2aq}$, is the concentration of dissolved CO$_2$ (μM); a and b are constants representative of the physiological, ecological, and environmental factors which influence the ε_p-CO$_2$ relationship.

Rau et al. (1992), Jasper et al. (1994) and Bidigare et al. (1997) proposed a simpler expression of the relationship:

$$\varepsilon_p = \varepsilon_f - \frac{b}{CO_{2aq}} \qquad (15)$$

where 'b' represents an integration of the physiological variables (a and b in Equation (14)) to include growth rate and cell geometry, affecting the total carbon isotope fractionation during photosynthesis (Jasper et al. 1994; Bidigare et al. 1997); ε_f, represents the carbon isotopic fractionation due to carbon fixation.

Popp et al. (1989) derived another estimate of ε_p by assuming the processes involved in photosynthetic fixation of carbon to be the transportation of inorganic carbon to C-fixing enzymes ('mass-transport') and the formation of chemical bonds ('fixation'). By assuming a steady state, mass-balancing allows the determination of ε_p as:

$$\varepsilon_p = ([(\delta_p + 1000) / (\delta_d + 1000)] - 1) \, 1000 \qquad (16)$$

where δ_p is the isotope composition of the total biomass of the primary photosynthate; δ_d, is the isotope composition of dissolved CO$_2$.

Jasper and Hayes (1990) used Equation 16 to reconstruct tropical surface water pCO$_2$ over the late Pleistocene assuming that the carbon isotope composition of the calcitic test of the planktonic foraminifera *Globigerinoides ruber* was representative of δ_d and the isotope composition of C$_{37}$ alkenones ($\delta_{37:2}$) was representative of δ_p. The ε_p-CO$_2$ relationship was subsequently calculated by estimating the constants a and b from measurements of particulate organic carbon in the modern ocean, derived primarily from highly productive waters.

The biggest problem with the estimation of pCO$_2$ is, however, the determination of the constants related to physiological variables. Pancost and Pagani (in press) suggested the available data provided evidence for a significant correlation between the physiological-dependent term 'b' in equation (15) and the concentration of reactive soluble phosphate.

This relationship likely stems from the influence of phosphate on growth rates which influences ε_p (Bidigare et al. 1997). Pancost and Pagani (in press) suggest it is unlikely that $[PO_4^{3-}]$ alone is responsible for this variability in growth rate inferred from variations in 'b'. Instead, Bidigare et al. (1997) and Laws et al. (2001) propose that the availability of specific trace elements, such as Se, Co and Ni, ultimately impact the growth characteristics of natural Haptophyte populations. Consequently, $[PO_4^{3-}]$ acts as a proxy for these growth-limiting trace elements that also exhibit phosphate-like distributions in the modern ocean.   Recent studies have applied estimates from alkenone $\delta^{13}C$ in conjunction with phosphate estimates to reconstruct Late Quaternary surface water pCO_2 values in the South Atlantic (Andersen et al. 1999) and global Miocene-age pCO_2 trends (Pagani et al. 1999a;b).

One drawback to this approach is that the alkenone-pCO_2 proxy rests on the assumption that intercellular CO_{2aq} fixed during photosynthesis arrives by diffusive flux, or that the proportion of diffusive and actively-transported carbon flux is constant under a variety of environmental conditions although there is support for this assumption from culturing experiments (Nimer et al. 1992; Nimer and Merrett 1996; Popp et al. 1998). In addition, Pagani et al. (2002) have demonstrated the accuracy of the alkenone-pCO_2 technique by comparing reconstructed pre-industrial water-column $[CO_{2aq}]$ values from sedimentary alkenone $\delta^{13}C$ values at 20 sites across a North Pacific transect to both observed water-column CO_{2aq} values and estimated pre-industrial concentrations. The results of Pagani et al. (2002) support the use of alkenone-based ε_p values as a proxy for palaeo-pCO_2 reconstruction and suggest that the alkenone approach can resolve relatively small differences in water column CO_2 when phosphate concentrations and temperatures are well constrained. Their results further suggest that light-limited growth effects and/or active carbon uptake are negligible or that these processes have negligible effects on the carbon isotope compositions of Haptophytes in the natural environment.

Terrestrial vegetation

Marine bulk organic matter $\delta^{13}C$ records have been used to investigate variations in terrestrial vegetation. This approach is, however, limited to areas where the majority of organic matter within the marine sediment is derived from terrestrial sources. An example of this are the records derived from the Amazon Fan by Kastner and Goñi (2003). They found no variation in the bulk organic matter $\delta^{13}C$ values from samples taken from glacial-aged Amazon Fan sediments and modern samples from the Amazon River, concluding there were no major changes in the extent of the Amazon rainforest between glacial and interglacial periods.

A more discerning approach uses particular biomarkers which originate only from terrestrial plants. For example long-chain n-alkyl compounds are major components of epicuticular waxes from the leaves of vascular plants (Eglinton et al 1962; Eglinton and Hamilton 1967).   These compounds, which include n-alkanes, n-alkanols, n-alkanoic acids and wax esters, are relatively resistant to degradation making them suitable for use as higher plant biomarkers (Cranwell 1981). n-alkanes have been most thoroughly characterised isotopically, occurring with carbon chain-lengths ranging from C_{25} to C_{35} (Eglinton et al 1962; Eglinton and Hamilton 1967; Kolattukudy 1976). Collister et al. (1994) observed that in C3 and C4 plant species the n-alkanes are depleted relative to biomass by c. 5.9‰ and 9.9‰ respectively, with similar isotope shifts believed to exist for *n*-aldehydes and other biosynthetically related *n*-alkyl compounds (e.g., *n*-acids and *n*-alkanols). For example, Chikarashi and Naraoka (2003) observed similar differences between biomass and *n*-alkane $\delta^{13}C$ values in a range of C3 angiosperms and gymnosperms and C4 plants.

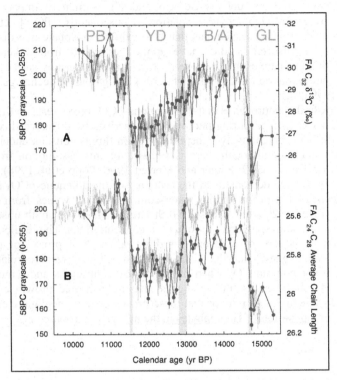

Figure 8. Comparison of grey scale and alkanic fatty acid leaf wax carbon isotope values and average chain length between 15.5 and 10 ka BP in the Cariaco Basin off the coast of Venezuela. PB = Preboreal, YD = Younger Dryas, B/A = Bølling/Allerød and GL = Glacial. By comparing the climate records and the alkanic fatty acid leaf wax carbon isotope values and average chain length, shifts between grassland and tropical forest can be observed in under 50 years (Hughen et al. 2004).

The majority of higher plant lipids found in marine or lacustrine sediments are thought to derive from sub-aerial land plants rather than submerged plants. Thus, atmospheric carbon dioxide is their sole carbon source causing controls on atmospheric CO_2 $\delta^{13}C$ to represent a major factor in determining higher plants' isotope compositions and consequently the $\delta^{13}C$ of higher plant biomarkers found in marine sediments. In the recent past, the primary control on atmospheric CO_2 $\delta^{13}C$ is the Suess effect. The Suess effect is caused by the burning of fossil fuels which have added lower $\delta^{13}C$ carbon to the atmosphere resulting in atmospheric carbon being c. 1.5‰ lower than pre-industrial times. Prior to that, atmospheric CO_2 $\delta^{13}C$ values appear to have varied little; for example since the LGM values have varied from –4‰ to –6.9‰ despite significant changes in CO_2 concentrations. On longer time scales, as discussed above, atmospheric carbon dioxide

$\delta^{13}C$ values have changed in-line with the entire ocean-atmosphere reservoir. In addition to global changes in atmospheric CO_2 $\delta^{13}C$ values, the carbon isotope composition of air in a plant's immediate growth environment can vary. An example of this is the canopy effect (e.g., Broadmeadow and Griffiths (1993)), wherein low $\delta^{13}C$ CO_2, generated by soil respiration, accumulates and is assimilated into plants growing below the forest canopy. Similarly, low $\delta^{13}C$ CO_2 derived from methane oxidation has been invoked to explain variations in peat bog vegetation isotope values.

The major effect on $\delta^{13}C$ values of higher plant lipids in marine sediments, however, is the relative contribution of lipids derived from C4 and C3 plant biomass. This has been demonstrated by Rommerskirchen et al. (2003) who showed a strong relationship between adjacent land vegetation type (C3 vs C4), aerosol transport, compound specific $\delta^{13}C$ terrestrial biomarkers (*n*-alkanes and *n*-alkanols) and pollen records from Holocene sediments in marine cores along the coast of South West Africa. Similarly, Hughen et al. (2004) used the $\delta^{13}C$ of *n*-alkanoic long chained (C_{32}) acid leaf waxes combined with their C_{24}-C_{28} average carbon chain length from the Cariaco Basin to demonstrate that vegetation in Venezuela can potentially switch from grassland to forest in less than 50 years (Figure 8).

Nitrogen isotopes in marine sediments

The nitrogen cycle plays a central role within marine biochemistry. As an essential nutrient, biologically available nitrogen has the potential to limit biological productivity over large regions of the oceans. While the magnitude of fixed nitrogen sources and sinks are poorly understood (Deutsch et al. 2004), with a residence time of 2-3 kyrs (Gruber and Sarmiento 1997; Brandes and Devol 2002) oceanic levels of nitrogen operate on a time-scale which could influence glacial-interglacial transitions (~10 kyrs). Consequently, hypotheses have even been put forward to suggest an indirect influence of the marine nitrogen budget on the climate (e.g., McElroy (1983); Ganeshram et al. (2000); Altabet et al. (2002)) since alterations in the oceanic inventory of fixed nitrogen, dominated by nitrate (NO_3^-), would be expected to impact the biological carbon pump in the low latitude ocean. In line with this, increases in the nitrogen inventory during the last glacial period have been highlighted as a possible cause of lower glacial atmospheric CO_2 concentrations (McElroy 1983).

Although the use of nitrogen isotopes in palaeoceanography is in its infancy, the approach also provides a unique means of investigating the highly complex oceanic nitrogen cycle and its long-term variations within the oceanic nitrogen budgets with alterations in the sources and sinks of nitrogen influencing the $\delta^{15}N$ in distinct ways. To date, nitrogen isotopes have been measured on bulk organic matter (Altabet et al. 1995; Farrel et al. 1995) and organic matter extracted from either diatoms (e.g., Shemesh et al. (1993); Robinson et al. (2004)) or foraminifera (Altabet and Curry 1989).

Palaeoproductivity and nutrient utilisation

The processes which cause the fractionation of nitrogen isotopes are similar to those of carbon isotopes (see above section). During nitrate assimilation, phytoplankton preferentially consumes ^{14}N-nitrate relative to ^{15}N-nitrate (Montoya and McCarthy 1995; Waser et al. 1998) leaving the surface nitrate pool higher in $\delta^{15}N$ (Sigman et al. 1999; Sigman and Haug 2004). This results in a correlation between the $\delta^{15}N$ of organic nitrogen and the degree of nitrate utilisation by phytoplankton in the surface waters. Hence a strong correlation exists between the $\delta^{15}N$ of the source nitrate consumed and organic matter exported from the euphotic zone. The degree of $\delta^{15}N$ fractionation, though, also depends on temporal variations in the amount and sources of nitrate as well as the rate of consumption of nitrate in the upper ocean (see Figure 9a). Changes in nutrient utilisation, surface water productivity and the amount of available nitrate can all influence the $\delta^{15}N$ signal present in marine sediments and need to be considered. For example, Altabet et al. (1991) observed seasonal 7‰ variations in $\delta^{15}N$ of sinking particles in the North Atlantic Ocean reflecting temporal shifts in surface water productivity. Similarly, large variations in the degree of fractionation have been observed in relation to distance from upwelling zones (Altabet and Francois 1994) and distance offshore (Holmes et al. 1998).

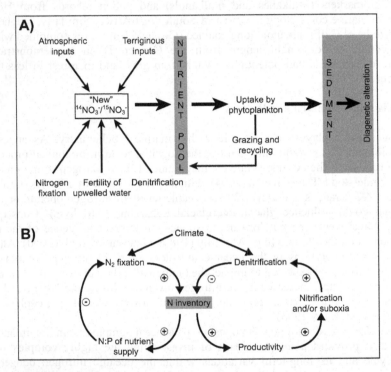

Figure 9. A) Factors influencing the nitrogen isotopes of organic matter found in marine sediments, detailed processes described within the text and in Table 3. B) Schematic hypothesised N budget feedbacks. Perturbation to the nitrogen (N) inventory could be regulated through feedbacks by either denitrification or N_2 fixation. The denitrification negative feedback loop also works for benthic denitrification which is not sensitive to bottom water oxygen content but is sensitive to sediment nitrification rates and thus to productivity driven fluxes of organic matter to the sediment. Each step in the feedback mechanism involves a direct proportionality (arrow with plus sign) or an inverse proportionality (arrow with minus sign) but the sum of the processes in each case constitutes a negative feedback. Redrawn and adapted from Deutsch et al. (2004)).

Table 3. Estimates of the expansion of carbon stored in the terrestrial biosphere since the LGM using different methods illustrating a range of glacial-Holocene $\delta^{13}C$ shifts in oceanic carbon isotope studies.

Cause	Process	Effect	Examples
Denitrification	^{14}N returned to the atmosphere by denitrifying bacteria	More ^{15}N enters marine NO_3^- pool, thereby limiting supply for phytoplankton	Altabet et al. 1995; Falkowski 1997; Farrell et al. 1995; Francoise et al. 1993; Ganeshram et al. 1995; Holmes et al. 1996; 1998; Naqvi et al. 1998; Pride et al. 1999
Nitrogen fixation	Cyanobacteria incorporates 'new' nitrogen into system	More ^{14}N enters marine NO_3^- pool, thereby increasing the amount of usable NO_3	Altabet and Francoise 1994; Falkowski 1997; Karl et al. 1997; Peters et al. 1978
Atmosphere input	e.g. rain incorporates more 'new' nitrogen into system	More ^{14}N enters marine NO_3^- pool, thereby increasing the amount of usable NO_3	Capone and Carpenter 1982; Seitzinger and Sanders 1999
Terriginous input	e.g. by rivers	Marine NO_3^- pool becomes lower/higher depending upon contaminant source	Capone and Carpenter 1982; Falkowski 1997; Holmes et al. 1996; 1998; Peters et al. 1978
Grazing and recycling	Increase in ^{15}N at each trophic level	More ^{15}N enters marine NO_3^- pool, thereby limiting supply for phytoplankton	Altabet et al. 1991; Bronk and Ward 1999; Montoya 1994
Diagenetic alteration	May occur both pre- and post- burial. Preferential removal of ^{14}N during oxidative degradation of organic matter, however, evidence has been found indicating a lowering of $\delta^{15}N$	Will cause increase/decrease in $\delta^{15}N$ in both the NO_3^- in the nitrate pool and in the sedimentary record	Altabet 1996; Altabet and Francoise 1994; Altabet et al. 1991; Farrell et al. 1995; Francoise et al. 1993; Holmes et al. 1996; 1998; Montoya 1994; Sigman et al. 1997

Down-core $\delta^{15}N$ analyses of bulk sediment, diatom intrinsic organic matter and foraminiferal organic matter display significant glacial-interglacial variations in the equatorial Pacific Ocean (e.g., Farrel et al. (1995)), Southern Ocean (e.g., Shemesh et al. (1993); Sigman et al. (1999); Crosta and Shemesh (2002)) and equatorial Atlantic Ocean (Altabet and Curry 1989) related to changes in glacial productivity relative to the modern ocean. In contrast, Kienast (2000) found no significant change in $\delta^{15}N$ in the South China Sea between glacial and interglacial periods; while Francois et al. (1997), using $\delta^{15}N$ records together with other palaeo-productivity proxies, suggested that increases in surface water stratification south of the Antarctic Polar Front may have contributed to lower glacial atmospheric CO_2 levels (see section below on silicon isotopes for the apparent anti-correlation of $\delta^{15}N$ and $\delta^{30}Si$ records in the Southern Ocean during glacial periods). At first sight $\delta^{15}N$ appears a promising and valuable proxy for past nutrient utilisation and surface water productivity. There are, however, but a number of key uncertainties exists which can complicate $\delta^{15}N$ interpretations including (see Sigman and Haug (2004)):

1) Variations in the isotope composition of oceanic nitrate through time. Both the amount and isotope composition of nitrate in the surface water can vary with inputs of nitrogen from the atmosphere, continents and upwelled nutrient rich sub-surface waters (Table 3). Each source contains a different isotopic signature while in addition, nitrogen fixation and denitrification can both influence the isotopic value of the surface water nitrate (see below).

2) Isotope effect of nitrate assimilation. Temporal variations in the relationship between nitrate utilisation and nitrogen isotopes in the surface ocean can impart a significant effect on the isotopic records. For example the fractionation between nitrate and organic nitrogen could be between 3-7‰. In addition, the amount of grazing and recycling of nutrients in the surface waters can initiate a fractionation affect of between 2-4‰ on the analysed $\delta^{15}N$ value, with an increase in $\delta^{15}N$ for each successive trophic stage the nitrogen passes through (Table 3).

3) Diagenetic alteration. The survival of the isotopic signal of sinking organic matter into the sedimentary record is an important consideration as diagenesis has been shown to have a significant effect at some locations (Table 3). Diagenesis may occur both pre- and post-burial, with a preferential removal of ^{14}N during oxidative degradation of organic matter. Some authors have suggested that the diagenetic effect at particular sites may be constant through time and can thus be ignored (e.g., Holmes et al. (1999)). Alternatively core sites characterised by bottom waters with a low oxygen content are analysed in order to limit the possible oxidation and thus diagenesis.

Palaeo-denitrification and nitrogen fixation

The biggest complication to interpreting nitrogen isotopes in terms of relative changes in nutrient utilisation are the processes of denitrification and nitrogen fixation. Denitrification is the heterotrophic reduction of nitrate to N_2 gas and is the dominant cause of the loss of bioavailable (fixable) nitrogen from the ocean. Denitrification is mediated by bacteria, usually in suboxic conditions, resulting in an increase in $\delta^{15}N$ in the subsurface nitrate and a return to the atmosphere of lower $\delta^{15}N$ nitrogen gas. Nitrogen isotope studies in currently active regions of denitrification provide compelling evidence that water column denitrification was reduced in these regions during glacial periods highlighting that the extent of denitrification can vary over time (Altabet et al. 1995; Ganeshram et al. 1995; 2002; Pride et al. 1999). Another important process that must be considered in nitrogen isotopic studies is N_2 fixation, the synthesis of new fixed nitrogen from atmospheric N_2 by cyanobacterial phytoplankton. This process incorporates nitrogen from the atmosphere, with relatively low $\delta^{15}N$ values, into the surface water nitrate reservoir. It has been suggested that during glacial periods N_2 fixation was greater due to the increased atmospheric supply of iron to the open oceans (Falkowski 1997); for, iron is a central requirement of the enzymes for N_2 fixation. Both the glacial decrease in water column denitrification and the increase in N_2 fixation would have increased the oceanic nitrate reservoir. It has been suggested that such a nitrate increase would lead to significantly higher export production of the open ocean during glacial periods potentially explaining part of the observed glacial/interglacial atmospheric CO_2 changes (McElroy

1983; Falkowski 1997; Broecker and Henderson 1998). However, there still remain some critical questions that need to be answered before this mechanism can be validated. The first: how strictly do marine organisms adhere to the Redfield ratios of both N/P and C/N (Pahlow and Riebesell 2000)? Unlike nitrogen there is no way for marine phytoplankton to control the availability of phosphorus which is fed into oceans via river discharge. The question, therefore, is whether increasing the availability of nitrate, through reduced denitrification and increased nitrogen fixation, can alone increase productivity (see Sigman and Haug (2004) for more detailed review).

Initial evidence from nitrogen isotope studies suggests that the nitrogen cycle is adequately dynamic to ensure large scale changes in the oceanic nitrate reservoir on glacial/interglacial timescales (Sigman and Haug 2004). For example, during the last glacial to Holocene transition a series of changes have been inferred. First, in regions of modern water column denitrification, nitrogen isotope records suggest denitrification in the thermocline increased during the last deglaciation but subsequently decreased through the early Holocene (Ganeshram et al. 2000; Altabet et al. 2002). Second, records from areas distant from denitrification zones show no significant change in sediment $\delta^{15}N$ from the last glacial period to the Holocene (Kienast 2000). Third, other areas outside the modern denitrification zones show a deglacial maximum in $\delta^{15}N$ similar to that observed in the modern denitrification zones. These observations suggest that during the last glacial period significant parts of the ocean suboxic zones experienced reduced denitrification rates. Subsequently, deglacial increases in water column denitrification appear to have occurred, possibly in response to climatic-related changes in intensity and/or extent of thermocline suboxia (Deutsch et al. 2004). However, the exact interpretation of these nitrogen isotope records is still debated and open to further investigation.

Interestingly, though, denitrification and nitrogen fixing both produce feedbacks which work to stabilise the nitrate content of the ocean (Figure 9b). Denitrification occurs in environments that are deficient in O_2. An increase in the ocean's nitrate content will drive higher export production, assuming the Redfield ratio is flexible, i.e., the ratio of C:P:N required by marine plants can vary significantly away from those proposed by the Redfield ratio. When this increased export production is oxidised in the ocean interior it will drive more extensive O_2 deficiency, which in turn leads to higher rates of global denitrification lowering the ocean's nitrate content. So it is the sensitivity of denitrification to the O_2 content of the ocean that generates the hypothetical negative feedback. A similar feedback exists for nitrogen fixing, with increased N_2 fixation increasing the ocean N-inventory which in turn increases the N:P ratio in supplied nutrients (Figure 9b). Even with a flexible Redfield ratio, at a threshold value nitrate ceases to be limiting resulting in N_2 fixation ceasing to be competitive (Deutsch et al. 2004; Sigman and Haug. 2004).

North Pacific Pliocene record

An example of the use of down-core sedimentary $\delta^{15}N$ data lies in the subarctic region of the North West Pacific across the onset of major Northern Hemisphere Glaciation (NHG) (c. 2.73 Ma) at ODP site 882. Analysis of $\delta^{15}N_{(bulk\ sediment)}$ from 2.8 Ma to 2.6 Ma reveals notably lower $\delta^{15}N_{(bulk\ sediment)}$ values prior to the onset of NHG with values increasing by c. 2‰ at c. 2.73 Ma (Haug et al. 1999; Sigman et al. 2004). A shift to higher $\delta^{15}N_{(bulk\ sediment)}$ likely indicates an increase in phytoplankton nutrient utilisation, which in turn suggests a possible increase in phytoplankton productivity. BSi data from the same site, however, reveals a three-fold decrease in sedimentary concentrations across the interval which, if dissolution rates are constant through time, equates to a marked reduction in surface water productivity with no evidence existing for a shift to carbonate organisms (Haug et al. 1999). A decrease in productivity without a change in nitrate supply would normally be expected to result in a reduction in nutrient utilisation rather than the c. 2‰ observed increase in $\delta^{15}N_{(bulk\ sediment)}$. Consequently, the inferred increase in nutrient utilisation and decrease in productivity must have occurred simultaneously with a decrease in nitrate/nutrient upwelling to the euphotic zone, causing phytoplankton to increasingly utilise the diminished available nitrate pool. With no observed major alteration in the global thermohaline circulation from c. 2.73 Ma, the reduced supply of nitrate to the surface waters can not be explained by whole water-column spatial shifts in the distribution and availability of nutrients within the Pacific Ocean. Complementary data provided by $\delta^{18}O_{(diatom)}$ and alkenone supersaturation ratios (U^{k}_{37} and U^{k}_{37}') (Haug et al. 2005) indicate that the North West Pacific was marked from c. 2.73 Ma by the development of a significantly warmer pool of surface water, indicating the presence of some form of stratification within the water column. The North West Pacific today also displays relatively high levels of nutrient utilisation with the presence of a halocline, salinity driven stratification, restricting the upwelling of nutrient enriched deep water to the euphotic zone. Consequently, the marked increase in $\delta^{15}N_{(bulk\ sediment)}$ in the post-2.73 Ma period, in combination with other palaeo proxy data, likely indicates the development of a stratification system which prevails to the modern day. This would have caused a reduction in the supply of nitrate to the surface waters, in turn initiating a reduction in surface water BSi productivity while causing a marked increase in nutrient utilisation by the remaining phytoplankton (Haug et al. 1999; Sigman et al. 2004). As highlighted through this example (other excellent studies include Sachs and Repeta (1999) and Thunell and Kepple (2004)), while records of $\delta^{15}N$ are more than suitable for palaeoceanographic reconstructions, $\delta^{15}N$ studies remain best utilised alongside other proxies to enable robust and accurate interpretations. Here in the North Pacific, tweaking out the environmental data within $\delta^{15}N$ records is a complex process with, for example, the observed increase in $\delta^{15}N$ potentially reflecting both an increase in nutrient utilisation and an increase or decrease in productivity. Without complementary evidence, fully understanding the implications of $\delta^{15}N$ records is fraught with uncertainties.

Silicon isotopes in marine sediments

Silicon represents the most common lithospheric element (c. 27%) after oxygen with over 95% of all marine sources of silicon occurring as silicic acid, predominantly in the form of $Si(OH)_4$. Within the marine environment, silicon or $Si(OH)_4$ represents a critical nutrient for organisms producing BSi. Consequently, the primary source of silicon cycling within the marine environment are diatoms, unicellular siliceous eukaryotic algae which, along with other siliceous organisms, uptake and convert $Si(OH)_4$ during biomineralisation into particulate silica (SiO_2). Sources of BSi preferentially take up the lighter ^{28}Si over the heavier ^{29}Si and ^{30}Si with silicon isotopes usually reported as $\delta^{30}Si$ ($^{30}Si/^{28}Si$) although occasionally as $\delta^{29}Si$ ($^{29}Si/^{28}Si$), see below. Analysis of the silicon isotope signal within BSi, therefore, provides a valuable insight into past variations within the marine silicon cycle including the availability of silicon for sources of BSi and the extent of past surface and intermediate water nutrient/silicic acid utilisation.

A major interest within studies of δ^{30}Si is the extraction of a palaeoenvironmental signal from diatoms. At present diatoms represent approximately 40% of all marine primary productivity increasing to 75%-90% in coastal and other high productivity waters (Nelson and Smith 1986; Nelson et al. 1995). Consequently, diatom export production represents an important mechanism in transferring carbon into the deep ocean storage leading to speculation that records of δ^{30}Si$_{(diatom)}$ may provide a unique insight into the role of the biological pump in controlling and instigating past changes in atmospheric CO_2 (e.g., Goldman (1993); Brzezinski et al. (2002); Matsumoto et al. (2002); Dugdale et al (2004)). Previous attempts to analyse and quantify the marine silicon cycle and the role of diatoms in exporting carbon to the deep ocean have primarily focused on analysing absolute sedimentary BSi concentrations. Such attempts, however, are inherently hindered by the presence of high, c. 97%, dissolution rates, which are subjected to significant temporal and spatial variability, and variability/decoupling of diatom Si/C ratios (Nelson et al. 1995; Tréguer et al. 1995; Boyle 1998; Dezileau et al. 2000; Ragueneau et al. 2000; Pondaven et al 2000). These issues do not influence the δ^{30}Si$_{(diatom)}$ signal with the isotope ratio assumed to have become firmly incorporated into the frustule following silica biomineralisation. Analysis of δ^{30}Si within both diatoms and other sources of BSi, therefore, potentially enables, in addition to information on the marine silicon cycle, a uniquely accurate and hitherto infeasible insight into the climatic implications of the biologically pump.

Until recently, measurement of silicon isotopes, reported relative to an NBS-28 standard, were restricted to analyses of terrestrial and extra-terrestrial material (e.g., Reynolds and Verhoogen (1953); Allenby (1954); Tilles (1961a; b); Epstein and Taylor (1970a; b); Taylor and Epstein (1970)). Technological advances, however, now permit silicon isotopes in marine sources of BSi to be reliably analysed through gas source fluorination, involving the conversion of purified silicon into SiF_4 for mass spectrometric analysis via a fluorine reagent (De La Rocha et al. 1996; Ding et al. 1996), ion microprobe analysis (e.g., Basile-Doelsch et al. (2005)) or through Multicollector Inductively Coupled Plasma Mass Spectrometer (MC-ICP-MS), operating in either wet or dry plasma mode (e.g., De La Rocha 2002; Cardinal et al. 2003). A present limitation of most MC-ICP-MS measurements, though, is the high intensity interference, predominantly originating from $^{14}N_2$, which prevents direct measurement of δ^{30}Si, requiring measurement to instead be reported as δ^{29}Si.

Marine silicic acid utilisation

Despite the widespread interest in using δ^{30}Si$_{(opal)}$ as a palaeoceanographic tracer, few examples of its application currently exist. Douthitt (1982) initially observed a 6‰ range in BSi silicon isotope values, while Spadaro (1983) reported the silicon isotope fractionation of a single diatom species. De La Rocha et al. (1997) determined the fractionation of silicon isotopes by three marine diatom species in a series of laboratory culture experiments demonstrating no inter-species vital effects and the lack of a temperature dependent signal on the diatom fractionation of δ^{30}Si between 12°C and 22°C, though further work is needed on a greater range of species and at colder temperatures to validate these observations.

One of the few studies to have applied δ^{30}Si within a palaeoceanographic context, De La Rocha et al. (1998), demonstrated markedly lower δ^{30}Si$_{(diatom)}$ values in three Southern

Ocean cores during the last glacial period relative to the Holocene. The Southern Ocean represents a non steady-state site south of the Antarctic Polar Front (APF). As such lower values of $\delta^{30}Si_{(diatom)}$ most likely indicate decreased silicic acid utilisation during the last glacial, which is in line with concordant decreases in $\delta^{13}C_{(organic)}$ (De La Rocha et al. 1998). This suggests the decreased usage of silicic acid by diatoms occurred during an overall lowering of marine productivity. Extended records of $\delta^{30}Si_{(diatom)}$ in the Southern Ocean over the last three glacial inter-glacial cycles (Brzezinski et al. 2002) reveal a similar trend with decreased silicic acid usage during glacial periods displaying a strong anti-correlation to records of $\delta^{15}N_{(bulk\ sediment)}$. The origin of this anti-correlation remains unclear but, given similar evidence from other cores, may reflect a decoupling of the silicon and nitrogen records and/or possible iron enrichment causing increased numbers of non-siliceous organisms which would in-turn lead to increased use of NO_3^- over $Si(OH)_4$ (see discussion in Brzezinski et al. (2002)). The increased supply of iron to the Southern Ocean during the last glacial, though, would have resulted in a marked reduction in the uptake of diatom $Si(OH)_4$:NO_3^- ratios (Hutchins and Bruland 1998; Takeda 1998). This may have caused a reduction in silicic acid utilisation, initiating the observed decrease in $\delta^{30}Si_{(diatom)}$, without an overall lowering of diatom productivity. If records of $\delta^{30}Si_{(diatom)}$ are reflecting decreased silicic acid usage, creating a pool of concentrated silicic acid, its northward migration during the last glacial may have caused diatoms to dominate at lower latitudes in place of coccolithophores (Brzezinski et al. 2002). Such an event has major implications leading to a marked net increase in the draw down of CO_2 into the deep ocean storage, thereby lowering atmospheric levels of CO_2 (Brzezinski et al. 2002). While as yet unconfirmed, this represents the basis of the so-called "silicic-acid leakage hypothesis" which is proposed to have caused significant changes in Southern Ocean productivity and consequently levels of pCO_2 during the last glacial period (Brzezinski et al. 2002; Matsumoto et al. 2002).

A potential uncertainty associated with these measurements and interpretations of $\delta^{30}Si$ is the possibility for shifts in $\delta^{30}Si_{(diatom)}$ to actually reflect variations in the isotope composition of silicic acid supplied to the euphotic zone as opposed to shifts in silicic acid utilisation; since the value of $\delta^{30}Si$ within the diatom frustule is entirely dependent upon the initial isotope composition of silicic acid. Whole ocean changes in the $\delta^{30}Si$ value of silicic acid, however, are at present believed to be minimal over long, glacial-interglacial, timescales. Furthermore, values of $\delta^{30}Si$ within sources of BSi are more likely to reflect the value of silicic acid entering the euphotic zone only in highly stratified regions and not in the largely un-stratified system of the Southern Ocean (De La Rocha et al. 1998).

Data within De La Rocha et al. (1998), however, indicates constant 0.3‰ regional offsets in $\delta^{30}Si_{(diatom)}$ in the Southern Ocean through time, likely reflecting regional differences in the value of dissolved $\delta^{30}Si$ supplied to the euphotic zone. Evidence suggests these regional offsets prevailed over glacial-interglacial cycles (De La Rocha et al. 1998) with similar regional offsets also observed in modern day values for Pacific and Atlantic deep waters (De La Rocha et al. 2002). Provided these assumptions hold true, at present such conclusions are based on limited observations, down-core analyses of $\delta^{30}Si_{(diatom)}$ enable at the very least a qualitative proxy of silicic acid utilisation, as demonstrated in De La Rocha et al. (1998). Further work is required, however, into expanding the available data on modern day silicic acid isotope composition if interpretations of $\delta^{30}Si_{(diatom)}$ are to be transformed into more detailed, quantitative based, reconstructions.

One possible solution for monitoring past $\delta^{30}Si$ secular variations in silicic acid involves the analysis of non-photosynthetic siliceous organisms. Diatoms due to their pre-requisite for light, live within the euphotic zone of the surface waters. With the silicon isotope composition of surface waters heavily affected by diatom uptake of silicic acid (De La Rocha et al 2002), analysis of $\delta^{30}Si_{(diatom)}$ alone is unsuitable to provide information on long/short-term secular changes in the marine $\delta^{30}Si$ pool. Analysis of the $\delta^{30}Si$ record in benthic sources of BSi may, however, provide a means of extracting the silicon isotope record of deep and intermediate waters and possible changes in the marine $\delta^{30}Si$ pool over time. Such records, in addition to indicating changes in the $\delta^{30}Si$ of silicic acid being supplied to the euphotic zone, may also enable the creation of correction factors, where necessary, to quantitatively correct for records of $\delta^{30}Si_{(diatom)}$. One type of benthic BSi potentially suitable for this approach are siliceous sponges, which demonstrate significantly greater silicon fractionation, c. −3.7‰, (Douthitt 1982; De La Rocha 2003) during biomineralisation when compared with the −1.1‰ range reported for diatoms (De La Rocha et al. 1997). In contrast to diatoms, however, significant inter-species vital effects have been detected within marine sponges (Vroon et al. 2002; De La Rocha 2003), emphasing the need for laboratory culture experiments before further work can be completed. Although initial results from benthic sponges over the Eocene-Oligocene boundary indicate considerable, 2‰, shifts in $\delta^{30}Si$, it remains to be seen what proportion of this signal is generated from secular shifts in silicic acid $\delta^{30}Si$ and what proportion from inter-species fractionation effects (De La Rocha 2003; Vroon et al 2004).

Although significant and rapid steps have been achieved over the past decade in developing protocols and techniques for measuring silicon isotopes, much work remains before measurements of $\delta^{30}Si_{(opal)}$ can be reliably used as a quantitative tracer of past nutrient dynamics. This includes the development of a greater understanding into the modern and past distribution of silicon isotopes in the water column and the functioning of the marine and terrestrial silicon isotope cycle (De La Rocha et al. 1998; 2000; De La Rocha 2003). Initial attempts at this have partially improved our understanding of the lithogenic/terrestrial controls and the marine distribution of silicon isotopes and their uptake into the diatom frustules. All existing studies, however, stress that the controls on silicic acid isotope composition are highly complex being a function of oceanic and terrestrial silicon inputs and outputs, biogenic cycling of silicic acid and the interaction of biogenic cycling with the global thermohaline circulation (e.g., De La Rocha et al. (2000), Ziegler et al. (2000; 2002), Wischmeyer et al. (2003); Ding et al. (2004), Milligan et al. (2004); Varela et al. (2004); Cardinal et al. (2005)). Such is the interest, though, in developing $\delta^{30}Si$ in order to examine the role of the biological pump in driving palaeoclimatic changes, that the next decade is liable to witness considerable developments in this field. Although recent work indicates the potential for variable $\delta^{30}Si_{(diatom)}$ values under different concentrations of silicic acid species (Mataliotaki et al. 2003), within the normal seawater pH range of c. 7.5 to c. 8.5, such variability is insignificant with $Si(OH)_4$ remaining by far the dominant species. In lacustrine environments, however, where the encountered range of pH values is considerably greater, such fractionation influences can potentially override any true environmental signal (Mataliotaki et al. 2003).

Boron isotopes in marine sediments

Boron stable isotopes (^{10}B and ^{11}B) represent another novel, but under-utilised, proxy within palaeoceanography that potentially enables palaeo-pH reconstructions. While this palaeo-pH tracer is present within both biogenic and abiogenic calcite, the extraction of δ^{11}B data from foraminifera represents the primary, though not solitary (e.g. Kasemann et al. (2005)), use of this palaeoceanographic proxy to date; particularly as the extraction of stable isotope data from different species enables the reconstruction of pH-depth profiles. The reconstruction of a palaeo-pH signal is all the more remarkable as it potentially provides an alternative measure of palaeo-pCO$_2$ that is independent of the δ^{13}C approach and its associated limitations (see above section on carbon isotopes). As demonstrated below, however, significant issues and uncertainties remain and require further attention before δ^{11}B can be adopted as a common and widespread technique in multi-proxy studies.

Boron palaeo-pH reconstructions

In seawater, two predominant species of dissolved boron coexist. Of these, B(OH)$_3$ is relatively more abundant in waters below a pH of 7 while B(OH)$_4^-$ in turn becomes the principal species at a pH of 11 of greater. At isotopic equilibrium B(OH)$_3$ is constantly higher in δ^{11}B, relative to B(OH)$_4^-$, by c. 19.5‰ (Kakihana et al. 1977). With no significant temperature, salinity or other environmental based fractionation, the δ^{11}B of the two species should vary with change in seawater pH given:

$$B(OH)_3 + H_2O = B(OH)_4^- + H^+ \tag{17}$$

Both biogenic and inorganic sources of CaCO$_3$ preferentially take up B(OH)$_4^-$ enabling the δ^{11}B signal precipitated within calcite to be regarded as a function of sea-water pH (Hemming and Hanson 1992; Hemming et al. 1995; Sanyal et al. 1996; 2000; 2001, Hönisch et al. 2004). Standard preparation and analysis of δ^{11}B$_{(foram)}$ follows that outlined in Sanyal et al. (1995) and Palmer et al (1998). Work has previously highlighted the presence of significant δ^{11}B$_{(foram)}$ shifts of up to c. 2.3‰, caused by variation in species shell sizes and shell partial dissolution, indicating the need for pristine similar sized foraminifera shells for isotope analysis (Hönisch and Hemming 2004). Knowledge of the apparent dissociation constants for boric acid (pK^*_B) enables the relative distribution of each boron species and hence the value of δ^{11}B$_{(foram)}$ to be explained as a function of pH. The accuracy of the reconstructed pH obtained from δ^{11}B$_{(foram)}$ remains, however, entirely dependent on the validity of additional assumptions made over palaeo-δ^{11}B$_{(seawater)}$, and the equilibrium isotope fractionation factor (α_{4-3}) which both contribute towards the construction of the δ^{11}B-pH curve. A detailed discussion of these individual parameters is contained within Pagani et al. (2005) and Zeebe (2005). Each of these parameters, though, contains a degree of uncertainty which have the potential to impart significant errors onto any palaeo-pH reconstruction. In particular, (pK^*_B) is temperature, salinity and pressure (calcification depth) dependent, requiring some form of estimate of these variables when reconstructing pH from δ^{11}B$_{(foram)}$.

Culture experiments have verified the presence of a strong relationship between pH and δ^{11}B in foraminifera and abiogenic calcite despite the additional occurrence of constant systematic inter-species vital effect over modern seawater pH values (Sanyal et al. 1996; 2000; 2001). Studies have demonstrated these offsets to be related to physiological processes, including calcification, photosynthesis and respiration, which influence the seawater micro-environment pH and thereby impact the $\delta^{11}B_{(foram)}$ signal (Hönisch et al. 2003; Zeebe et al. 2003). Such processes and their overall influence on the $\delta^{11}B_{(foram)}$ signal, however, are believed to be negligible when deriving overall palaeo-pH reconstructions provided that mono-specific foraminiferal samples are analysed (Sanyal et al. 2000; 2001; Hönisch et al. 2003; Zeebe et al. 2003). Using $\delta^{11}B$ to reconstruct palaeo-pH assuming the presence of a palaeo-$\delta^{11}B_{(seawater)}$ value of +39.5‰, Spivack et al. (1993) reconstructed pH in the surface waters of the equatorial Pacific Ocean over the last 21 Ma inferring a 0.8 pH increase between 21 Ma and 6 Ma. Sanyal et al. (1995; 1997) subsequently demonstrated changes in surface and deep ocean water pH in the Atlantic and Pacific Ocean between the last glacial and the Holocene and over other glacial-interglacial cycles by analysing both planktonic and benthic foraminiferal $\delta^{11}B$.

A contentious issue surrounding these earlier studies involves the validity of the values used for palaeo-$\delta^{11}B_{(seawater)}$. Residence times of oceanic boron are estimated to lie in the range of tens of millions of years with no/minimal changes over glacial-interglacial cycles (Spivack and Edmond 1987; Lemarchand et al. 2000; 2002). Consequently, a modern $\delta^{11}B_{(seawater)}$ value of +39.5‰, as applied in Spivack et al. (1993) and Sanyal et al. (1995; 1997), has been widely regarded as a reliable estimate of palaeo-$\delta^{11}B_{(seawater)}$. $\delta^{11}B_{(foram)}$ results in Palmer et al. (1998), however, suggest minor variations in $\delta^{11}B_{(seawater)}$ may have occurred through time. Error analysis examining the impact of a potential 1‰ shift in $\delta^{11}B_{(seawater)}$ reveals that such potential changes in the oceanic boron pool impart a significant degree of uncertainty onto the final reconstructed pH value (Palmer et al. 1998). To account for such uncertainties, Pearson and Palmer (1999; 2000) modelled pH through time as a function of different values of $\delta^{11}B_{(seawater)}$. Within the Eocene, values of $\delta^{11}B_{(seawater)}$ were constrained between +38‰ and +41‰ based on modern-day oceanic seawater pH ranges and water column pH-depth gradients resulting in the reconstruction of an upper, lower and 'best estimate' of palaeo-pH, constituting a total pH range of 0.42 units (Pearson and Palmer, 1999). More conservative values of possible palaeo-$\delta^{11}B_{(seawater)}$ values have been estimated over the last 60 Ma by Pearson and Palmer (2000). Other work though has suggested that the range of plausible $\delta^{11}B_{(seawater)}$ in the past was considerably greater than those suggested by (Pearson and Palmer 1999; 2000) with changes in $\delta^{11}B_{(seawater)}$ controlled by both continental input and the weathering of the oceanic crust (Lemarchand et al. 2000; 2002). Consequently, the current poor constraints on possible past variations in $\delta^{11}B_{(seawater)}$, significantly affect the reliability of existing studies which have reconstructed past variations in oceanic pH.

Boron pCO₂ reconstructions

Provided the uncertainties and errors associated with palaeo-pH reconstructions from $\delta^{11}B_{(foram)}$ data can be minimised, analysis of $\delta^{11}B_{(foram)}$ permits the simultaneous reconstruction of pCO_2. Use of $\delta^{11}B$ to infer palaeo-pCO_2 assumes that an increase in atmospheric CO_2 results in increased dissolved CO_2 within the ocean surface waters; in turn causing a reduction in oceanic pH. Analysed planktonic foraminifera must additionally be assumed to have inhabited waters either in or close to equilibrium with respect to atmospheric levels of CO_2. Inferring pCO_2 from pH, however, is problematic with levels of both pCO_2 and pH in the modern ocean subjected to unrelated spatial and temporal variations in line with changes in upwelling and freshwater inflows. As such, to minimise the propagation of errors, analysed core sites need to be situated away from continental locations and areas of significant deep water upwelling.

Five primary environmental variables affect the carbonate chemistry of seawater: pH, concentration of dissolved CO_2, amount of dissolved inorganic carbon ($\sum CO_2$), carbonate ion concentration CO^{2-}_3 and alkalinity. A major limitation of the $\delta^{11}B_{(foram)}$ approach to reconstructing pCO_2 is, therefore, the lack of the necessary complementary data with either $\sum CO_2$ or alkalinity needed to calculate pCO_2 from pH. To date, the principal exponents of $\delta^{11}B_{(foram)}$ derived pCO_2-reconstruction are represented by a series of papers based on sediments within a thermally stratified region of the equatorial tropical Pacific, which is believed to have experienced no significant changes in productivity over time. Initial work at this site resulted in the construction of pH water-column profiles at five time slices between the Miocene and Pleistocene using $\delta^{11}B$ from individual foraminiferal species that lived at different depth habitats (Palmer et al. 1998). These all demonstrated similar pH-depth profiles to the modern day. Using a similar approach, Pearson and Palmer (1999) repeated this for the Eocene before attempting to reconstruct pCO_2. To circumvent the uncertainty over $\sum CO_2$, reconstructed values of palaeo-pCO_2 were modelled as a function of modern day values of $\sum CO_2$ (Pearson and Palmer 1999). In an attempt to resolve the remaining uncertainties associated with this approach, Pearson and Palmer (2000) utilised existing knowledge of the Carbonate Compensation Depth (CCD) to obtain estimates of past levels of $\sum CO_2$ and alkalinity before re-calculating pCO_2 (Figure 10). While the reconstructed record is remarkable, many inherent, but at present unavoidable, errors and uncertainties remain over the reliability of the data used to calculate pCO_2; including the assumption that the waters are at equilibrium with atmospheric CO_2 and that the use of CCD to calculate alkalinity and $\sum CO_2$ is both accurate and reliable.

Errors introduced into the final pCO_2 value through the use of CCD may be negligible though when compared to the analytical errors originating from the original analysis of $\delta^{11}B_{(foram)}$, which are comparatively high at c. 0.1-0.6‰ (Pearson and Palmer 2000) (Figure 10), and from the initial $\delta^{11}B_{(foram)}$-pH calculations. For example, assuming oceanic DIC remains steady, decreasing pH by 0.2 results in a doubling of atmospheric CO_2 (Pagani et al. 2005). On top of this, further errors originate from the uncertainties around (pK^*_B), α_{4-3} and $\delta^{11}B_{(seawater)}$ which also need to be considered. For example, altering the value of α_{4-3} from the value used in Pearson and Palmer (2000) to a value of 0.974, which now appears more valid for carbonates, significantly alters the reconstructed values of pCO_2 (see Pagani et al. (2005)). While an independent data-set of alkenone based carbon isotopic variations appears to verify much of the $\delta^{11}B$ inferred pCO_2 records between 16 Ma and 4 Ma, diverging trends between the two records appear during the expansion of the Antarctic ice sheet at c. 15 Ma (Pagani et al. 1999b). Issues over the value of palaeo-$\delta^{11}B_{(seawater)}$ used when calculating changes in pH and pCO_2 can largely be neglected over more recent timescales, e.g., over the last 23,000 years (Palmer and Pearson 2003), when a modern $\delta^{11}B_{(seawater)}$ value of +39.5‰ can be used with greater confidence. Over this time frame, the likelihood of oceanic boron having varied significantly is minimal. However, over longer periods, values of $\delta^{11}B_{(seawater)}$ require more in-depth modelling (e.g., Pearson and Palmer (1999; 2000)) to resolve some the current uncertainties that exists for the reconstructed estimates of pCO_2.

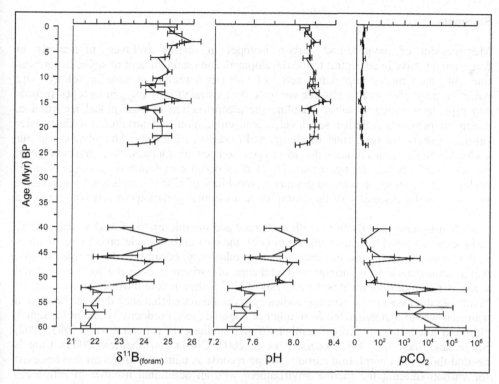

Figure 10: $\delta^{11}B$ from planktonic foraminifera with 'best estimate' values of pH and pCO_2 over the last 60 Ma (Pearson and Palmer 2000). Error margins reflect the analytical uncertainty from measuring $\delta^{11}B$. Further (not displayed) errors also originate over the assumptions and values used for $\delta^{11}B_{(seawater)}$, (pK^*_B), α_{4-3} and $\sum CO_2$ when calculating pH and pCO_2. Re-analysis of the data using a a_{4-3} value of 0.974, which may be more accurate than the value of 0.981 used in Pearson and Palmer (2000), and alternative values of $\delta^{11}B_{(seawater)}$ significantly alters the reconstructed pH and pCO_2 (see Pagani et al. (2005)). Data from Pearson and Palmer (2003).

Much work remains to refine and harness the huge potential for $\delta^{11}B_{(foram)}$ before it can be employed to provide robust palaeoceanographic and palaeoclimatic information. However, provided pristine, single species, samples of similar shell size are analysed (e.g., Pearson and Palmer (2000); Palmer and Pearson (2003)), $\delta^{11}B_{(foram)}$ potentially provides a unique source of oceanic palaeo-pH and pCO_2. Before this, however, a better understanding is required on species vital effects, the values of (pK^*_B), α_{4-3}, $\sum CO_2$ and the temporal variations, both seasonally and at longer time-scales, of $\delta^{11}B$ concentrations in seawater. Currently such uncertainties, particularly over potential shifts in the marine boron budget, induce significant and unavoidable errors onto the reconstructed palaeo-pH and pCO_2. Assuming these uncertainties can be resolved, the limits to using $\delta^{11}B$ as a palaeoceanographic proxy lie only at the lower range of seawater pH values where the sensitivity of the $\delta^{11}B$-pH relationship falls off as $B(OH)_3$ increases to represent over c. 98% of all boron species.

Summary

Measurement of oxygen and carbon isotopes in marine archives, particularly in foraminifera, have been central to the development and establishment of stable isotopes as one of the most important sets of proxies within palaeoclimatology and palaeoceanography. In this chapter we have demonstrated how oxygen isotopes records can provide stratigraphy while enabling the reconstruction of past global ice volume, ocean temperatures, relative sea level, ocean circulation and structure, surface water salinity, iceberg melting, river discharge and monsoonal intensity. The advances in our understanding of past climates due to oxygen isotopes are fundamental: confirmation of the theory of orbital forcing, variability of deep ocean circulation on a range of time-scales from millions of years to centuries, variability of ENSO, monsoonal and riverine systems and the discovery of large amplitude millennial variations in relative global sea level.

Carbon isotope records from both carbonate and organic matter found within marine sediments can provide crucial information on the past carbon cycle on a range of time-scales, enabling an insight into past marine productivity, ocean circulation, past surface water, atmospheric CO_2, storage and exchange of carbon on both the local and global scale. A recent development is the analysis of $\delta^{13}C$ within diatom intrinsic organic matter. While not discussed here, existing studies have demonstrated that such data can be used in a similar way to that generated from other proxies, i.e., as a productivity and pCO_2 signal, although at present only limited numbers of such studies exist (Shemesh et al. 1993; 2002; Singer and Shemesh 1995; Rosenthal et al. 2000; Crosta and Shemesh 2002). Care is needed though in interpreting carbon isotope records as numerous different sources exist for carbon entering the marine environment with an additional number of influences affecting the isotope values of each source. The rewards are great, however, if carbon isotope records can be interpreted as they provide one of the primary ways of investigating the carbon cycle which is highly relevant considering the continued debate over the future impacts of global warming. Recent years have also witnessed the development of a variety of other stable isotope proxies, in particular $\delta^{15}N$ and $\delta^{11}B$. While still in their infancy, great steps forward have been achieved in our understanding of them although significant difficulties remain in ensuring robust and accurate interpretations from the generated data. The potential of both isotopes, however, to provide a more significant insight into past oceanic conditions and palaeo-pCO_2 levels is notable. As such, their use are likely to expand over the next decade as a greater understanding of the marine nitrogen cycle, controls on $\delta^{15}N$ and the errors associated with $\delta^{11}B$ based pCO_2 reconstructions are achieved.

Outside foraminiferal isotope based studies, there is great scope to generate isotope results from other biological sources. A current major limitation with foraminifera studies is the void of data from regions depleted in $CaCO_3$ preservation, particularly in the Southern Ocean and other high latitude regions. Future development and use of $\delta^{18}O_{(diatom)}$ and $\delta^{30}Si_{(diatom)}/\delta^{29}Si_{(diatom)}$ alongside $\delta^{13}C$ and $\delta^{15}N$ analysis of intrinsic organic matter within diatom frustules will ultimately open the gateway for the extension of detailed, isotope based, palaeoceanographic reconstructions to a greater swarth of the world's oceans. However, further research is first required into the systematics of oxygen and silicon uptake into diatoms, in particular addressing the potential issues of inter-species and size related fractionation effects.

References

Adams J.M. and Faure H. 1998. A new estimate of changing carbon storage on land since the last glacial maximum, based on global land ecosystem reconstruction. Global Planet. Change 16-17: 2–24.

Adams J.M., Faure H., Faure-Denard L., McGlade J.M. and Woodward F.I. 1990. Increases in terrestrial carbon storage from the Last Glacial maximum to the present. Nature 348: 711–714.

Adkins J.F. and Schrag D.P. 2001. Pore fluid constraints on deep ocean temperature and salinity during the last glacial maximum. Geophys. Res. Lett. 28: 771–774.

Adkins J.F., Cheng H., Boyle E.A., Druffel E.R.M. and Edwards R.L. 1998. Deep-sea coral evidence for rapid change in ventilation of the deep North Atlantic 15,400 years ago. Science 280: 725–728.

Alegret L., Molina E. and Thomas E. 2001. Benthic foraminifera at the Cretaceous/Tertiary boundary around the Gulf of Mexico. Geology 29: 891–894.

Alegret L., Arenillas I., Arz J.A., Liesa C., Meléndez A., Molina E., Soria A.R. and Thomas E. 2002. The Cretaceous/Tertiary boundary: sedimentology and micropaleontology at El Mulato section, NE Mexico. Terra Nova 14: 330–336.

Alegret L., Molina E. and Thomas E. 2003. Benthic foraminiferal turnover across the Cretaceous/Paleogene boundary at Agost (southeastern Spain): paleoenvironmental inferences. Mar. Micropaleontol. 48: 251–279.

Allenby R.J. 1954. Determination of the isotopic ratios of silicon in rocks. Geochim. Cosmochim. Ac. 5: 40–48.

Altabet M.A. 1991. Nitrogen isotopic evidence for the source and transformation of sinking particles in the open ocean. Abstr. Pap. Am. Chem. S. 201: 1, 41–GEOC.

Altabet M. 1996. Nitrogen and carbon isotope tracers of the source and transformation of particles in the deep sea. In: Ittekkot V., Schaefer P., Honjo S. and Depetris, P.J. (eds), Particle Flux in the Ocean. SCOPE Report, J. Wiley and Sons, London, pp.155–184.

Altabet M.A. and Curry W.B. 1989. Testing models of past ocean chemistry using foraminifera 15N/14N. Paleoceanography 3: 107–119.

Altabet M.A. and Francois R.L. 1994. Sedimentary nitrogen isotope ratio as a recorder for surface ocean nitrate utilization. Global Biogeochem. Cy. 8: 103–116.

Altabet M.A., Deuser W.G., Honjo S. and Stienen C. 1991. Seasonal and depth related changes in the source of sinking particles in the North Atlantic. Nature 354: 136–139.

Altabet M.A., Francois R., Murray D.W. and Prell W.L. 1995. Climate related variations in denitrification in the Arabian Sea from sediment ^{15}N/^{14}N ratios. Nature 373: 506–509.

Altabet M.A., Higginson M.J. and Murray D.W. 2002. The effect of millennial-scale changes in Arabian Sea denitrification on atmospheric CO_2. Nature 415: 159–162.

Andersen N., Müller P.J., Kirst G. and Schneider R.R. 1999. The δ^{13}C signal in $C_{37:2}$ alkenones as a proxy for reconstructing Late Quaternary pCO$_2$ in surface waters from the South Atlantic. In: Fischer G. and Wefer G. (eds), Use of proxies in paleoceanography: examples from the South Atlantic. Springer-Verlag, New York, pp. 469–488.

Anderson T. and Arthur M. 1983. Stable isotopes of oxygen and carbon and their application to sedimentological and paleoenvironmental problems. In: Arthur M. and Anderson T. (eds) SEPM Short Course No. 10: Stable isotopes in Sedimentary Geology. Dallas, pp. 1-1 – 1-1.151.

Anderson O.R., Spindler M., Bé A.W.H. and Hemleben Ch. 1979. Tropic activity of planktonic foraminifera. J. Mar. Biol. Assoc. UK. 59: 791–799.

Bard E., Hamelin B., Arnold M., Montaggioni L., Cabioch G., Faure G. and Rougerieet F. 1996. Deglacial sea-level record from Tahiti corals and the timing of global meltwater discharges. Nature 382: 241–244.

Bard E., Hamelin B., Fairbanks R. and Zindler A. 1999. Calibration of the ^{14}C timescale over the past 30,000 years using mass spectrometric U-Th ages from Barbados corals. Nature 345: 405–409.

Barrera E. and Johnson C. (eds). 1999. Evolution of the Cretaceous ocean-climate system. Geol. Soc. Am. Special Paper 332, pp. 1–445.

Basile-Doelsch I., Meunier J.D. and Parron C. 2005 Another continental pool in the terrestrial silicon cycle. Nature 433: 399–402.

Bé A.W.H. 1977. An ecological, zoogeographic and taxonomic review of recent planktonic foraminifera. In: Ramsay A.T.S. (ed.), Oceanic micropaleontology. Academic Press, London, pp. 1–88.

Beerling D.J. 1999. New estimates of carbon transfer to terrestrial ecosystems between the last glacial maximum and the Holocene. Terra Nova 11: 162–167.

Berger W.H. and Vincent E. 1986. Deep-sea carbonates: reading the carbon isotope signal. Geol. Rundschau. 75: 249–269.

Berger W.H., Killingley J.S. and Vincent E. 1978. Stable isotopes in deep sea carbonates: Box core ERDC-92 western equatorial Pacific. Oceanol. Acta. 1: 203–216.

Berger W.H., Lange C.B. and Weinheinmer A. 1997. Silica depletion of the thermocline in the eastern North Pacific during glacial conditions. Geology 25: 619–622.

Bidigare R.R., Fluegge A., Freeman K.H., Hanson K.L., Hayes J.M., Hollander D., Jasper J.P., King L.L., Laws E.A., Millero F.J., Pancost R.D., Popp B.N., Steinberg P.A. and Wakeham S. G. 1997. Consistent fractionation of 13C in nature and in the laboratory: Growth-rate effects in some haptophyte algae. Global Biogeochem. Cy.11: 279–292.

Bird M.I., Lloyd J. and Farquhar G.D. 1994. Terrestrial carbon storage at the LGM. Nature 371: 566.

Brovkin V., Hofmann M., Bendtsen J. and Ganopolski A. 2002. Ocean biological could control atmospheric $\delta^{13}C$ during glacial-interglacial cycles. Geochem. Geophys. Geosyst. 3: doi: 10.1029/2001GC000270.

Boyle E.A. 1995. Limits on benthic foraminiferal chemical analyses as precise measures of environmental properties. J. Foramin. Res. 25: 4–13.

Boyle E. 1998. Pumping iron makes thinner diatoms. Nature 393: 733–734.

Brandes J.A. and Devol A.H. 2002. A global marine fixed nitrogen budget: implications for Holocene nitrogen cycling. Global Biogeochem. Cy. 16: 1120. doi: 10.1029/2001GB001856.

Brandriss M.E., O'Neil J.R., Edlund M.B. and Stoermer E.F. 1998. Oxygen isotope fractionation between diatomaceous silica and water. Geochim. Cosmochim. Ac. 62: 1119–1125 .

Broadmeadow M.S.J. and Griffiths H. 1993. Carbon isotope discrimination and the coupling of CO_2 fluxes within forest canopies. In: Ehleringer J.R., Hall A.E. and Farquhar G.D. (eds), Stable isotopes and plant carbon-water relations. Academic Press, San Diego, pp. 109–129.

Broecker W.S. 1982. Glacial to interglacial changes in ocean chemistry. Prog. Oceanogr. 11: 151–197.

Broecker W.S. and Peng T-H. 1982. Tracers in the Sea. Eldigino Press, New York. pp. 690

Broecker W.S. and Denton G.H. 1989. The role of ocean-atmosphere reorganisation in glacial cycles. Geochim. Cosmochim. Ac. 53: 2465–2501.

Broecker W.S. and Peng T-H. 1993. What caused the glacial to interglacial CO_2 change. In: Heimann M. (ed.), The global carbon cycle, NATO ASI Series, Vol. 15. Springer, Berlin. pp. 95–115.

Broecker W. and Henderson G.M. 1998. The sequence of events surrounding Termination II and their implications for the cause of glacial-interglacial CO_2 changes. Paleoceanography 13: 352–364.

Bronk D.A. and Ward B.B. 1999. Gross and net nitrogen uptake and DON release in the euphotic zone of Monterey Bay, California. Limnol. Oceanogr. 44: 573–585.

Brook E.J., Harder S., Severinghaus J. and Bender M. 1999. Atmospheric methane and millennial-scale climate change. In: Clark P.U., Webb R.S. and Keigwin L.D. (eds), Mechanisms of global climate change at millennial time scales. AGU Geophysical Monographs 112: 165–175.

Brook E.J., Harder S., Severinghaus J., Steig E.J. and Sucher C.M. 2000. On the origin and timing of rapid changes in atmospheric methane during the last glacial period. Global Biogeochem. Cy. 14: 559–572.

Brzezinski M.A., Pride C.J., Franck V.M., Sigman D.M., Sarmiento J.L., Matsumoto K., Gruber N., Rau G.H. and Coale K.H. 2002. A switch from $Si(OH)_4$ to NO_3^- depletion in the glacial Southern Ocean. Geophys. Res. Lett. 29: 1564. doi: 10.1029/2001GL014349.

Buffett B. A. 2000. Clathrate Hydrates. Annu. Rev. Earth Pl. Sc. 28: 477–507.

Burns S. and Maslin M.A. 1999. Composition and circulation of bottom water in the western Atlantic Ocean during the last glacial, based on pore-water analyses from the Amazon Fan. Geology 27: 1011–1014.

Calder J.A. and Parker P.L. 1973. Geochemical implications of induced changes in 13C fractionation of blue-green algae. Geochim. Cosmochim. Ac. 37: 133–140.

Capone D.G. and Carpenter E.J. 1982. Nitrogen fixation in the marine environment. Science 217: 1140–1142.

Cardinal D., Alleman L.Y., de Jong J., Ziegler K. and André L. 2003. Isotopic composition of silicon measured by multicollector plasma source mass spectrometry in dry plasma mode. J. Anal. Atom. Spectrom. 18: 213–218.

Cardinal D., Alleman L.Y., Dehairs F., Savoye N., Trull T.W. and André L. 2005. Relevance of silicon isotopes to Si-nutrient utilization and Si-source assessment in Antarctic waters. Global Biogeochem. Cy. 18: GB2007. doi: 10.1029/2004GB002364.

Cerling T.E., Harris J.M., MacFadden B.J., Leakey M.G., Quade J., Eisenmann V. and Ehleringer J.R. 1997. Global vegetation change through the Miocene/Pliocene boundary. Nature 389: 153–158.

Chapman M. and Maslin M.A. 1999. Low latitude forcing of meridional temperature and salinity gradients in the North Atlantic and the growth of glacial ice sheets. Geology 27: 875–878.

Charles C. and Fairbanks R. 1992. Evidence from Southern Ocean sediments for the effects of North Atlantic deep-water flux on climate. Nature 355: 416–419.

Chikaraishi Y. and Naraoka H. 2003. Compound-specific δD-$\delta^{13}C$ analyses of n-alkanes extracted from terrestrial and aquatic plants. Phytochemistry 63: 361–371.

Clark P.U. and Mix A.C. 2002. Ice sheet and sea level of the Last Glacial Period. Quaternary Sci. Rev. 21: 1–8.

Coccioni R., Fabbrucci L. and Galeotti S. 1993. Terminal Cretaceous deep-water benthic foraminiferal decimation, survivorship and recovery at Caravaca (SE Spain). Paleopelagos 3: 3–24.

Cole J. 2001. A slow dance for El Niño. Science 291: 1496–1497.

Cole J. 2003. Holocene coral records: windows on tropical climate variability. In: Mackay A., Battarbee R., Birks J. and Oldfield F. (eds), Global Change in the Holocene. Arnold, London, pp. 168–184.

Collister J.W., Rieley G., Stern B., Eglinton G. and Fry B. 1994. Compound-specific $\delta^{13}C$ analyses of leaf lipids from plants with differing carbon dioxide metabolisms. Org. Geochem. 21: 619–627.

Craig H. and Gordon L.I. 1965. Isotope oceanography: deuterium and oxygen 18 variations in the ocean and the marine atmosphere. Symposium on Marine Geochemistry, University of Rhode Island, Occasional Publications 3: 277–374.

Cranwell P.A. 1981. Diagenesis of free and bound lipids in terrestrial detritus deposited in a lacustrine sediment. Org. Geochem. 3: 79–89.

Criss R.E. 1999. Principles of stable isotope distribution. Oxford University Press, New York, 254 pp.

Crosta X. and Shemesh A. 2002. Reconciling down core anticorrelation of diatom carbon and nitrogen isotopic ratios from the Southern Ocean. Paleoceanography 17: 1010. doi: 10.1029/2000PA000565.

Crowley T.J. 1995. Ice age terrestrial carbon changes revisited. Global Biogeochem. Cy. 9: 377–389.

Culver S.J. 2003. Benthic foraminifera across the Cretaceous-Tertiary (K-T) boundary: a review. Mar. Micropaleontol. 47: 177–226.

Curry W.B., Duplessy J-C., Labeyrie L.D. and Shackleton N.J. 1988. Changes in the distribution of $\delta^{13}C$ of deep water $\sum CO_2$ between the last glaciation and the Holocene. Paleoceanography 3: 327–337.

Curry W.B., Ostermann, D.R., Guptha, M.V. and Ittekkot, V. 1992. Foraminiferal production and monsoonal forcing in the Arabian Sea: evidence from sediment traps. In: Summerhayes C.P., Prell W.L., Emeis K-C. (eds), Upwelling Systems: evolution since the early Miocene. Geological Society Special Publication 64, pp 93–106.

Dansgaard W. and Tauber H. 1969. Glacial oxygen-18 content and Pleistocene ocean temperatures. Science 166: 499–502.

Day S. and Maslin M.A. in press. Linking large impacts, gas hydrates and carbon isotope excursions through widespread sediment liquifaction and continental slope failure: the example of the K-T boundary: event. In: Kenkmann T., Horz F. and Deutsch A. (eds), Large meteorite impacts III. Geol. Soc. Am. Special Paper 384, pp. 239–259.

de Garidel-Thoron T., Beaufort L., Bassinot F. and Henry P. 2004. Evidence for large methane releases to the atmosphere from deep-sea gas-hydrate dissociation during the last glacial episode. P. Natl. Acad. Sci. USA. 101: 9187–9192.

De Freitas A.S.W., McCulloch A.W. and McInnes A.G. 1991. Recovery of silica from aqueous silicate solutions via trialkyl or tetraalkylammonium silicomolybdate. Can. J. Chemistry. 69: 611–614.

De La Rocha C.L. 2002. Measurement of silicon stable isotope natural abundances via multicollector inductively coupled plasma mass spectrometry (MC-ICP-MS). Geochem. Geophys. Geosyst. 3: 1045. doi: 10.1029/2002GC000310.

De La Rocha C.L. 2003. Silicon isotope fractionation by marine sponges and the reconstruction of the silicon isotope composition of ancient deep water. Geology 31: 423–426.

De La Rocha C.L., Brzezinski M.A. and De Niro M.J. 1996. Purification, recovery and laser-driven fluorination of silicon from dissolved and particulate silica for the measurements of natural stable isotopes abundances. Anal. Chem. 68: 3746–3750.

De La Rocha C.L., Brzezinski M.A. and DeNiro M.J. 1997. Fractionation of silicon isotopes by marine diatoms during biogenic silica formation. Geochim. Cosmochim. Ac. 61: 5051–5056.

De La Rocha C.L., Brzezinski M.A., DeNiro M.J. and Shemesh A. 1998. Silicon-isotope composition of diatoms as an indicator of past oceanic change. Nature 395: 680–683.

De La Rocha C.L., Brzezinski M.A. and DeNiro M.J. 2000. A first look at the distribution of the stable isotopes of silicon in natural waters. Geochim. Cosmochim. Ac. 64: 2467–2477.

Degens E., Guillard R., Sackett W. and Hellebust J. 1968. Metabolic fractionation of carbon isotopes in marine plankton 1, temperature and respiration experiments. Deep-Sea Res. 15: 1–9.

Descolas-Gros C. and Fontugne M.R. 1985. Carbon fixation in marine phytoplankton: carboxylase activities and stable carbon isotope ratios; physiological and paleoclimatological aspects. Mar. Biol. 87: 1–6.

Deutsch C., Sigman D.M., Thunell R.C., Meckler A.N. and Haug G.H. 2004. Isotopic constraints on glacial/interglacial changes in the oceanic nitrogen budget. Global Biogeochem. Cy. 18: GB4012. doi: 10.1029/2003GB002189.

Dezileau L., Bareille G., Reyss J.L. and Lemoine F. 2000. Evidence for strong sediment redistribution by bottom currents along the southeast Indian ridge. Deep-Sea Research Pt. I. 47: 1899–1936.

Dickens G.R. 2001. The potential volume of oceanic methane hydrates with variable external conditions. Org. Geochem. 32: 1132–1193.

Dickens G.R., Paull C.K., Wallace P. and the ODP Leg 164 Scientific Party. 1997. Direct measurements of in situ methane quantities in a large gas-hydrate reservoir. Nature 385: 426–428.

Ding T., Jiang S., Wan D., Li Y., Li J., Song H., Liu Z. and Yao X. 1996. Silicon Isotope Geochemistry. Geological Publishing House, Beijing, China, 125 pp.

Ding T., Wan D., Wang C. and Zhang F. 2004. Silicon isotope compositions of dissolved silicon and suspended matter in the Yangtze River, China, Geochim. Cosmochim. Ac. 68: 205–216.

Douthitt C.B. 1982. The geochemistry of the stable isotopes of silicon. Geochim. Cosmochim. Ac. 46: 1449–1458.

Dugdale R.C., Lyle M., Wilkerson F.P., Chai F., Barber R.T. and Peng T.-H. 2004. Influence of equatorial diatom processes on Si deposition and atmospheric CO_2 cycles at glacial/interglacial timescales. Paleoceanography 19: PA3011. doi: 10.1029/2003PA000929.

Duplessy J.-C., Shackleton N.J., Matthews R.K., Prell W., Ruddiman W.F., Caralp M. and Hendy C.H. 1984. 13C record of benthic foraminifera in the last interglacial ocean: implications for the carbon cycle and the global deep water circulation. Quaternary Res. 21: 225–243.

Duplessy J-C., Shackleton N.J., Fairbanks R.J., Labeyrie L.D., Oppo D. and Kallel N. 1988. Deepwater source variation during the last climatic cycle and their impact on the global deepwater circulation. Paleoceanography 3: 343–360.

Duplessy J-C., Labeyrie L., Juillet-Leclerc A., Maitre F., Duprat J. and Sarnthein M. 1991. Surface salinity reconstruction of the North Atlantic Ocean during the last glacial maximum. Oceanol. Acta. 14: 311–324.

Duplessy J-C, Labeyrie L., Arnold M., Paterne M., Duprat J. and van Weering T. 1992 Changes in surface water salinity of the North Atlantic Ocean during the last deglaciation. Nature 358: 485–488.

Duplessy J-C., Labeyrie L. and Waelbroeck C. 2002. Constraints on the ocean oxygen isotopic enrichment between the Last Glacial Maximum and the Holocene. Quaternary Sci. Rev. 21: 315–330.

Edwards R.L., Chen J.H., Ku T-L. and Wasserburg G.L. 1987. Precise timing of the last interglacial period from mass spectrometric determination of thorium-230 in corals. Science 236: 1547–1553.

Eglinton G. and Hamilton R.J. 1967. Leaf epicuticular waxes. Science 156: 1322–1334.

Eglinton G., Hamilton R.J., Raphael R.A. and Gonzalez A.G. 1962. Hydrocarbon constituents of the wax coatings of plant leaves: A taxonomic survey. Phytochemistry 1: 89–102.

Eglinton G., Bradshaw S.A., Rosell A., Sarnthein M., Pflaumann U. and Tiedemann R. 1992. Molecular record of secular sea surface temperature changes on 100 year timescales for glacial terminations I, II, and IV. Nature 356: 423–425.

Ehleringer J.R., Cerling T.E. and Helliker B. 1997. C_4-photosynthesis, atmospheric CO_2 and climate. Oecologia 112: 285–299.

Elderfield H. and Ganssen G. 2000. Past temperature and delta18O of surface ocean waters inferred from foraminiferal Mg/Ca ratios. Nature 405: 442–445.

Emiliani C. 1955. Pleistocene temperatures. J. Geol. 63: 538–578.

Emiliani C. 1971. The amplitude of Pleistocene climatic cycles at low latitudes and the isotopic composition of glacial ice. In: Turehian K.K. (ed.), The late Cenozoic glacial ages. Yale University, New Haven, Connecticut, pp. 183–197.

Epstein S. and Taylor H.P. 1970a. The concentration and isotopic composition of hydrogen, carbon and silicon in Apollo 11 lunar rocks and minerals. Proceedings of the Apollo 11 lunar science conference. 2: 1085–1096.

Epstein S. and Taylor H.P. 1970b. Stable isotopes, rare gases, solar wind and spallation products. Science 167: 533–535.

Epstein S., Buchsbaum R., Lowenstam H.A. and Urey H.C. 1953. Revised carbonate-water isotopic temperature scale. Geol. Soc. Am. Bull. 64: 1315–1325.

Erez J. and Honjo S. 1981 Comparison of isotopic composition of planktonic foraminifera in plankton tows, sediment traps and sediments. Palaeogeogr. Palaeocl. 33: 129–156.

Ettwein V. 2005. South American palaeoclimate reconstruction. PhD, thesis University College London, London.

Fairbanks R.G. 1989. A 17,000 year glacio-eustatic sea level record: influence of glacial melting rates on the Younger Dryas event and deep-ocean circulation. Nature 342: 637–642.

Fairbanks R., Sverdlove M., Free R., Wiebe P. and Bé A. 1982. Vertical distribution and isotopic fractionation of living planktonic foraminifera from the Panama Basin. Nature 298: 841–843.

Falkowski P.G. 1997. Evolution of the nitrogen cycle and its influence on the biological sequestration of CO_2 in the ocean. Nature 387: 272–275.

Farrell J.W., Pedersen T.F., Calvert S.E. and Nielsen B. 1995. Glacial-interglacial changes in nutrient utlization in the equatorial Pacific Ocean. Nature 377: 514–517.

Faure H., Adams J.M., Debenay J.P., Faure-Denard L., Gant D.R., Pirazzoli P.A., Thomassin B., Velichko A.A. and Zazo C. 1996. Carbon storage and continental land surface change since the last glacial maximum. Quaternary Sci. Rev. 15: 843–849.

Fillon R.H. and Williams D.F. 1984. Dynamics of meltwater discharge from Northern Hemisphere ice sheet during the last deglaciation. Nature 310: 674–676.

Fontugne M.R. and Duplessy J-C. 1986. Variations of the monsoon regime during the upper Quaternary: evidence from carbon isotopic record of organic matter in North Indian ocean sediment cores. Palaeogeogr. Palaeocl. 56: 69–88.

Francois L. and Godderis Y. 1998. Isotopic constraints on the Cenozoic evolution of the carbon cycle. Chem. Geol. 145: 177–212.

Francois L.M., Delire C., Warnant P. and Munchoven G. 1998. Modelling the glacial-interglacial changes in continental biosphere. Global Planet. Change 16-17: 37–52.

Francoise R., Bacon P. and Altabet M. 1993. Glacial/interglacial changes in sediment rain rate in the SW Indian Sector of Subantarctic waters as recorded by 230Th, 231Pa, U, and 15N. Paleoceanography 8: 611–629.

Francoise R., Altabet M.A., Yu E-F., Sigman D.M., Bacon M.P., Frank M., Bohrmann G., Bareille G. and Labeyrie L. 1997. Contribution of Southern Ocean surface water stratification to low atmospheric CO_2 concentrations during the last glacial period. Nature 389: 929–935.

Freeman K.H., Hayes J.M., Trendel J.M. and Albrecht P. 1990. Evidence from carbon isotope measurements for diverse origins of sedimentary hydrocarbons. Nature 343: 254–256.

Friedlingstein C., Delire C., Muller J.F. and Gerard J.C. 1992. The climate induced variation of the continental biosphere: a model simulation of the last glacial maximum. Geophys. Res. Lett. 19: 897–900.

Friedlingstein C., Prentice K.C., Fung I.Y., John J.G. and Brasseur G.P. 1995. Carbon-biosphere-climate interaction in the last glacial maximum climate. J. Geophys. Res. 100: 7203–7221.

Gagan M.K., Ayliffe L.K., Hopley D., Cali J.A., Mortimer G.E., Chappell J., McCulloch M.T., and Head M.J. 1998. Temperature and surface water balance of the mid-Holocene tropical western Pacific. Science 279: 1014–1018.

Gajewski K., Viau A., Sawada M., Atkinson D. and Wilson S. 2001. Sphagnum peatland distribution in North America and Eurasia during the past 21,000 years. Global Biogeochem. Cy. 15: 297–310.

Ganeshram R.S., Pedersen T.F., Calvert S.E. and Murray J.W. 1995. Large changes in oceanic nutrient inventories from glacial to interglacial periods. Nature 376: 755–758.

Ganeshram R.S., Pedersen T.F., Calvert S.E., McNeill G.W. and Fontugne M.R. 2000. Glacial-interglacial variability in denitrification in the world's ocean: Causes and consequences. Paleoceanography 15: 316–376.

Ganeshram R.S., Pedersen T.F., Calvert S.E. and Francois R. 2002. Reduced nitrogen fixation in the glacial ocean inferred from changes in marine nitrogen and phosphorus inventories. Nature 415: 156–159

Ganssen G. 1983. Dokumentation von küstennahem auftrieb anhand stabiler Isotopen in rezenten foraminiferen vor Nordwest-Afrika. Meteor Forsch. Ergebnisse C 37: 1–46.

Ganssen G. and Sarnthein M. 1983. Stable isotope composition of foraminifera: the surface and bottom waters record of coastal upwelling. In: Suess A.E. and Thiede J. (eds), Coastal Upwelling: its sediment record, Part A. Plenum, New York, pp. 99–121.

Garlick G.D. 1974. The stable isotopes of oxygen, carbon, hydrogen in the marine environment. In: Goldberg G.D. (ed.), The Sea. John Wiley & Sons, New York, pp. 393–425.

GEOSECS. 1987. Atlantic, Pacific and Indian Ocean expeditions. Shorebased data and graphics. Vol. 7. In: Östlund H.G., Craig H., Broecker W.S. and Spencer D. (eds), I.D.O.E. National Science Foundation, Washington, USA, 230 pp.

Goldman J.C. 1993. Potential role of large oceanic diatoms in new primary production. Deep-Sea Res. Pt. I. 40: 159–168.

Grootes P.M. 1993. Interpreting continental oxygen isotopes records. In: Swart P.K., Lohmann K.C., McKenzie J. and Savin S. (eds), Climate change in continental isotope records. AGU Geophysical Monograph. 78: 37–46.

Gruber N. and Sarmiento J.L. 1997. Global patterns of marine nitrogen fixation and denitrification. Global Biogeochem. Cy. 11: 235–266.

Haimson M. and Knauth L.P. 1983. Stepwise fluorination – a useful approach for the isotopic analysis of hydrous minerals. Geochim. Cosmochim. Ac. 47: 1589–1595.

Haq B. 1998. Natural gas hydrates: searching for the long-term climate and slope stability records. In: Henriet J.P. and Mienert J. (eds), Gas hydrates: relevance to world margin stability and climate change. Geological Society of London Special Publication 137, The Geological Society, London, pp. 303–318.

Haug G.H. and Tiedemann R. 1998. Effect of the formation of the Isthmus of Panama on Atlantic Ocean thermohaline circulation. Nature 393: 673–675.

Haug G.H., Sigman D.M., Tiedemann R., Pedersen T.F. and Sarnthein M. 1999. Onset of permanent stratification in the subarctic Pacific Ocean. Nature 401: 779–782.

Haug G.H., Ganopolski A., Sigman D.M., Rosell-Mele A., Swann G.E.A., Tiedemann R., Jaccard S, Bollmann J., Maslin M.A., Leng M.J. and Eglinton G. 2005. North Pacific seasonality and the glaciation of North America 2.7 million years ago. Nature 433: 821–825.

Haupt B. and Seidov D. 2001. Warm deep-water ocean conveyor during Cretaceous times. Geology 29: 295–298.

Hays P.D and Grossman E.L. 1991. Oxygen isotopes in meteoric calcite cements as indicators of continental paleoclimate. Geology 19: 441–444.

Hays J.D., Imbrie J. and Shackleton N.J. 1976. Variations in the Earth's orbit: Pacemaker of the Ice Ages. Science 194: 1121–1132.

Hemleben Ch., Spindler M. and Anderson O.R. 1988. Modern Planktonic Foraminifera. Spinger-Verlag, New York, 363 pp.

Hemming N.G. and Hanson G.N. 1992. Boron isotopic composition and concentration in modern marine carbonates. Geochim. Cosmochim. Ac. 56: 537–543.

Hemming N.G., Reeder R.J. and Hanson G.N. 1995. Mineral-fluid partitioning and isotopic fractionation of boron in synthetic calcium carbonate. Geochim. Cosmochim. Ac. 59: 371–379.

Hendy E.J., Gagan M.K., Alibert C.A., McCulloch M.T., Lough J.M. and Isdale P.J. 2002. Abrupt decreased in tropical Pacific Sea surface salinity at the end of the Little Ice Age. Science 295: 1511–1514.

Henriet J-P. and Mienert J. (eds) 1998. Gas Hydrates: Relevance to world margin stability and climate change. Geological Society of London Special Publication 137, The Geological Society, London, 338 pp.

Hoefs J. 1997. Stable isotope geochemistry. Spinger-Verlag, Berlin, 213 pp.

Holmes E.M., Muller P.J., Schneider R.R., Segl M., Patzold J. and Wefer G. 1996. Stable nitrogen isotopes in Angola Basin surface sediments. Mar. Geol. 134: 1–12.

Holmes M.E., Muller P.J., Schneider R.R., Segl M. and Wefer G. 1998. Spatial variations in euphotic zone nitrate utlization based in $\delta^{15}N$ in surface sediments. Geo-Marine Letter. 18: 58–65.

Holmes M.E., Eichner C., Struck U. and Wefer G. 1999. Reconstruction of surface water nitrate utilization using stable nitrogen isotopes in sinking particles and sediments. In: Fischer G. and Wefer G. (eds), Use of proxies in paleoceanography. Springer, Berlin, pp. 447–468.

Hönisch B. and Hemming N.G. 2004. Ground-truthing the boron isotope-paleo-pH proxy in planktonic foraminifera shells: Partial dissolution and shell size effects. Paleoceanography 19: PA4010. doi: 10.1029/2004PA001026.

Hönisch B., Bijma J., Russell A.D., Spero H.J., Palmer M.R., Zeebe R.E. and Eisenhauer A. 2003. The influence of symbiont photosynthesis on the boron isotopic composition of foraminifera shells. Mar. Micropaleontol. 49: 87–96.

Hönisch B., Hemming N.G., Grottoli A.G., Amat A., Hanson G.N. and Buma J. 2004. Assessing scleractinian corals as recorders for paleo-pH: Empirical calibration and vital effects. Geochim. Cosmochim. Ac. 68: 3675–3685.

Houghton R.A. 2004. The contemporary carbon cycle. In: Schlesinger W.H. (ed.), Treatise on Geochemistry, Volume 8: Biogeochemistry. Elsevier, Amsterdam, pp. 473–513.

Hughen K., Eglinton T.E., Xu L. and Makou M. 2004. Abrupt Tropical Vegetation Response to Rapid Climate Changes. Science 304: 1955–1959.

Hutchins D.A. and Bruland K.W. 1998. Iron-limited diatom growth and Si:N uptake ratios in a coastal upwelling zone. Nature 393: 561–564.

Imbrie J. and Kipp N.G. 1971. A new micropaleontological method for quantitative paleoclimatology. In: Turekian K.K. (ed.), Late Cenozoic Glacial Ages. Yale University Press, New Haven, Connecticut, pp. 71–182.

Imbrie J., Hays J.D., Martinson D.G., McIntyre A., Mix A.C., Morley J.J., Pisias N.G., Prell W.L. and Shackleton N.J. 1984. The orbital theory of Pleistocene climate: support from a revised chronology of the marine $\delta^{18}O$ record. In: Berger A., Imbrie J., Hays J., Kugla G. and Satzmann B. (eds), Milankovitch and climate. Reidel Publishing Company, Dordrecht, pp. 269–305.

Ivany L.C. and Salawitch R.J. 1993. Carbon isotopic evidence for biomass burning at the K-T boundary. Geology 21: 487–490.

Jasper J.P. and Hayes J. 1990. A carbon isotope record of CO_2 level during the late Quaternary. Nature 347: 462–464.

Jasper J.P. and Gagosian R.B. 1990. The sources and deposition of organic matter in Late Quaternary Pigmy Basin, Gulf of Mexico. Geochim. Cosmochim. Ac. 54: 1117–1132.

Jasper J.P. Mix A.C., Prahl F.G. and Hayes J.M. 1994. Photosynthetic fractionation of 13C and concentrations of dissolved CO_2 in the central equatorial Pacific during the last 255,000 years. Paleoceanography 6: 781–798.

Jennings A.E. and Weiner N.J. 1996. Environmental change on eastern Greenland during the last 1300 years: Evidence from formainifera and lithofacies in Nansen Fjord. Holocene 6: 179–191.

Juillet-Leclerc A. and Labeyrie L. 1987. Temperature dependence of the oxygen isotopic fractionation between diatom silica and water. Earth Planet. Sc. Lett. 84: 69–74.

Kasemann S.A., Hawkesworth C.J., Prave A.R., Fallick A.E. and Pearson P.N. 2005. Boron and calcium isotope composition in Neoproterozoic carbonate rocks from Namibia: evidence for extreme environmental change. Earth Planet. Sc. Lett. 231: 73–86.

Kaiho H., Kajiwara Y., Tazaki K., Ueshima M., Takeda N., Kawahata H., Arinobu T., Ishiwatari R., Hirai A. and Lamolda M.A. 1999. Oceanic primary productivity and dissolved oxygen levels at the Cretaceous/Tertiary boundary: their decrease, subsequent warming, and recovery. Paleoceanography 14: 511–524.

Kakihana H., Kotaka M., Satoh S., Nomura M. and Okamoto M. 1977. Fundamental studies on the ion-exchange of boron isotopes. B. Chem. Soc. Jpn. 50: 158–163.

Kaplan J., Prentice I.C., Knorr W. and Valdes P.J. 2002. Modelling the dynamics of terrestrial carbon storage since the LGM. Geophys. Res. Lett. 29: 2074. doi: 10.1029/2002GL015230.

Karl D., Letelier R., Tupas L., Dore J., Christian J. and Hebel D. 1997. The role of nitrogen fixation in biogeochemical cycling in the subtropical North Pacific Ocean. Nature 388: 533–539.

Kastner T.P. and Goñi M.A. 2003. Constancy in the vegetation of the Amazon basin during the late Pleistocene. Geology 31: 291–294.

Kennett J.P. and Fackler-Adams B.N. 2000. Relationship of clathrate instability to sediment deformation in the upper Neogene of California. Geology 28: 215–215.

Kennett J.P, Cannariato K.G., Hendy I.L. and Behl R.J. 2000. Carbon isotopic evidence for methane hydrate instability during Quaternary interstadials. Science 288: 128–133.

Kennett J., Cannariato K.G., Hendy I.L. and Behl R.J. 2003. Methane hydrates in Quaternary climate change: The clathrate gun hypothesis: American Geophysical Union, 216 pp.

Kienast M. 2000. Unchanged nitrogen isotope composition of organic matter in the South China Sea during the last climatic cycle: Global implications. Paleoceanography 15: 244–253.

Kolattukudy P.E. 1976. The Chemistry and Biochemistry of Natural Waxes. Elsevier, Netherlands, 459 pp.

Kroon D. and Ganssen G. 1989. Northern Indian ocean upwelling cells and the stable isotope composition of living foraminifera. Deep-sea Res. 36: 1219–1236.

Kucera M., Weinel M., Kiefer T., Pflaumann U., Hayes A., Weinelt M., Chen M-T., Mix A.C., Barrows T.T., Cortijo E., Duprat J., Juggins S. and Waelbroeck C. 2005a. Reconstruction of sea-surface temperatures from assemblages of planktonic foraminifera: multi-technique approach based on geographically constrained calibration data sets and its application to glacial Atlantic and Pacific Oceans. Quaternary Sci. Rev. 24: 951–998.

Kucera M., Rosell-Melé A., Schneider R., Waelbroeck C. and Weinelt M. 2005b. Multiproxy approach for the reconstruction of the glacial ocean surface (MARGO) Quaternary Sci. Rev. 24: 813–819.

Kump L.R. and Arthur M.A. 1999. Interpreting carbon-isotope excursions: carbonates and organic matter. Chem. Geol. 161: 181–198.

Kuroyanagi A. and Kawahata H. 2004. Vertical distribution of living planktonic foraminifera in the seas around Japan. Mar. Micropaleontol. 53: 173–196.

Kvenvolden K.A. 1993. Gas hydrates – geological perspective and global change. Rev. Geophys. 31: 173–187.

Kvenvolden K.A. 1998. A primer on the geological occurrence of gas hydrate. In: Henriet J.P. and Mienert J. (eds), Gas hydrates: relevance to world margin stability and climate change. Geological Society of London Special Publication 137, The Geological Society, London, pp. 9–30.

Labeyrie L.D. 1974. New approach to surface seawater paleotemperatures using (18)O/(16)O ratios in silica of diatom frustules. Nature 248: 40–42.

Labeyrie L.D. and Juillet A. 1982. Oxygen isotopic exchangeability of diatom valve silica; interpretation and consequences for palaeoclimatic studies. Geochim. Cosmochim. Ac. 46: 967–975.

Labeyrie L.D., Duplessy J-C., Duprat J., Juillet-Leclerc A., Moyes J., Michel E., Kallel N. and Shackleton N.J. 1992. Changes in the vertical structure of the North Atlantic Ocean between glacial and modern times. Quaternary Sci. Rev. 11: 401–413.

Lamb A.L., Leng M.J., Sloane H.J. and Telford R. J. In press. A comparison of the palaeoclimatic signals from diatom oxygen isotope ratios and carbonate oxygen isotope ratios from a low latitude crater lake. Palaeogeogr. Palaeocl.

Lambeck K. and Chappell J. 2001. Sea level change through the last glacial cycle. Science 292: 679–686.

Laws E.A., Popp B.N., Bidigare R.R., Riebesell U., Burkhardt S. and Wakeham S. 2001. Controls on the molecular distribution and carbon isotopic composition of alkenones in certain haptophyte algae. Geochem. Geophys. Geosyst. 2: doi.2000GC000057.

Lemarchand D., Gaillardet J., Lewin E. and Allegre C.J. 2000. The influence of rivers on marine boron isotopes and implications for reconstructing past ocean pH. Nature 408: 951–954.

Lemarchand D., Gaillardet J., Lewin E. and Allegre C.J. 2002. Boron isotope systematics in large rivers: implications for the marine boron budget and paleo–pH reconstruction over the Cenozoic. Chem. Geol. 190: 123–140.

Leng M., Barker P., Greenwood P., Roberts N. and Reed J. 2001. Oxygen isotope analysis of diatom silica and authigenic calcite from Lake Pinarbasi, Turkey. J. Paleolimnol. 25: 343–349.

Leng M.J., Lamb A.L, Heaton T.H.E, Marshall J.D., Wolfe B.B., Jones M.D., Holmes J.A. and Arrowsmith C. This volume. Isotopes in lake sediments. In: Leng M.J. (ed.), Isotopes in palaeoenvironmental Research. Springer, Dordrecht.

Lücke A., Moschen R. and Schleser G.H. 2005. High-temperature carbon reduction of silica: a novel approach for oxygen isotope analysis of biogenic opal. Geochim. Cosmochim. Ac. 69: 1423–1433.

Lynch-Stieglitz J. 2004. Tracers of past ocean circulation. In: Elderfield H. (ed.), Treatise on Geochemistry, Volume 6: The Oceans and Marine Geochemistry. Elsevier, Amsterdam, pp. 433–453.

Lynch-Stieglitz J., Curry W., Slowey N. and Schmidt G. 1999a. The overturning of the glacial Atlantic: A view from the top. In: Abrantes F. and Mix A. (eds), Reconstructing Ocean History: A window into the future. Kluwer Academic, New York, pp. 7–32.

Lynch-Stieglitz J., Curry W. and Slowey N. 1999b. Weaker Gulf Stream in the Florida Straits during the Last Glacial Maximum. Nature 402: 644–648.

MacDonald G.J. 1990. Role of methane clathrates in past and future climates. Climatic Change 16: 247–281.

Mackensen A., Hubberten H.W., Bickert T., Fischer G. and Fütterer D.K. 1993. The $\delta^{13}C$ in benthic foraminifera tests of *Fontbotia wuellerstorfi* relative to the $\delta^{13}C$ of dissolved inorganic carbon in Southern Ocean deep Water: Implications for ocean circulation models. Paleoceanography 8: 587–610.

McCorckle D.C., Keigwin L.D., Corliss B.H. and Emerson S.R. 1990. The influence of microhabitats on the carbon isotopic composition of deep-sea benthic foraminifera. Paleoceanography 5: 161–185.

McElroy M.B. 1983. Marine biological controls on atmospheric CO_2 and climate. Nature 302: 328–329.

Martinson D.G., Pisias N.G., Hays J.D., Imbrie J., Moore T.C. Jr. and Shackleton N.J. 1987. Age dating and the orbital theory of ice ages: development of a high-resolution 0 to 300,000 years chronostratigraphy. Quaternary Res. 27: 1–29.

Maslin M.A. and Burns S.J. 2000. Reconstruction of the Amazon Basin effective moisture availability over the last 14,000 years. Science 290: 2285–2287.

Maslin M.A. and Thomas E. 2003. Balancing the deglacial global carbon budget: the hydrate factor. Quaternary Sci. Rev. 22: 1729–1736.

Maslin M.A., Shackleton N.J. and Pflaumann U. 1995a. Temperature, salinity and density changes in the Northeast Atlantic during the last 45,000 years: Heinrich events, deep water formation and climatic rebounds. Paleoceanography 10: 527–544.

Maslin M.A., Adams J., Thomas E., Faure H. and Haines-Young R. 1995b. Estimating the carbon transfer between the oceans, atmosphere and the terrestrial biosphere since the last glacial maximum. Terra Nova 7: 358–366.

Maslin M.A., Haug G.H., Sarnthein M. and Tiedemann R. 1996. The progressive intensification of northern hemisphere glaciation as seen from the North Pacific. Geologische Rundschau. 85: 452–465.

Maslin M.A., Mikkelsen N., Vilela C. and Haq B. 1998. Sea-level and gas hydrate controlled catastrophic sediment failures of the Amazon Fan. Geology 26: 1107–1110.

Maslin M.A., Pike J., Stickley C. and Ettwein V. 2003. Evidence of Holocene climate variability from marine sediments. In: Mackay A.W., Battarbee R., Birks J. and Oldfield F. (eds), Global Change in the Holocene. John Wiley, London, pp. 185–209.

Maslin M.A., Vilela C., Mikkelsen N. and Grootes P. In press. Causation of the Quaternary catastrophic failures of the Amazon Fan deduced from stratigraphy and benthic foraminiferal assemblages. Quaternary Sci. Rev.

Mataliotaki I., De La Rocha C.L., Passow U. and Wolf-Gladrow D. 2003. Impact of the pH-dependent speciation of silicic acid on the silicon isotope composition of diatoms. Geophysical Research Abstracts. 5: www.cosis.net/abstracts/EAE03/09375/EAE03-J-09375.pdf

Matheney R.K. and Knauth L.P. 1989. Oxygen-isotope fractionation between marine biogenic silica and seawater. Geochim. Cosmochim. Ac. 53: 3207–3214.

Matsumoto R., Watanabe Y., Satoh M., Okada H., Hiroki Y., Kawasaki M. and the ODP Leg 164 Shipboard Scientific Party. 1996. Distribution and occurrence of marine gas hydrates - preliminary results of ODP Leg 164: Blake Ridge Drilling. Journal of the Geological Society of Japan. 102: 932–944.

Matsumoto K., Sarmiento J.L. and Brzezinski M.A. 2002. Silicic acid leakage from the Southern Ocean: A possible explanation for glacial atmospheric pCO$_2$. Global Biogeochem. Cy. 16: 1031. doi: 10.1029/2001GB001442.

Melosh H.J., Schneider N.M., Zahnle K. and Latham D. 1990. Ignition of global wildfires at the Cretaceous/Tertiary boundary. Nature 350: 251–254.

Mikkelsen N., Labeyrie L. and Berger, W.H. 1978. Silica oxygen in diatoms: a 20,000 yr record in deep-sea sediments. Nature 271: 536–538.

Milankovitch M.M. 1949. Kanon der Erdbestrahlung und seine Anwendung auf das Eiszeitenproblem. Royal Serbian Sciences, Special publication 132, Section of Mathematical and Natural Sciences, Belgrade, 633 pp.

Milligan A.J., Varela D.E., Brzezinski M.A. and Morel F.M.M. 2004. Dynamics of silicon metabolism and silicon isotopic discrimination in a marine diatom as a function of pCO_2. Limnol. Oceanogr. 49: 322–329.

Mix A. and Ruddiman W.F. 1984. Oxygen isotope analyses and Pleistocene Ice Volumes. Quaternary Res. 21: 1–20.

Mix A. and Fairbanks R. 1985. North Atlantic surface-ocean control of Pleistocene deep-ocean circulation. Earth Planet. Sc. Lett. 73: 231–243.

Montoya J.P. 1994. Nitrogen isotope fractionation in the modern ocean: implications for sedimentary record. NATO ASI Series, Vol 17, Springer, Berlin, pp. 259–279.

Montoya J.P. and McCarthy J.J. 1995. Isotope fractionation during nitrate uptake by marine phytoplankton growth in continuous culture. J. Plankton. Res. 17: 439–464.

Mopper K. and Garlick G.D. 1971. Oxygen isotope fractionation between biogenic silica and ocean water. Geochim. Cosmochim. Ac. 35: 1185–1187.

Morley D.W., Leng M.J., Mackay A.W., Sloane H.J., Rioual P. and Battarbee R.W. 2004. Cleaning of lake sediment samples for diatom oxygen isotope analysis. J. Paleolimnol. 31: 391–401.

Morris I. 1980. Paths of carbon assimilation in marine phytoplankton. In: Falkowski P. (ed.), Primary Productivity of the Sea. Plenum, New York, pp. 139–159.

Mortlock R.A., Charles C.D., Froelich P.N., Zibello M.A., Saltzman J., Hays J.D. and Burckle L.H. 1991. Evidence for lower productivity in the Antarctic Ocean during the last glaciation. Nature 351: 220–223.

Mulitza S., Durkoop A., Hale W., Wefer G. and Niebler H.S. 1997. Planktonic foraminifera as recorders of past surface water stratification. Geology 25: 335–338.

Müller P.J., Erenkeuser H. and Grafenstein R. 1983. Glacial interglacial cycles in ocean productivity inferred from organic carbon content in eastern North Atlantic sediment cores. In: Thiede J. and Suess E. (eds), Coastal Upwelling: Its sediment record (Part B). Plenum Press, New York, pp. 365–398.

Naqvi S.W.A., Yoshinari T., Brandes J.A., Devol A.H., Jayakumar D.A., Narvekar P.V., Altabet M.A. and Codispoti L.A. 1998. Nitrogen isotopic studies in the suboxic Arabian Sea. P. Indian As-Earth. 107: 367–378.

Nelson D.A. and Smith W.O. Jr 1986. Phytoplankton bloom dynamics of the western Ross Sea Ice edge II. Mesoscale cycling of nitrogen and silicon. Deep-Sea Res. 33: 1389–1412.

Nelson D.M., Tréguer P., Brzezinski M.A., Leynaert A. and Quéguiner B. 1995. Production and dissolution of biogenic silica in the ocean: revised global estimates, comparison with regional data and relationship to biogenic sedimentation. Global Biogeochem. Cy. 9: 359–372.

Niebler H-S., Hubberten H-W. and Gersonde R. 1999. Oxygen isotope values of planktonic foraminifera: a tool for the reconstruction of surface water stratification. In: Fischer G. and Wefer G. (eds), Use of proxies in Paleoceanography. Springer, Berlin, pp. 165–189.

Nimer N.A. and Merrett M.J. 1996. The development of a CO_2-concentrating mechanism in Emiliania huxleyi. New Phytol. 133: 383–389.

Nimer N.A., Dixon G.K. and Merrett M.J. 1992. Utilization of inorganic carbon by the coccolithophorid Emiliania huxleyi (Lohmann) Kamptner. New Phytol. 120: 153–158.

Nisbet E.G. 1990. The end of the ice age. Can. J. Earth Sci. 27: 148–157.

Nisbet E.G. 1992. Sources of atmospheric CH4 in early postglacial time. J. Geophys. Res. 97: 12,859–12,867.

Norris R.D., Klaus A. and Kroon D. 2001. Mid-Eocene deep water, the late Paleocene thermal maximum and continental slope mass wasting during the Cretaceous-Palaeogene impact. In: Kroon D., Norris R.D. and Klaus A. (eds), Western North Atlantic Paleogene and Cretaceous Paleoceanography. Geological Society of London Special Publication 183, The Geological Society, London, pp. 23–28.

O'Neil J., Clayton R. and Mayeda T. 1969. Oxygen isotope fractionation in divalent metal carbonates. J. Chem. Phys. 51: 5547–5558.

Otto D., Rasse D., Kaplan J., Warnant P. and Francois L. 2002. Biospheric carbon stocks reconstructed at the Last Glacial maximum. Global Planet. Change 33: 117–138.

Pagani M., Freeman K.H. and Arthur M.A. 1999b. Late Miocene atmospheric CO_2 concentrations and the expansion of C_4 gasses. Science 185: 876–879.

Pagani M., Arthur M.A. and Freeman K.H. 1999a The Miocene evolution of atmospheric carbon dioxide. Paleoceanography 14: 273–292.

Pagani M., Freeman K.H., Ohkouchi N. and Caldeira K. 2002. Comparison of water column CO_{2aq} with sedimentary alkenone-based estimates: A test of the alkenone-CO_2 proxy. Paleoceanography 17: 1069. doi: 10.1029/2002PA000756.

Pagani M., Lemarchand D., Spivack A. and Gaillardet J. 2005. A critical evaluation of the boron isotope-pH proxy: The accuracy of ancient ocean pH estimates. Geochim. Cosmochim. Ac. 69: 953–961.

Pahnke K., Zahn R., Elderfield H. and Schulz M. 2003. 340,000-year centennial-scale marine record of Southern Hemisphere climatic oscillation. Science 301: 948–952.

Palmer M.R., Pearson P.N. and Cobb S.J. 1998. Reconstructing past ocean pH-depth profiles. Science 282: 1468–1471.

Palmer M.R. and Pearson P.N. 2003 A 23,000 year record of surface water pH and pCO_2 in the Western Equatorial Pacific Ocean. Science 300: 480–482.

Pahlow M. and Riebesell U. 2000. Temporal changes in deep ocean Redfield ratios. Science 287: 831–833.

Pancost R. and Pagani, M. in press. Controls on the Carbon Isotopic Compositions of Lipids in Marine Environments. In Volkman et al (eds).

Paull C.K., Ussler W. and Dillon W.P. 1991. Is the extent of glaciation limited by marine gas hydrates? Geophys. Res. Lett. 18: 432–434.

Paull C.K., Brewer P.G., Ussler W. III, Peltzer E.T., Rehder G. and Clague D. 2003. An experiment demonstrating that marine slumping is a mechanism to transfer methane from seafloor gas-hydrate deposits into the upper ocean and atmosphere. Geo-Mar. Lett. 22: 198–203.

Pearson P.N. and Palmer M.R. 1999. Middle Eocene seawater pH and atmospheric carbon dioxide concentrations. Science 284: 1824–1826.

Pearson P.N. and Palmer M.R. 2000. Atmospheric carbon dioxide concentrations over the past 60 million years. Nature 406: 695–699.

Pearson P.N. and Palmer M.R. 2003. Reconstructed 60 Million Year Atmospheric Carbon Dioxide Concentration Data. IGBP PAGES/World Data Center for Paleoclimatology: Data Contribution Series # 2003-069.

Peltier W.R. 2002. On eustatic sea level history: Last Glacial Maximum to Holocene. Quaternary Sci. Rev 21: 1–8.

Peng C.H., Guiot J. and Van Campo E. 1995. Reconstruction of the past terrestrial carbon storage of the Northern Hemisphere from the Osnabrueck Biosphere Model and palaeodata. Climate Res. 5: 107–118.

Peng C.H., Guiot J. and van Campo E. 1998. Estimating changes in terrestrial vegetation and carbon storage using palaeoecological data and models. Quaternary Sci. Rev. 17: 719–735.

Peters K.E., Sweeny R.E. and Kaplan I.R. 1978. Correlation of carbon and nitrogen stable isotopes in sedimentary organic matter. Limnol. Oceanogr. 23: 598–604.

Pflaumann U., Sarnthein M., Chapman M., d'Abreu L., Funnell B., Huels M., Kiefer T., Maslin M.A., Schulz H., Swallow J., van Kreveld S., Vautravers M., Vogelsang E. and Weinelt M. 2003. The Glacial North Atlantic: Sea-surface conditions reconstructed by GLAMAP-2000. Paleoceanography 18: 1065. doi: 10.1029/2002PA000774.

Pike J. and Kemp A.E.S. 1997. Early Holocene decadal-scale ocean variability recorded in Gulf of California laminated sediments. Paleoceanography 12: 227–238.

Pondaven P., Ragueneau O., Treguer P., Hauvespre A., Dezileau L. and Reyss J.L. 2000. Resolving the 'opal paradox' in the Southern Ocean. Nature 405: 168–172.

Popp B., Takigiku R., Hayes J., Louda J. and Baker E. 1989. The past paleozoic chronology and mechanism of ^{13}C depletion in primary marine organic matter. Am J. Sci. 289: 436–454.

Popp B.N., Laws E.A., Bidigare R.R., Dore J.E., Hanson K.L. and Wakeham S.G. 1998. Effect of phytoplankton cell geometry on carbon isotopic fractionation. Geochim. Cosmochim. Ac. 62: 69–77.

Prell W.L., Imbrie J., Martinson D., Morley J., Shackleton N. and Streeter H. 1986. Graphic correlation of oxygen isotope stratigraphy application to the Late Quaternary. Paleoceanography 1: 137–162.

Prentice I.C. and Fung I.Y. 1990. The sensitivity of terrestrial carbon storage to climate change. Nature 346: 48–51.

Prentice I.C., Sykes M.T., Lautenschlager M., Harrison S.P., Denissenki O. and Bartlein P.J. 1993. Modelling the increase in terrestrial carbon storage after the last glacial maximum. Global Ecol. Biogeogr. 3: 67–76.

Pride C., Thunell R., Sigman D.M., Keigwin L., Altabet M. and Tappa E. 1999. Nitrogen isotope variations in the Gulf of California since the last deglaciation: Response to global climate change. Paleoceanography 14: 397–409.

Ragueneau O., Treguer P., Leynaert A., Anderson R.F., Brzezinski M.A., DeMaster D.J., Dugdale R.C., Dymond J., Fischer G., Francois R., Heinze C., Maier-Reimer E., Martin-Jezequel V., Nelson D.M. and Queguiner B. 2000. A review of the Si cycle in the modern ocean: recent progress and missing gaps in the application of biogenic opal as a paleoproductivity proxy. Global Planet. Change 26: 317–365.

Raubitschek S., Lücke A. and Schleser G.H. 1999. Sedimentation patterns of diatoms in Lake Holzmaar, Germany - on the transfer of climate signals to biogenic silica oxygen isotope proxies. J. Paleolimnol. 21: 437–448.

Rau G.H., Takahashi T. and Des Marais D. 1989. Latitudinal variation in plankton δ^{13}C: implications for CO_2 and productivity in past oceans. Nature 341: 516–518.

Rau G.H., Takahashi T., Des Marais D.J., Repeta D.J. and Martin J.H. 1992. The relationship between δ^{13}C of organic matter and $[CO_{2(aq)}]$ in ocean surface water: data from a JGOFS site in the northeast Atlantic Ocean and a model. Geochim. Cosmochim. Ac. 56: 1413–1419.

Ravelo A. and Andreasen D. 1999. Using planktonic foraminifera as monitors of tropical surface ocean. In: Abrantes F. and Mix A. (eds), Reconstructing Ocean History: A window into the future. Kluwer Academic, New York, pp. 217–243.

Raymo M., Oppo D. and Curry W. 1997. The mid-Pleistocene climate transition: A deep sea carbon isotope perspective. Paleoceanography 12: 546–559.

Reynolds L. and Thunell R.C. 1985. Seasonal succession of planktonic foraminifera in the subpolar north Pacific. J. Foramin. Res. 15: 282–301.

Reynolds L.A. and Thunell R.C. 1986. Seasonal production and morphological variation of Neogloboquadrina pachyderma (Ehrenberg) in the Northeast Pacific. Micropaleontology 32: 1–18.

Reynolds J.H. and Verhoogen J. 1953. Natural variations in the isotopic constitution of silicon. Geochim. Cosmochim. Ac. 3: 224–234.

Rings A., Lücke A. and Schleser G.H. 2004. A new method for the quantitative separation of diatom frustules from lake sediments. Limnol. Oceanogr. Methods. 2: 25–34.

Robinson R.S., Brunelle B.G. and Sigman D.M. 2004. Revisiting nutrient utilization in the glacial Antarctic: Evidence from a new method for diatom-bound N isotopic analysis. Paleoceanography 19: PA3001. doi: 10.1029/2003PA000996.

Rohling E.J. 2000. Paleosalinity: confidence limits and future applications. Mar. Geol. 163: 1–11.

Rohling E.J. and Cooke S. 1999. Stable oxygen and carbon isotope ratios in foraminiferal carbonate. In: Sen Gupta B.K. (ed.), Modern Foraminifera. Kluwer Academic, Dordrecht, The Netherlands, pp. 239–258.

Rohling E.J, Sprovieri M., Cane T., Casford J.S.L., Cooke S., Bouloubassi I., Emeis K.C., Schiebel R., Rogerson M., Hayes A., Jorissen F.J. and Kroon D. 2004. Reconstructing past planktic foraminiferal habitats using stable isotope data: a case history for Mediterranean sapropel S5. Mar. Micropaleontol. 50: 89–123.

Rommerskirchen F., Eglinton G., Dupont L., Gunter U., Wenzel C. and Rullkotter J. 2003. A north to south transect of Holocene southeast Atlantic margin sediments: Relationship between aerosol transport and compound-specific $\delta^{13}C$ land plant biomarker and pollen records. Geochem. Geophys. Geosyst. 4: 1101. doi: 10.1029/2003GC000541.

Rosenthal Y., Dahan M. and Shemesh A. 2000. Southern Ocean contributions to glacial-interglacial changes of atmospheric pCO_2: an assessment of carbon isotope record in diatoms. Paleoceanography 15: 65–75.

Rostek F., Ruhland G., Bassinot F.C., Müller P.J., Labeyrie L.D., Lancelot Y. and Bard E. 1993. Reconstructing sea surface temperature and salinity using $\delta^{18}O$ and alkenone records. Nature 364: 319–321.

Ruhlemann C., Mulitza S., Muller P.J., Wefer G. and Zahn R. 1999. Tropical Atlantic warming during conveyor shut down. Nature 402: 511–514.

Sackett W.M. 1964. The depositional history and isotope organic composition of marine sediments. Mar. Geol. 2: 173–185.

Sackett W.H., Eckelmann W.R., Bender M.L. and Bé A.W.H. 1965. Temperature dependence of carbon isotopic composition in marine plankton and sediments. Science 148: 235–237.

Sachs J.P. and Repeta D.J. 1999. Oligotrophy and nitrogen fixation during eastern Mediterranean sapropel events. Science 286: 2485–2488.

Sanyal A., Hemming N.G., Hanson G.N. and Broecker W.S. 1995. Evidence for a higher pH in the glacial ocean from boron isotopes in foraminifera. Nature 373: 234–236.

Sanyal A., Hemming N.G., Broecker W.S., Lea D.W., Spero H.J. and Hanson G.N. 1996. Oceanic pH control on the boron isotopic composition of foraminifera: Evidence from culture experiments. Paleoceanography 11: 513–517.

Sanyal A., Hemming N.G., Broecker W.S. and Hanson G.N. 1997. Changes in pH in the eastern equatorial Pacific across stage 5-6 boundary based on boron isotopes in foraminifera. Global Biogeochem. Cy. 11: 125–133.

Sanyal A., Nugent M., Reeder R.J. and Buma J. 2000. Seawater pH control on the boron isotopic composition of calcite: Evidence from inorganic calcite precipitation experiments. Geochim. Cosmochim. Ac. 64: 1551–1555.

Sanyal A., Bijma J., Spero H. and Lea D.W. 2001. Empirical relationship between pH and the boron isotopic composition of *Globigerinoides sacculifer*: Implications for the boron isotope paleo-pH proxy. Paleoceanography 16: 515–519.

Sarnthein M. and Tiedemann R. 1990. Younger Dryas-style cooling events at glacial terminations I-VI at ODP Site 685: associated benthic $\delta^{13}C$ anomalies constrain meltwater hypothesis. Paleoceanography 5: 1041–1055.

Sarnthein M., Winn K., Duplessy J-C. and Fontugne M. 1988. Global variations of surface ocean productivity in low and mid latitudes. Paleoceanography 3: 361–398.

Sarnthein M., Winn K., Jung S.J.A., Duplessy J.C., Labeyrie L., Erlenkreuser H. and Ganssen G.M. 1994. Changes in east Atlantic deepwater circulation over the last 30,000 years: eight time slice reconstructions. Paleoceanography 10: 1063–1094.

Sautter L.R. and Thunell. R.G. 1991. Seasonal varibility in the $\delta^{18}O$ and $\delta^{13}C$ of planktonic foraminifera from an upwelling environment. Paleoceanography 3: 307–334.

Schmidt G.A. 1999. Error analysis of paleosalinity calculations. Paleoceanography 14: 422–429.

Schmidt M., Botz R., Stoffers P., Anders T. and Bohrmann G. 1997. Oxygen isotopes in marine diatoms: A comparative study of analytical techniques and new results on the isotope composition of recent marine diatoms. Geochim. Cosmochim. Ac. 61: 2275–2280.

Schmidt M., Botz R., Rickert D., Bohrmann G., Hall S.R. and Mann S. 2001. Oxygen isotopes of marine diatoms and relations to opal-A maturation. Geochim. Cosmochim. Ac. 65: 201–211.

Schneider R., Muller P.J. and Wefer G. 1994. Late Quaternary paleoproductivity changes off the Congo deduced from stable isotopes of planktonic foraminifera. Palaeogeogr. Palaeocl. 110: 255–274.

Schrag D.P., Hampt G. and Murray D.W. 1996. Pore fluid constraints on the temperature and oxygen isotope composition of the glacial ocean. Science 272: 1930–1932.

Schrag D.P., Adkins J.F., McIntyre K., Alexander J.L., Hodell D.A., Charles C.D. and McManus J.F. 2002. The oxygen isotope composition of seawater during the Last Glacial Maximum. Quaternary Sci. Rev. 21: 331–342.

Seidov D., Haupt B.J. and Maslin M.A. (eds). 2001. The Oceans and Rapid Climate Change: Past, Present and Future. AGU Geophysical Monograph Series Volume 126, Washington DC, 293 pp.

Seitzinger S.P. and Sanders R.W. 1999. Atmospheric inputs of dissolved organic nitrogen stimulate estuarine bacteria and phytoplankton. Limnol. Oceanogr. 44: 721–730.

Severinghaus J.P. and Brook E.J. 1999. Abrupt climate change at the end of the last glacial period inferred from trapped air in polar ice. Science 286: 930-934.

Shackleton N.J. 1967. Oxygen isotope analyses and Pleistocene temperatures re-assessed. Nature 215: 15–17.

Shackleton N.J. 1974. Attainment of isotopic equilibrium between ocean water and the benthic foraminifera Genus Uvigerina: isotope changes in the ocean during the last glacial. In: Les méthodes quantitatives d'étude des variations due climat au cours du Pleistocene, 219. Colloques Internationaux de Central National de la Recherce Scientifique CNRS, Paris, pp. 203-209.

Shackleton N.J. 1977. The oxygen isotope stratigraphic records of the Pleistocene. Philos. T. Roy. Soc. B. 280: 169–179.

Shackleton N.J. 2000. The 100,000 year Ice Age cycle identified and found to lag temperature, carbon dioxide and orbital eccentricity. Science 289: 1897–1902.

Shackleton N.J. and Opdyke N.D. 1973. Oxygen isotope and paleomagnetic stratigraphy of equatorial Pacific core V28-238. Quaternary Res. 3: 39–55.

Shackleton N.J., Imbrie J. and Hall M.A. 1983. Oxygen and carbon isotope record of East Pacific core V19-30: implications for deep water in the late Pleistocene North Atlantic. Earth Planet. Sc. Lett. 65: 233–244.

Shackleton N.J., Hall M.A. and Pate D. 1995. Pliocene stable isotope stratigraphy of Site 846. In: Pisias N.G., Mayer L.A., Janecek T.R., Palmer-Julson A. and van Andel T.H. (eds), Proc. ODP Scientific Results: 138. College Station, Texas, pp 337–357.

Shemesh A., Charles C.D. and Fairbanks R.G. 1992. Oxygen isotopes in biogenic silica: global changes in ocean temperature and isotopic composition. Science 256: 1434–1436.

Shemesh A., Macko S.A., Charles C.D. and Rau G.H. 1993. Isotopic evidence for reduced productivity in the glacial Southern Ocean. Science 262: 407–410.

Shemesh A., Burckle L.H. and Hays J.D. 1994. Meltwater input to the Southern Ocean during the Last Glacial Maximum. Science 266: 1542–1544.

Shemesh A., Burckle L.H. and Hays J.D. 1995. Late Pleistocene oxygen-isotope records of biogenic silica from the Atlantic sector of the Southern-Ocean. Paleoceanography 10: 179–196.

Shemesh A., Hodell D., Crosta C., Kanfoush S., Charles C. and Guilderson T. 2002. Sequence of events during the last deglaciation in Southern Ocean sediments and Antarctic ice cores. Paleoceanography 17: 1056. doi: 10.1029/2000PA000599.

Skinner L., Shackleton N.J. and Elderfield H. 2003. Millennial-scale variability of deep water temperature and $\delta^{18}O_{dw}$ indicating deep-water source variations in the Northeast Atlantic, 0-34 cal. Ka BP. Geochem. Geophys. Geosyst. 4: 1098. doi: 10.1029/2003GC000585.

Siddall M., Rohling E.J., Almogi-Labin A., Hemleben C., Meischner D., Schmelzer I. and Smeed D.A. 2003. Sea-level fluctuations during the last glacial cycle. Nature 423: 853–858.

Sigman D.M. and Haug G.H. 2004. The biological pump in the past. In: Elderfield H. (ed.),

Treatise on Geochemistry, Volume 6: The Oceans and Marine Geochemistry. Elsevier, Amsterdam, pp. 491–528.

Sigman D.M., Altabet M.A., Francoise R. and Whelan J. 1997. Diatom microfossil N isotopes support the hypothesis of higher nitrate utilisation in the Southern Ocean during the last ice age. Abstr. Pap. Am. Chem. S. 214: 69-GEOC.

Sigman D.M., Altabet M.A., Francois R., McCorkle D.C. and Fischer G. 1999. The $\delta^{18}O$ of nitrate in the Southern Ocean: Consumption of nitrate in surface water. Global Biogeochem. Cy. 13: 1149–1166.

Sigman D.M., Altabet M.A., Francois R., McCorkle D.C. and Gaillard J-F. 1999. The isotopic composition of diatom-bound nitrogen in Southern Ocean sediments. Paleoceanography 14: 118–134.

Sigman D.M., Jaccard S.L. and Haug G.H. 2004. Polar ocean stratification in a cold climate. Nature 428: 59–63.

Simstich J., Sarnthein M. and Erlenkeuser H. 2003. Paired $\delta^{18}O$ signals of *Neogloboquadrina pachyderma* (s) and *Turborotalita quinqueloba* show thermal stratification structure in Nordic Seas. Mar. Micropaleontol. 48: 107–125.

Singer A.J. and Shemesh A. 1995. Climatically linked carbon isotope variation during the past 430,000 years in Southern Ocean sediments. Paleoceanography 10: 171–177.

Smith B.M. and Epstein S. 1971. Two categories of $^{13}C/^{12}C$ ratios for higher plants. Plant Physiol. 47: 380–384.

Spadaro P.A. 1983. Silicon isotope fractionation by the marine diatom *Phaeodactylum tricornutum*. MSc thesis. University of Chicago, Chicago.

Spivack A.J. and Edmond J.M 1987. Boron isotope exchange between seawater and the oceanic crust. Geochim. Cosmochim. Ac. 51: 1033–1043.

Spivack A.J., You C.F. and Smith H.J. 1993. Foraminifera boron isotope ratios as a proxy for surface ocean pH over the past 21 Myr. Nature 363: 149–151.

Sundquist E.T. and Visser K. 2004. The geologic history of the carbon cycle. In: Schlesinger W.H. (ed.), Treatise on Geochemistry, Volume 8: Biogeochemistry. Elsevier, Amsterdam, pp. 425–472.

Takeda S. 1998. Influence of iron availability on nutrient consumption ratio of diatoms in oceanic waters. Nature 393: 774–777.

Taylor H.P. and Epstein S. 1970. Oxygen and silicon isotope ratios of lunar rock. Earth Planet. Sc. Lett. 9: 208–210.

Tiedemann R., Sarnthein M. and Shackleton N.J. 1994. Astronomic timescale for the Pliocene Atlantic $\delta^{18}O$ and dust flux records of ODP Site 659. Paleoceanography 9: 619–638.

Tilles D. 1961a. Variations of silicon isotope ratios in a zoned pegmatite, J. Geophys. Res. 66: 3015–3020.

Tilles D. 1961b Natural variations in isotopic abundances of silicon, J. Geophys. Res. 66: 3003–3014.

Thunell R.C. and Kepple A. 2004. Glacial-Holocene $\delta^{15}N$ record from the Gulf of Tehuantepec, Mexico: Implications for denitrification in the eastern equatorial Pacific and changes in atmospheric N_2O. Global Biogeochem. Cy. 18: GB1001, doi:10.1029/2002GB002028.

Tolderlund D.S. and Bé A.W.H. 1971. Seasonal distributions of planktonic foraminifera in the western North Atlantic. Micropaleontology. 30: 241–260.

Tréguer P., Nelson D.M., Van Bennekom A.J., DeMaster D.J., Leynaert A. and Queguiner B. 1995. The silica balance in the world ocean – a re-estimate. Science 268: 375–379.

Tudhope A.W., Chilcott C.P., McCulloch M.T., Cook E.R., Chappell J., Ellam R.M., Lea D.W., Lough J.M. and Shimmield G.B. 2001. Variability in the El Niño-Southern Oscillation through a glacial-interglacial cycle. Science 291: 1511–1517.

Turchyn A.V. and Schrag D. 2004. Oxygen isotope constraints on the sulphur cycle over the last 10 million years. Science 303: 2004–2007.

Uchida M., Shibata Y., Ohkushi K., Ahagon N. and Hoshiba M. 2004. Episodic methane release events from Last Glacial marginal sediments in the western North Pacific Geochem. Geophys. Geosyst. 5: Q08005, doi: 10.1029/2004GC000699.

Uliana E., Lange C.B. and Wefer G. 2002. Evidence for Congo River freshwater load in Late Quaternary sediments of ODP Site 1077 (5°S, 10°E). Palaeogeogr. Palaeocl. 187: 137–150.

Urey H.C. 1947. The thermodynamic properties of isotopic substances. J. Chem. Soc. 152: 190–219.

Urey H.C. 1948. Oxygen isotopes in nature and in the laboratory. Science 108: 489–496.

van Campo E., Guiot J. and Peng C.H. 1993. A data-based re-appraisal of the terrestrial carbon budget at the Last Glacial Maximum. Global Planet. Change 8: 189–201.

Varela D.E., Pride C. J. and Brzezinski M. A. 2004. Biological fractionation of silicon isotopes in Southern Ocean surface waters. Global Biogeochem. Cy. 18: GB1047, doi: 10.1029/2003GB002140.

Vroon P.Z., Beets C.J., Van Soest R.W.M., Schwieters J., Troelstra S.R., Van Belle J.C., Davies G.R. and Andriessen P.A.M. 2002. Silicon isotope composition of sponge spicules determined by MC-ICPMS. Geochim. Cosmochim. Ac. 67: A812 supplement.

Vroon P.Z., Beets C.J., Soenarjo D.H., Van Soest R.W.M., Troelstra S.R., Schwieters J., Van Belle J.C. and Van der Wagt B. 2004. Species dependent fractionation of silicon isotopes by present-day demosponges. Geochim. Cosmochim. Ac. 68: A213 supplement.

Waelbroeck C., Labeyrie L., Michel E., Duplessy J.C., McManus J.F., Lambeck K., Balbon E. and Labracherie M. 2002. Sea level and deep water temperature changes derived from benthic foraminifera isotope records. Quaternary Sci. Rev. 21: 295–305.

Wang L.J., Sarnthein M., Erlenkeuser H., Grootes P., Grimalt J., Pelejero C. and Linck G. 1999a. Holocene variations in Asian monsoon moisture: a bidecadal sediment record from the South China Sea. Geophys. Res. Lett. 26: 2889–2892.

Wang L.J, Sarnthein M., Erlenkeuser H., Grimalt J., Grootes P., Heilig S., Ivanova E., Kienast M., Pelejero C. and Pflaumann U. 1999b. East Asian monsoon climate during the late Pleistocene: High resolution sediment records from the South China Sea. Mar. Geol. 156: 245–284.

Waser N.A.D., Turpin D.H., Harrison P.J., Nielsen B. and Calvert S.E. 1998. Nitrogen isotope fractionation during the uptake and assimilation of nitrate and urea by marine diatom. Limnol. Oceanogr. 43: 215–224.

Wefer G., Berger W.H., Bijma J. and Fischer G. 1999. Clues to Ocean History: A brief overview of proxies: In: Fischer G. and Wefer G. (eds), Uses of proxies in Paleoceanography: Examples from the South Atlantic. Springer, Berlin, pp. 1–68.

Wischmeyer A.G., De La Rocha C.L., Maier-Reimer E. and Wolf-Gladrow D.A. 2003. Control mechanisms for the oceanic distribution of silicon isotopes. Global Biogeochem. Cy. 1083. doi: 10.1029/2002GB002022.

Yokoyama Y., Lambeck K., De Deckker P.P.J. and Fifield L.K. 2000. Timing of the last glacial maximum from observed sea-level minima. Nature 406: 713–716.

Zachos J.C., Pagani M.N., Sloan L., Thomas E. and Billups K. 2001. Trends, rhythms and aberrations in global climate 65 Ma to present. Science 292: 686–693.

Zahn R., Winn K. and Sarnthein M. 1986. Benthic foraminifera Episodic methane release events from Last Glacial marginal sediments $\delta^{13}C$ and accumulation rates of organic carbon. Paleoceanography 1: 27–42.

Zeebe R.E. 2005. Stable boron isotope fractionation between dissolved $B(OH)_3$ and $B(OH)^-_4$. Geochim. Cosmochim. Ac. 69: 2753–2766.

Zeebe R.E., Wolf-Gladrow D.A. 2001. CO_2 in seawater: equilibrium, kinetics, isotopes. Elsevier Oceanography Series, 65, Amsterdam, 346 pp.

Zeebe R.E., Wolf-Gladrow D.A., Bijma J. and Hönisch B. 2003. Vital effects in foraminifera do not compromise the use of $\delta^{11}B$ as a paleo-pH indicator: Evidence from modelling. Paleoceanography 18: 1043. doi: 10.1029/2003PA000881.

Ziegler K., Chadwick O.A., Kelly E.F., Brzezinski M.A. and DeNiro M.J. 2000. Silicon isotope fractionation during weathering and soil formation: experimental results. Journal of Conference Abstracts 5: 1135.

Ziegler K., Chadwick O.A., Kelly E.F. and Brzezinski M.A. 2002. The δ^{30}Si values of soil weathering profiles: Indicators of Si pathways at the lithosphere/hydro(bio)sphere interface. Geochim. Cosmochim. Ac. 66: A881 supplement.

GLOSSARY, ACRONYMS AND ABBREVIATIONS

α-cellulose: one of a number of forms of cellulose having the highest molecular weight and being insoluble in both water and aqueous sodium hydroxide (also termed 'alkali-resistant cellulose').

δ: the delta notation, a relative measure of the difference, typically ratios of molar concentrations of stable isotopes between a substance and an agreed-upon standard material.

$\delta^{11}B$: notation used to express the value of the boron isotope ratio ($^{11}B/^{10}B$) in a sample relative to that in a standard. Thus $\delta^{11}B = (((^{11}B/^{10}B_s / ^{11}B/^{10}B_{std})-1)*1000)$ where s is the sample (unknown) and std a standard that has been calibrated relative to NBS boric acid SRM 951.

$\delta^{13}C$: notation used to express the value of the carbon isotope ratio ($^{13}C/^{12}C$) in a sample relative to that in a standard. Thus $\delta^{13}C = (((^{13}C/^{12}C_s / ^{13}C/^{12}C_{std})-1)*1000)$ where s is the sample (unknown) and std a standard that has been calibrated relative to the Pee Dee Bellemnite (PDB), usually expressed as Vienna Pee Dee Bellemnite (VPDB).

$\delta^{13}C_{carb}$: carbon isotope ratio of carbonate.

$\delta^{13}C_{TDIC}$: carbon isotope ratio of Total Dissolved Inorganic Carbon.

$\Delta^{14}C$: $^{14}C/^{12}C$ ratio of the atmosphere represented as ‰ variations. Reconstructed using the dendrochronologically constrained tree-ring record of radiocarbon ages.

δ^2H: also known as δ^2H. Notation used to express the value of the hydrogen isotope ratio (D/H) in a sample relative to that in a standard. Thus $\delta^2H = (((D/H_s / D/H_{std})-1)*1000)$ where s is the sample (unknown) and std is a standard that has been calibrated relative to Vienna Standard Mean Ocean Water (VSMOW).

δD: also known as δD. Notation used to express the value of the hydrogen isotope ratio ($^2H/^1H$) in a sample relative to that in a standard. Thus $\delta^2H = (((^2H/^1H_s / ^2H/^1H_{std})-1)*1000)$ where s is the sample (unknown) and std is a standard that has been calibrated relative to Vienna Standard Mean Ocean Water (VSMOW).

δ^2H_{org}: hydrogen isotope ratio of the organic compound measured.

δ^2H_{precip}: hydrogen isotope ratio of precipitation.

$\delta^{15}N$: notation used to express the value of the nitrogen isotope ratio ($^{15}N/^{14}N$) in a sample relative to that in a standard. Thus $\delta^{15}N = (((^{15}N/^{14}N_s / ^{15}N/^{14}N_{std})-1)*1000)$ where s is the sample (unknown) and std is a standard that has been calibrated relative to atmospheric nitrogen (described as AIR).

$\delta^{18}O$: notation used to express the value of the oxygen isotope ratio ($^{18}O/^{16}O$) in a sample relative to that in a standard. Thus $\delta^{18}O = (((^{18}O/^{16}O_s / ^{18}O/^{16}O_{std})-1)*1000)$ where s is the sample (unknown) and std is a standard that has been calibrated relative to the Pee Dee Belemnite (PDB) (usually expressed as Vienna Pee Dee Bellemnite (VPDB)) or Vienna Standard Mean Ocean Water (VSMOW).

$\delta^{29}Si$: notation used to express the value of the silicon isotope ratio ($^{29}Si/^{28}Si$) in a sample relative to that in a standard. Thus $\delta^{29}Si = (((^{29}Si/^{28}Si_s / ^{29}Si/^{28}Si_{std})-1)*1000)$ where s is the sample (unknown) and std is a standard that has been calibrated relative to NBS28.

$\delta^{30}Si$: notation used to express the value of the silicon isotope ratio ($^{30}Si/^{28}Si$) in a sample relative to that in a standard. Thus $\delta^{30}Si = (((^{30}Si/^{28}Si_s / ^{30}Si/^{28}Si_{std})-1)*1000)$ where s is the sample (unknown) and std is a standard that has been calibrated relative to NBS28.

$\delta^{18}O_{carb}$: oxygen isotope ratio of carbonate.

$\delta^{18}O_{evaporation}$: oxygen isotope ratio of evaporated water.

$\delta^{18}O_{lakewater}$: oxygen isotope ratio of lake water.

$\delta^{18}O_{lw}$: oxygen isotope ratio of lake water.

$\delta^{18}O_{org}$: oxygen isotope ratio of the organic compound measured.

$\delta^{18}O_{precip}$: oxygen isotope ratio of precipitation.

δp: oxygen isotope ratio of precipitation.

δss: the contemporary terminal basin steady-state delta value.

ε^{18}_{hydro}: oxygen isotope enrichment between water and meteoric water due to hydrological factors.

$\varepsilon^{18}_{org-lakewater}$: the oxygen isotope enrichment that occurs during synthesis of the organic compound.

$\varepsilon^{2}_{org-lakewater}$: the hydrogen isotope depletion that occurs during synthesis of the organic compound.

Adenonsine diphosphate (ADP): a nucleotide derived from adenosine that is converted to ATP for energy storage.

Adenosine triphosphate (ATP): a nucleotide derived from adenosine that occurs in muscle tissue and used as the major source of chemical energy for cellular reactions.

amino acids: organic molecules containing both basic (amino) and acidic (carboxylic) functional groups that, when linked together, make proteins.

Antarctic Bottom Water (AABW): extremely cold and dense ocean bottom water formed around Antarctica.

Antarctic Polar Front (APF): An oceanographic term to describe the southern front of the Antarctic Circumpolar Current which separates the cold waters in the Antarctic Zone to the south from the relatively warmer surface waters within the polar frontal zone to the north.

aquifer: geological formation containing water, especially one that supplies water for wells, springs, etc.

authigenic: sediment component(s) formed within the lake/ocean. Sometimes also described as endogenic.

benthic foraminifera: a single celled protoza which produces a calcium carbonate test and lives on or just within the sediment at the bottom of the ocean.

biogenic: being of a biological origin.

bioturbation: the mixing of sediment at the bottom of the ocean due to the movement and feeding of benthic (bottom dwelling) organisms.

C3 plants: plants that employ the Calvin pathway during photosynthesis. C3 plants are so named because they employ an intermediate compound, phosphoglyceric acid, that contains three carbon atoms. Most higher plants and algae use the C3 pathway. These plants strongly discriminate (20‰) against ^{13}C during carbon incorporation.

C4 plants: plants that employ the Hatch-Slack pathway during photosynthesis. C4 plants are named such because they employ an intermediate compound, oxaloacetic acid, that contains four carbon atoms. These plants moderately discriminate (8 to 14‰) against ^{13}C during carbon incorporation. C4 plants are usually higher land plants that are specially adapted to dry and low-$pCO2$ conditions.

C/N: total organic carbon divided by total nitrogen within organic material. Normally the weight ratio of organic carbon divided by total nitrogen (sometimes the atomic weight is used).

calcite: the carbonate mineral $CaCO_3$.

carbohydrate: a class of organic molecule embracing sugars and also polymers such as cellulose and starch, and the main metabolites in the conversion of energy in plants and animals.

carbonates: CO_3 minerals including calcite and aragonite ($CaCO_3$) and dolomite $CaMg(CO_3)_2$.

Carbonate Compensation Depth (CCD): the depth in the ocean/lake when calcium carbonate becomes under-saturated in the water so that any calcium carbonate entering this zone will start to dissolve.

carbon fixation: the biological process whereby inorganic carbon is converted to a form useful to living organisms. Usually this is by photosynthesis.

carboxylation: the introduction of a carboxyl group into a compound or molecule.

cellulose: (1) a group of polymeric carbohydrates with the formula $C_n(H_2O)_n$ that are major components of the tissue of terrestrial macrophytes. It is the most abundant natural organic compound. (2) A polysaccharide ($C_6H_{10}O_5$) of linked glucose units and is the principal component of cell walls of higher and lower plants.

chitin: long-chain, nitrogen-containing polysaccharide compound found in the exoskeletons of arthropods and cell walls of fungi, etc. Resistant to deterioration when buried in lake sediments.

chironomids: non-biting midges (Diptera family) that are sensitive to temperature changes.

closed lakes: occur particularly in arid regions where water loss is mainly through evaporation. See also open lakes.

co-variation: when two sets of data vary together, this can be positive (direct) or negative (inverse).

coccolith n-alkenone U^k_{37}: biomarker found within coccolithophores which can be used to estimate the sea surface temperature in which the organisms inhabited.

coccolithophores: nannofossils, plants which produce a carbonate test.

cyanobacteria (formerly cyanophyceae = blue-green algae): a large group of prokaryotes that possess chlorophyll 'a' and carry out photosynthesis with the concomitant production of oxygen. Many species fix atmospheric nitrogen. Colonies sometimes occur in lake sediments.

$d\delta^{18}O/dT$: rate of change of $\delta^{18}O$ as a function of temperature.

$d\delta p/dT$: rate of change of $\delta^{18}O_{precipitation}$ as a function of temperature.

dV: Change in volume.

Dansgaard relationship: at intermediate and high latitudes the oxygen isotope composition of mean annual precipitation correlates directly with changes in temperature with a gradient of approximately +0.6‰/°C.

Dansgaard/Oeschger events: millennial scale warming (interstadial) events during the last glaciation. Temperature records based on isotope data from the Greenland ice cores show more than 20 interstadial (Dansgaard/Oeschger) events between 110 and 15 kyr BP.

dendrochronology: the analysis and dating of annual growth layers in wood.

dendroclimatology: the study of climate and environmental change through analysis of trees and annual growth layers in wood. A sub-discipline of dendrochronology.

denitrification: the process by which nitrate is reduced into gaseous nitrogen. The opposite of nitrogen fixation.

dentine: the main structural part of a tooth, beneath the enamel and surrounding the pulp chamber and root canals. It consists, like bone, of a composite of mineral (hydroxyapatite) and organic (collagen) material.

deuterium: minor, 'heavy' stable isotope of hydrogen, ^2H, present at low concentration in all natural waters.

diatoms: unicellular, eukaryotic algae containing a highly silicified cell wall, found in both freshwater and marine environments, which play a major role in regulating the marine silicon cycle.

disequilibrium: commonly known as 'vital' effects in the case of biogenic carbonates, include a variety of rate effects and micro-environment induced changes that cause the mineral to have an isotope composition that is different from that predicted purely by thermodynamics.

Dissolved Inorganic Carbon (DIC): (1) carbon contained in carbon dioxide, bicarbonate and carbonate ions dissolved in water. (2) The sum of all dissolved inorganic carbon species ($CO_2 + HCO_3^- + CO_3^{2-} + H_2CO_3$) in an aqueous solution.

Dissolved Inorganic Nitrogen (DIN): the sum of the dissolved inorganic nitrogen species ($NO_3^- + NO_2^- + NH_4^+$) in an aqueous solution. DIN normally excludes dissolved N_2.

Dissolved Organic Nitrogen (DON): the sum of the dissolved organic nitrogen species (e.g., humic acids, fulvic compounds).

dolomite: the carbonate mineral $CaMg(CO_3)_2$.

enamel: the hard, mineralised substance covering the exposed portion of a tooth, composed mostly of hydroxylapatite.

ephemeral water: periodic surface water, often only occurs as a direct result of heavy rainfall.

epikarst zone: zone of partially weathered limestone just beneath the soil-rock interface in karst regions.

epilimnion: the warm, relatively less dense upper layer of water in a stratified lake, subject to mixing by wind action.

equilibrium: condition in which isotope exchange reactions between co-existing chemical species (e.g., compounds, minerals, phases or individual molecular species) achieve no net reaction. An equilibrium constant (K) can be written to describe the ratio of the light to heavy isotopes in the coexisting species and this is typically temperature dependent.

Equivalent Sea Level (ESL): a term for changes in relative global sea level.

fluid inclusions: generally small voids formed during the development of minerals such as calcite in speleothems and quartz in hydrothermal veins, containing traces of the depositing fluids (possibly modified by diagenesis).

foraminifera: single celled protozoa which produces a calcium carbonate test. Can be abbreviated to foram.

fractionation factor: ratio of two isotopes in one chemical compound divided by the same ratio in another compound (α). If X and Y are two chemical compounds then the fractionation factor $\alpha_{X-Y} = R_X/R_Y$.

fractionation: partitioning of isotopes between two substances or between two phases that results in a preferential uptake or release of one isotope (i.e., lighter or heavier) relative to another is termed 'fractionation'.

frustule: the siliceous component of the diatom cell wall.

FTIR: Fourier Transform Infrared Spectroscopy. Radiation is separated into its component wave numbers by Fourier transformation of the interferogram produced by an interferometer. The technique has proved useful for the identification of different forms of water trapped in speleothem calcite.

gastropod: a member of the class Gastropoda (Mollusca).

General Circulation Models (GCMs): numerical (computer) model that uses physical principles to model the behaviour of the atmosphere through time, usually coupled with a model for the upper layer of the ocean.

geostrophic flow: ocean movement due to the topography of the ocean basin.

GISP2: Greenland Ice Sheet Project 2, refers to ice-cores completed by a US led drilling programme at Summit, Greenland on July 1st 1993 after five years of drilling (see http://www.gisp2.sr.unh.edu/GISP2).

Glacial-Interglacial cycles: the cycles between ice ages (glacials) and warm periods (interglacials) such as the Holocene.

global ice volume: the amount of ice in ice sheets and sea ice on the planet.

Global Meteoric Water Line (GMWL): a line on the $\delta^{18}O$-$\delta^{2}H$ plane formed when the $\delta^{18}O$ and $\delta^{2}H$ values of meteoric precipitation from all parts of the globe are plotted together. For precipitation collected at any one locality, the slope or the intercept may deviate from this line.

Heinrich events: millennial-scale ice rafting events first found in the North Atlantic during the last glacial period.

hydroxyapatite: a microcrystalline material containing calcium, phosphorus, oxygen and hydrogen, and constituting the main component in bone and tooth, to which it gives compressive strength.

hypolimnion: non-circulating, cooler, denser layer of water below the *epilimnion* (see above), separated by the thermocline.

Ice-Rafted Debris (IRD): material carried by icebergs or sea ice out to sea where is it is dropped once the ice has melted.

Intertropical Convergence Zone (ITCZ): low-latitude zone where the northern and southern hemispheric trade winds converge, and which tends to follow the Sun's annual movements north and south of the equator.

isotope fractionation: processes that result in a change in the isotope ratio of a compound.

isotopically depleted: a substance that is depleted in the heavy (usually less abundant) stable isotope over a light (usually more abundant) isotope of the same element is said to be isotopically depleted. In the context of carbon isotopes for example an isotopically depleted substance is relatively enriched in ^{12}C and therefore has a relatively low $\delta^{13}C$ value.

isotopically enriched: a substance that is enriched in the heavy (usually less abundant) stable isotope over a light (usually more abundant) isotope of the same element is said to be isotopically enriched. In the context of carbon isotopes for example an isotopically enriched substance is relatively depleted in ^{12}C and therefore has a relatively high $\delta^{13}C$ value.

ITCZ: *see* Intertropical Convergence Zone

K/T boundary: a period of major transition between the Cretaceous and Tertiary at 65.5 Ma when large numbers of genera died out, including dinosaurs.

karst systems: karst refers to dissolution phenomena in soluble bedrocks such as limestones. Examples of karst systems include dissolution fissures and caverns.

kerogen: the least reactive form of organic matter. Often retained after hydrofluoric and hydrochloric digestion.

kinetic fractionation: non-equilibrium fractionation, found for example in diffusion-controlled evaporation, or biologically-controlled processes.

Krebs cycle: a series of enzymatic reactions in aerobic organisms involving oxidative metabolism of acetyl units and producing high-energy phosphate compounds, which serve as the main source of cellular energy.

Late Glacial: short climatic fluctuation at the end of the last glacial comprising the Late Glacial (or Windermere) Interstadial and the Younger Dryas (or Loch Lomond) Stadial (c. 13,000-10,000 BP).

Late Glacial Interstadial: short warm interval at the end of the last glacial. In the U.K. this is known as the Windermere Interstadial, c. 13,000-11,000 BP.

Laurentide ice sheet: the North Eastern America ice sheet which existed during the last glacial period.

lignin: the substance that binds together fibres in higher plants to stiffen their structures. Wood may contain as much as 50% lignin. It is a high-molecular-weight polyphenolic compound that varies in composition with different plant types.

lipid: a class of organic molecules, of a generally oily nature, whose chemical constitution renders them largely insoluble in water but soluble in organic solvents.

lithogenic: being of a rock or soil origin.

Little Ice Age: a cool period recognised in several parts of the world occurring from about 1550 – 1850 AD, the timing varying from area to area.

Local Evaporation Line (LEL): the isotope composition of surface waters within catchments or within the same climatic region commonly plot on lines with slopes ranging from about 4 to 7 in $\delta^{18}O$-δD space, due to equilibrium and fractionation effects associated with evaporation.

Local Meteoric Water Line (LMWL): a line on the $\delta^{18}O$-$\delta^2 H$ plane formed when the $\delta^{18}O$ and $\delta^2 H$ values of meteoric precipitation from a particular region are plotted together. Each region may have its own slope or intercept. The LMWL may deviate from the GMWL.

Maunder and Dalton Minima: refer to well-documented minima in sunspot numbers (and reduced solar irradiance) in the time intervals between about 1650 and 1710 AD (Maunder minimum) and between about 1790-1830 AD (Dalton minimum).

meromixis: stratification of a water body, this can result in low oxygen concentrations and reduced nutrient supply.

methionine: an amino acid containing sulphur found in most proteins and essential for the nutrition of higher organisms.

mycorrhiza: the symbiotic association of the mycelium of a fungus with the roots of certain plants.

microbial: product of bacterial mediation: bacteria typically prefer to utilise the light isotope of a particular compound, e.g., ^{12}C rather than ^{13}C.

Mid-Pleistocene Revolution (MPR): the transitional period when the glacial-interglacial cycles changed from a periodicity of c. 41,000 years to c. 100,000 years between 900 and 650 kyr BP.

n-alkanes/n-alkanols/n-alkanoic acid: biomarker found in higher plants which can be found in marine/lake sediments.

nitrogen fixing: the fixing of atmospheric nitrogen by organisms either in the surface ocean or on land.

North Atlantic Deep Water (NADW): deep water which is predominantly formed by cooling of the surface waters in the North Atlantic and Greenland Sea.

ontogenetic: the origin and development of an individual organism from embryo to adult.

open lakes: occur where there is both an inflow and outflow of water, often at high altitudes and latitudes.

ostracods: a millimetre-sized crustacean with the soft body enclosed in a carapace comprised of a pair of kidney-shaped, spherical or lenticular shells made of chitin and low-Mg calcite.

pCO₂: partial pressure of carbon dioxide, a measurement of the concentration of CO_2 in the atmosphere and water.

palaeotemperatures: temperatures of past environments (e.g., past air or water temperatures).

palaeotemperature equations: equations derived from equilibrium fractionation, which is related directly to the thermodynamics of the mineral precipitation.

Palaeocene-Eocene Thermal Maximum (PETM): the rapid increase in global temperatures observed at the Paleocene-Eocene boundary approximately 55 million years ago.

per mille (also per mil, permil and ‰): literally parts per thousand, the conventional unit for expressing isotope delta (δ) values.

phloem water: water stored and transported within the phloem of the tree. In trees the phloem is the region containing food conducting tissues in the bark. Phloem water may return from the leaf to the site of cellulose synthesis its composition having been modified through photosynthetic reactions in the leaf.

PHREEQC (PH (pH), RE (redox), EQ (equilibrium), C (program written in C)): a widely used geochemical modelling program that uses C isotope mass balance.

phytoplankton: free floating flora which convert inorganic compounds into complex organic compounds.

planktonic foraminifera: a single celled protozoa which produces a calcium carbonate test and lives in the surface ocean.

protein: macromolecules composed of chains of amino acids, and vital to living organisms both for their structural and catalytic properties.

Quaternary: the second and last period in the Cenozoic Era. Normally, it is subdivided into the Holocene (Recent) and Pleistocene Epochs. The short chronology places its beginning at the traditionally defined boundary at roughly 1.78 Ma, while the long chronology places it at the start of the Matuyama Magnetochron at approximately 2.48 Ma.

radiolaria: form of phytoplankton found through the marine environment from the Cambrian onwards which produce siliceous skeletons.

proxy: a measurement that provides indirect evidence of some process or a participant in a process. Various proxies exist for palaeoenvironmental reconstructions. They range from fossil remains to isotope ratios that provide evidence of past conditions.

"Segment length effect": the phenomenon by which the minimum signal frequency that may be recovered from a palaeorecord (e.g., a tree ring series) is limited by its temporal span.

SEM: scanning electron microscope or microscopy.

silicates: silica bearing minerals, including diatom.

silicic acid: the predominant form of silicon within water $(SiOH)_4$.

soil $pCO2$: the partial pressure of CO_2 in a soil gas.

SPECMAP (Spectral Mapping Project Timescale): the stack of benthic foraminifera oxygen isotope records which provided the first composite stratigraphy for the last 600,000 years.

speleothems: secondary carbonate accumulations such as stalagmites, stalactites and flowstones deposited in caves by degassing of cave drip-waters that are supersaturated with respect to calcium carbonate.

stoichiometry: quantitative relationships among the reactants and products in a chemical reaction as expressed by the equation for the reaction.

stomata: a minute pore in the epidermis of a leaf through which gas (CO_2 and water vapour) may pass.

stratosphere: part of the upper atmosphere, extending from the *tropopause* (see below) to about 50 km above the Earth's surface, and having a low water vapour content compared to the troposphere. Characterised by increasing temperature with increasing altitude.

summer insolation: solar radiation incident on the Earth's surface, usually at a specified latitude (e.g., 45 °N).

teleconnections: observation that spatially separated (on inter-continental scales) climate proxies record simultaneous, possibly related climate change events.

Thermal Ionisation Mass Spectrometry (TIMS): mass spectrometry in which a chemically purified form of the element of interest (e.g., uranium, thorium) is loaded onto a solid metal filament (usually Re or Ta) that is then heated by an electrical current in a high vacuum. The element of interest is ionised, producing positively charged ions that are then accelerated in a flight tube and separated by mass in a magnetic field. Ion beams are measured using a Faraday collector or other ion-counting circuitry.

Total Dissolved Inorganic Carbon (TDIC): (1) carbon contained in carbon dioxide, and bicarbonate and carbonate ions dissolved in water. (2) The sum of all dissolved inorganic carbon species ($CO_2 + HCO_3^- + CO_3^{2-} + H_2CO_3$) in an aqueous solution.

travertine: calcium carbonate deposited by hot or cold water, including speleothems and accumulations at springs.

tritium: radioactive isotope of hydrogen, ^3H, produced mainly during the testing of thermonuclear devices and by nuclear reactors.

tropopause: boundary layer between the *troposphere* (see below) and the *stratosphere* (see above), characterised by little change in temperature with increasing altitude.

troposphere: lowest region of the atmosphere, extending to 10-15 km above the Earth's surface, characterised by clouds and weather, and decreasing temperature with increasing altitude.

U-series: refers to the chain of nuclides produced by radioactive decay of uranium. The decay chain of interest in the context of speleothem dating is $^{238}U \rightarrow {}^{234}U \rightarrow {}^{230}Th$. As time elapses the concentration of ^{230}Th builds up in the speleothem calcite as a result of uranium decay.

wax esters: biomarker found in higher plants which can be found in marine/lake sediments.

XRD: x-ray diffraction, diffractometer or diffractometry.

xylem water: the water stored and transported within the xylem of a plant. In trees the xylem is the water conducting tissue which forms the woody cylinder of the tree (see also phloem water).

Younger Dryas: a cold reversal in the Late-Glacial warming trend during the Glacial/Holocene transition (equivalent to the Loch Lomond Stadial in the U.K.). It occurred between c. 12,700 and 11,500 cal BP (c. 11,000-10,000 ^{14}C BP). The climatic changes occurred rapidly in decades or less. They are registered most strongly around the North Atlantic region, in ice cores and in the biological and sedimentary records in marine and terrestrial sediments. The term 'Younger Dryas' is now more precisely defined as Greenland Stadial 1 (GS-1) in the GRIP ice core. Named after the plant *Dryas octopetala*, which now has an arctic-alpine distribution.